American Seacoast Forts

A Directory with Period Military Maps
1890-1950

Volume 3

The Pacific Coast
San Diego to Puget Sound

PREPARED BY

TERRANCE C. MCGOVERN
MARK A. BERHOW
GLEN M. WILLIFORD

Published by the Coast Defense Study Group Press
2025

PLEASE DIRECT ANY COMMENTS OR CORRECTIONS TO THE PUBLISHER - INFO@CDSG.ORG

IBSN 978-0-9748167-8-4 (Hardcover B&W)
LIBRARY OF CONGRESS CATALOG CARD NUMBER 2025941598

Library of Congress Cataloging-in-Publication data
American Seacoast Forts: A Directory / Terry McGovern, Mark Berhow and Glen Williford
p.   cm.
Includes bibliographical references and index.
Library of Congress Control Number: 2025491598
ISBN 978-0-9748167-8-4 (h.c.)
1. Military History, 2. Artillery I. Terry McGovern, Mark Berhow and Glen Williford

First Edition: August 2025
Printed in the USA by Ingram Spark

Cover Photographs
Front cover: Battery Spencer, Fort Baker and the Entrance to San Francisco Bay, California (Terry McGovern)

Rear cover (clockwise from upper left): Map of the Entrance to San Francisco Bay CA 1915 (National Archives); Battery Mendell, Marin Headlands, CA (Terry McGovern); The Fort MacArthur Upper Reservation batteries 1937 (National Archives); Battery Worth, Fort Casey State Park, WA (Terry McGovern)

THE COAST DEFENSE STUDY GROUP, INC.
CDSG.ORG

The Coast Defense Study Group, Inc. (CDSG) is a tax-exempt corporation dedicated to the study of seacoast fortifications. The purposes of the CDSG include educational research and documentation, preservation and interpretation of historic sites, and assistance to other organizations dedicated to the preservation and interpretation of coast defense sites. Membership is open to any person or organization interested in the study or history of coast defenses and fortifications. Membership in the CDSG will allow you to attend annual conferences, special tours, and receive quarterly newsletter and journal. To find our more about the CDSG, please visit the CDSG website at **cdsg.org.**

Acknowledgments

This book is dedicated to three key members of the CDSG who began research on this subject in the 1970s and provided much of the initial information for those that followed in this study. Robert Zink (member number 1) and Glen Williford (member number 2) were both dedicated in their studies and unfailing supportive to those that sought more information on American seacoast defenses. The third key CDSG member was Bolling W. Smith who did a large amount of research work in obtaining copies of documents from the National Archives of which many were used in this compilation.

Sighting a 10-inch gun at Fort Standish, Massachusetts
(Leslie Jones, Digital Commonwealth, Massachusetts Online Collection, Boston Public Library)

The CDSG Press

Coast Defense Study Group Press is a division of Coast Defense Study Group (cdsg.org), which publishes books of historical interest, especially concerning seacoast fortifications. The CDSG Press also offers an ever-expanding number of key reprints reports and manuals in electronic PDF format on compact disks. To order these books and other **CDSG Press** publications, please access the **CDSG Press** pages on the **CDSG web site** at **cdsg.org.**

CDSG Press is interested in new titles, especially those dealing with fortifications, please contact Terry McGovern at 703/538-5403 or at tcmcgovern@att. net if you have a title that you are seeking to have published. Visit www.cdsg.org/press.

Under the CDSG Press label, our organization has published:

Notes on Seacoast Fortification Construction by Col. Eben E. Winslow, 1920, 428 pp. 1994 reprint HC with bound drawings
Seacoast Artillery Weapons Technical Manual (TM) 9-210 by U.S. War Dept. 1944, 202 pp. 1995 reprint PB
The Service of Coast Artillery by F. Hines & F. Ward, 1910, 736 pp. 1997 reprint HC
Permanent Fortifications & Sea-Coast Defenses by U.S. Congress, 1862, 544 pp. 1998 reprint HC
American Coast Artillery Material Ordnance Dept. Doc#2042 by U.S. War Dept., 1922, 528 pp., 2001 Reprint HC
American Seacoast Defenses: A Reference Guide (3rd Edition) by Mark A. Berhow, (2015) 732 pp. HC
The Endicott & Taft Board Reports, reprint of original reports of 1886 and 1905 by U.S. Congress, 525 pp. 2007 HC
Artillerists and Engineers: The Beginnings of US Fortifications 1794-1815 by Col. Wade, U.S. Army, 226 pp. 2011 PB
World War II Harbor Defenses of San Diego by Commander (Ret.) Everett, U.S. Navy, 226 pp. 2020 HC

The CDSG Presss offers these Hole in the Head Press Books

Artillery at the Golden Gate by Brian B. Chin (Hole-in-the-Head Press), 176 pp. 2019 PB
Fort Baker Through the Years by Kristin L Baron and John A. Martini (Hole-in-the-Head Press), 99 pp. 2013 PB
Rings of Supersonic Steel (3rd Edition) by Mark Morgan and Mark Berhow (Hole-in-the-Head Press), 358 pp. 2010 PB
The Last Missile Site by Stephen Hailer and John A. Martini (Hole-in-the-Head Press), 158 pp. 2010 PB
To Defend and Deter by John Lonnquest and David Winker (Hole-in-the-Head Press), 432 pp. 2014 PB

The CDSG Press
1700 Oak Lane
McLean, VA 22101-3322 USA

AMERICAN SEACOAST DEFENSES
A DIRECTORY OF AMERICAN SEACOAST DEFENSES 1890-1950

Table of Contents

U.S. Coast Artillery 1890-1945 Harbor Defense locations

Directory to American Seacoast Forts 1885-1950 in 4 volumes
    (21 Continental US Harbor Defenses and Overseas Harbor Defenses)
• Volume 1 North Atlantic Coast: New England – Long Island Sound - New York
• Volume 2 Mid-Atlantic, South Atlantic, and Gulf Coasts: Cheasapeake Bay to Galveston
• Volume 3 Pacific Coast San Diego to Puget Sound
• Volume 4 Alaska and Overseas Bases: Hawaii - Philippines - Panama - Caribbean - Newfoundland
    - Bermuda

Volume 3: The Pacific Coast: San Diego to the Puget Sound

VOLUME 1: THE NORTH ATLANTIC COAST: PORTLAND TO NEW YORK

Portland & Kennebec River, ME-—Fort and Gun Battery Descriptions
Portsmouth, NH—Fort and Gun Battery Descriptions
Boston, MA—Fort and Gun Battery Descriptions
New Bedford, MA—Fort and Gun Battery Descriptions
Narrangansett Bay, RI—Fort and Gun Battery Descriptions
Long Island Sound, CT & NY—Fort and Gun Battery Descriptions
New York, East —Fort and Gun Battery Descriptions
West Point Military Academy practice batteries
New York, NY & NJ—Fort and Gun Battery Descriptions

VOLUME 2: THE MID-ATLANTIC, SOUTH ATLANTIC, AND GULF COASTS: DELAWARE BAY TO GALVESTON

Delaware Bay, DE & NJ—Fort and Gun Battery Descriptions
Chesapeake Bay (Baltimore, Potomac River, James River, Hampton Roads)
Baltimore, MD—Fort and Gun Battery Descriptions
Potomac River, MD & VA—Fort and Gun Battery Descriptions
Entrance to the Chesapeake Bay—Fort and Gun Battery Descriptions
Cape Fear River, NC—Fort and Gun Battery Descriptions
Charleston, SC—Fort and Gun Battery Descriptions
Port Royal Sound, SC—Fort and Gun Battery Descriptions
Savannah, GA—Fort and Gun Battery Descriptions
St. Mary's River - Fort Clinch—Fort and Gun Battery Descriptions
St. John's River, FL—Fort and Gun Battery Descriptions
Key West, FL—Fort and Gun Battery Descriptions
Tampa Bay, FL—Fort and Gun Battery Descriptions
Pensacola, FL—Fort and Gun Battery Descriptions
Mobile Bay, AL—Fort and Gun Battery Descriptions
Mississippi River, LA—Fort and Gun Battery Descriptions
Galveston, TX—Fort and Gun Battery Descriptions

VOLUME 4: ALASKA AND THE OVERSEAS BASES

The Harbor Defenses of Sitka, Alaska—WWII Program sites
The Harbor Defenses of Seward, Alaska—WWII Program sites
The Harbor Defenses of Kodiak, Alaska—WWII Program sites
The Harbor Defenses of Dutch Harbor, Alaska—WWII Program sites

The Harbor Defenses of Honolulu—Fort and Gun Battery Descriptions
The Harbor Defenses of Pearl Harbor—Fort and Gun Battery Descriptions
The Harbor Defenses of Kaneohe Bay and the North Shore of Oahu—Fort and Gun Battery Descriptions

The Harbor Defenses of Manila Bay—Fort and Gun Battery Descriptions
The Harbor Defenses of Subic Bay—Fort and Gun Battery Descriptions

The Harbor Defeses of Cristobal—Fort and Gun Battery Descriptions
The Harbor Defenses of Balboa—Fort and Gun Battery Descriptions

The Harbor Defenses of Vieques Sound, Puerto Rico, Virgin Islands (Roosevelt Roads)
    —WWII Program sites
The Harbor Defenses of San Juan, Puerto Rico—WWII Program sites
Planned Defenses in Guantanamo Bay, Cuba
Planned Defenses in Jamaica
Planned defenses in Trinidad

Harbor Defenses in Newfoundland, Canada—WWII Program sites

Defenses in Bermuda —WWII Program sites

Fire control stations at Fort Casey State Park (Terry McGovern)

Horseshoe Bay and Fort Baker, Marin Headlands, Golden Gate National Recreation Area (Terry McGovern)

INTRODUCTION

The United States has long focused on defending its seacoasts against overseas enemies due to its geopolitical situation with its long coastlines and generally peaceful borders with Canada and Mexico. Earlier American fortification efforts resulted in the First System, the Second System, and the Third System of coastal defenses. The great brick and stone forts built or remodeled during 1820 to 1860 are well known from their military importance and use during the American Civil War. For many years after that great internal conflict most U.S. fortification efforts languished. After 1885, due to the great advances in military technology and America's increasing worldwide economic presence that the United States embarked on a new round of fortification building to protect its shores. The U.S. Army expended much of its limited manpower and resources to protecting America's coast from 1890 to 1950.

The technical development and tactical objectives of the coast defenses of the modern era (1890 to 1950) is a product of America's earlier policies and experiences. Until the advent of air power and the missile age, the defense of the United States has primarily been one of defending our shores from naval attack. Only during the nation's early years did the threat of land invasion exist. Accordingly, the United States relied on the ships of its Navy to provide the first line of defense. Its U.S. Army was called upon to provide the second line of defense by building forts at key points along its coastline to defend major harbors. This defense policy of denying an enemy fleet access to its major harbors and anchorages developed into the array of former coastal fortifications that remain today.

Based on concerns over external threats and on internal politics, the United States government has built coastal fortification in a series of construction programs. After inheriting the remains of the fortifications from the colonial era and the revolutionary war, the first in a series of national fortification construction programs began in 1794 and these programs continued into the late 1940s. For the ease of use in this book, these fortification periods have been organized into distinct groups and have been named the following: First System (1794-1801), Second System (1802-1815), Third System including the American Civil War (1816-1867), 1870s Period (1868-1879), Endicott Program (1885-1904), Taft Program (1905-1916), World War I & Interwar Period (1917-1939), and the 1940 Modernization Program including World War II (1940-1950).

The legacy of these seacoast defenses is a series of concrete structures scattered along America's coastline, many now in public shoreline parks. The nature of fortifications, the fact that they were designed to withstand the pounding of naval artillery, has allowed these massive structures to withstand the attack of both the natural elements and economic development. That these structures are still standing many years after their effective use ended draws our attention to them. They captivate us regardless of whether it's a large brick and stone multi-story structure surrounded by a dry ditch or an odd shaped, concrete structure covered with thick vines or bright graffiti and surrounded by worn fences sporting weathered warning signs. Visitors to our seashores are curious about the nature of these structures. Some of the questions that they ask are: What are these structures? Why are these structures here? When were these structures built? This directory provides answers to many of these questions. Many of these parks have visitor's centers and gift shops selling range of books and other items, but few have any books which explain the fortifications that once existed at that site. This directory will fill that void.

This directory is a companion work to the CDSG Press's *American Seacoast Defenses – A Reference Guide* by Mark Berhow. This directory is to aid students of American seacoast fortifications to locate and visit the key defenses of the "Modern Era" of coastal defense (1890 to 1950), which is defined by its use of concrete, steel, and breech-loading rifles. The following pages provide a brief review of the function and history of the development of coastal fortifications in the United States. This history reflects the politics of changing external threats to our nation and rapid advancement of military technology. The directory's focus is a guide to the "Modern Era" of fortifications along the Atlantic, Pacific, and Gulf coasts, as well as U.S. overseas bases by providing key maps, plans, photographs, and short description of their history.

The directory is organized into several sections. The first section is brief history of the modern era of American coastal defenses, including background on the U.S. coast artillery material, organization, armament, design of military reservations, and garrison life. The second section is a brief description of the history and its current status (ownership, public access, remaining assets, things to see, etc.) of the major military reservations that had seacoast artillery and a short history of each major concrete gun battery. This history includes each battery is described as to its rational, authorization, construction and transfer dates, engineering cost, naming citation, armament, service history, ultimate disarming, and current status. These battery histories do not include railway and mobile artillery sites, including those with Panama mounts. Also excluded are temporary batteries, especially those using loaned naval guns, and anti-aircraft batteries that did not also serve a seacoast role. The directory is organized by Harbor Defense around the United States clockwise from Portland Maine to the Puget Sound in Washington State, followed by the Alaskan defenses of World War II, and the defenses in Hawaii, the Philippines, Panama, the Caribbean, Newfoundland and Bermuda.

Supporting the directory of defense sites is a compilation of maps for all the harbor defense reservations utilized during the period of 1900 to 1946. The map collection includes general maps of the location of elements (sites) for each defended harbor and the individual location site maps showing buildings, gun emplacements, fire control stations and other elements. Each Harbor Defense section has an overall 1920s-30s period map of the defense sites and a selected set of Confidential Blueprint Maps for each military reservation.

The maps are arranged more or less in order from the south to the north. A series of map symbol and abbreviation keys from 1921 and 1945 are included in the introduction. Some harbors (Baltimore, Potomac River, Cape Fear, Port Royal, Savannah, Tampa, Mobile, and the Mississippi River) did not receive new defenses during World War II.

12-inch Rifle on a barbette mount, Battery Godfrey, Fort Winfield Scott, California,
(Golden Gate National Recreation Area Collection, NPS)

CHRONOLOGY
Key events during the Modern Era of American coast defenses

1875   – Funding for new construction of coast defenses is stopped by the U.S. Congress

1883   – The U.S. Navy begins the first new construction program since the Civil War

1885   – President Cleveland appoints a joint army, navy, and civilian board headed by the Secretary of War, William Endicott, to evaluate the threats and needs for U.S. coastal defenses (Endicott Board)

1886   – The Endicott Board reports on state of the U.S. defenses and recommends a $126 million construction program of breech-loading cannons and mortars, floating batteries, warships, and submarine mines in 29 locations around the nation

1888   – Congress creates the Board of Ordnance and Fortifications to test weapons and implement the Endicott Program
         – Dynamite guns were developed to fire high explosive shells using compressed air

1890   – Congress approves funding for the construction of first Endicott Program batteries

1892   – The first Endicott Program battery is completed (Gun Lift Battery Potter)

1893   – First group of controlled mine casemates are completed

1894   – Buffington-Crozier disappearing carriage for 8-inch and 10-inch guns developed

1896   – Development of the Buffington-Crozier disappearing carriage for 12-inch guns

1898   – The Spanish-American War – 150 coast artillery pieces mounted
         – U.S. adds the Philippines, Guam, Puerto Rico as colonies; establishes military bases in Cuba; annexes Hawaiian Is.

1899   – 288 heavy coast artillery guns, 154 rapid-fire guns, and 312 mortars have been mounted

1901   – Reorganization of U.S. Army artillery corps to 30 batteries of field artillery and 126 companies of coast artillery

1902   – Work begins on the fortifications of Corregidor, Philippines

1904   – Work begins on the Panama Canal
         – First specially built mine planters constructed

1905   – President Roosevelt appointed a joint army, navy, and civilian board headed by the Secretary of War William Taft, to review the Endicott Program and to bring it up to date

1906   – The Taft Board reports on state of the U.S. defenses and recommends improvement in existing defenses by adding searchlights, electrification of defenses, and a modern system of fire control, as well as new defenses for newly acquired overseas bases

1907   – Establishment of the separate U.S. Army Coast Artillery Corps

1914   –World War I begins; Panama Canal opens

1915   – Report of the Board of Review on the coast defenses of the U.S., Panama Canal, and the Insular Possessions

1916   – First Coast Artillery anti-aircraft units formed

1917   – U.S. enters the World War II
         – Construction begins on the first long-range barbette batteries using existing 12-inch gun barrels

1918   – End of the World War I

1920   – Construction of the first Panama mount for 155m GPF guns in the Canal Zone

1922   – Washington Naval Treaty limits naval construction and Pacific fortifications
         – 16-inch gun and howitzer barbette batteries are constructed

1925   – Ten U.S. Harbor Defenses on active status and 15 are on caretaker status

1937   – Construction begins on first 16-inch casemated gun battery (Battery Davis)

1939   – Outbreak of World War II; U.S. Coast Artillery Corps has 4,200 troops

1940   – Congress approves the 1940 Modernization Program for 19 harbors in the U.S.
         – U.S. draft begins, coast artillery units brought up to wartime strength, national guard units federalized

1941   – U.S. enters the World War II
         – Establishment of the Harbor Entrance Control Posts (HECP)

1942   – U.S. Coast Artillery Corps has 70,000 troops

1945   – End of the World War II

1948   – All construction efforts cease, the coast defenses are abandoned, and armament salvaged

1950   – Disestablishment of the U.S. Coast Artillery Corps; remaining units reunited with Field Artillery

DESIGN AND FUNCTION OF AMERICAN SEACOAST DEFENSES IN THE MODERN ERA 1890-1950

Historical Development of American Coast Defenses during the Modern Era

The key development that led to the new American coast defense era was the development of new heavy rifled breech-loading guns that had a longer range, were more accurate and delivered a heavier projectile than the muzzle-loading smoothbore cannons of the Civil War. These new guns were made of high-quality steel that were lighter and stronger, which took advantage of new propellants that replaced gunpowder. Equally important was the development of effective breech mechanisms that could withstand the high pressures and temperatures generated by the new guns and allow for the gun to be loaded from the rear instead of the muzzle, which increased the rate of fire and allowed for improved protection of the gun crew. The new guns and mortars could accurately fire projectiles at effective ranges that were two to three times farther than the muzzle-loading smoothbore cannons used during the American Civil War. These developments coincided with the building of the new steel naval vessels that featured these new big guns starting in 1875. However, between 1875 and 1890 the U.S. Congress did not appropriate any funds for the construction of new coastal fortifications.

Seacoast Defenses built after the Endicott Board Report 1886-1904

As U.S. coastal fortifications were allowed to deteriorate in 1870s and early 1880s, new steam powered, ocean navigating iron warships were being built by foreign navies. As the U.S. Navy embarked on its new construction program, it required protected bases for its operations. The military began to lobby to overhaul the obsolete existing defenses. In 1885, a board was created by U.S. Congress to examine and report upon the state of U.S. coastal defenses. The board headed by the U.S. Secretary of War, William C. Endicott, was comprised of four officers from U.S. Army, two officers from the U.S. Navy, and two civilians. This joint board made an extensive study of fortifications, type of armament, and defense that would be needed, by evaluating current European developments. In 1886, the Endicott Board published its recommendations for new coastal fortifications to be built at 29 key harbors, along with floating batteries, torpedo boats, and submarine mines. The board's original plan called for over 1,300 guns and mortars of 8-inch or larger of the newest design to be installed. The costs of board's recommendations were estimated to be $126 million dollars (in 1886-dollar value). While the U.S. Congress took no immediate action on the board's report, the estimates provided in the report would be cited for the next 20 years as a measure of the construction progress of this new generation of U.S. coastal fortifications.

In 1888, Congress established an U.S. Army board for ordnance and fortifications who as charged with testing new weapons and to design new coastal fortifications. In 1890 Congress made the first appropriations for the first new construction of coastal fortifications in 16 years with an initial funding of $1.2 million dollars. This funding was for the first of new defenses: a 12-inch barbette battery at San Francisco; an 8-inch disappearing battery at New York Harbor; a 12-inch gun lift battery at Sandy Hook, New Jersey; and for 12-inch mortar batteries at Sandy Hook and at the Presidio, San Francisco. The design of these new coastal batteries would set the pattern of coastal defenses that would be duplicated at all of America's major harbors, and this was the beginning of what would become known as the "Endicott Program" of American seacoast defenses.

The designs used for the Endicott Program coastal fortifications demonstrated the shift in importance from the large multi-tiered multi-gunned "fortresses" to weapons emplaced in dispersed concrete "batteries" protected by earthen embankments. The "fort" became a defined reservation of land that contained guns of a range of calibers, along with the housing for the men required to man these defenses, and supply and maintenance buildings. The weapons were grouped into batteries containing from one to sixteen guns. The batteries were located along the shoreline to maximize their range and field of fire and were designed to blend into the landscape as not to be seen from the sea. The armament of these batteries ranged from weapons to engage enemy capital ships to small-caliber rapid fire guns to knock out fast moving torpedo boats, as well as to protect the fields of electrically controlled submarine mines from minesweepers. The dominance of the armament during this period is reflected by the dramatic increase in the time and cost in constructing a gun barrel and breech mechanism along with its carriage over that of the weapons of the earlier periods.

The primary weapons of the Endicott Program were 8-inch, 10-inch and 12-inch rifled breech-loading guns, a growth in size that reflected the need to match the increase in the size of opposing naval guns. These guns were mounted on both barbette and disappearing carriages that had a maximum elevation of 15 degrees and range of about seven to eight miles. The relatively unique American "disappearing" carriage allowed the gun to be raised over the parapet by using a counterweight to fire. The energy from the recoil caused the gun to drop back down behind the parapet into the emplacement to be reloaded while being protected from direct fire from attacking warships. These heavy weapons were mounted in large concrete emplacement with thick frontal walls that were in turn protected by many feet of earthen fill. Located below or adjacent to the firing platform were support areas that included the ammunition magazines containing projectiles and powder propellants. About three hundred of these heavy guns were installed around the United States during the Endicott Program in batteries of from one to six guns. It was a less expensive alternative to the armored turret mount favored by several European nations.

10-inch disappearing guns in Battery Hale, Fort Greble, Rhode Island (C.T. Gardner Collection)

The other large caliber weapon installed during the Endicott Program was the short-barreled 12-inch mortar. The mortar was designed to fire a shell in a high arc that descended down onto the lightly armored decks of warships of that era. To increase the opportunity of making a hit, these mortars were emplaced in groups of eight to sixteen mortars in square concrete pits that were protected by earthen hills. The use of these pits would give maximum protection to the mortars and their magazines from the flat trajectory of naval gunfire of the era. About four hundred of these mortars were installed around the United States during the Endicott Program.

The secondary smaller caliber weapons were installed to protect the controlled submarine mine fields from small craft that could sweep paths through the mines for larger warships, and to protect from attack by newly developed fast torpedo boats that could potentially penetrate the harbor and torpedo the shipping within. These threats called for guns that could be aimed, loaded and fired very rapidly. While not

12-inch mortar firing at Battery Alexander, Fort Barry, California (B.W. Smith Collection)

specified by the Endicott Report, several new gun and carriage systems were developed for this role rang-ing from 3-inch to 6-inch is size. These guns were generally mounted on either disappearing carriages or on pedestal carriages with simple steel shields. The concrete emplacements for these guns had low parapets and magazines below the guns. A rapid-fire battery had between two and six guns per battery. Over five hundred rapid-five weapons were installed during the Endicott Program.

While the use of submarine mines or "torpedoes," as well as channel obstructions or barriers, has a long history in defense of harbors, it was during the Endicott Program that a widespread and a structured use of submarine mines occurred. The U.S. Army developed a system of controlled submarine mines; sta-tionary explosive devices located below the surface of water where ships were likely to pass. The submarine mines used from 1890 to 1930 were the buoyant type (floating but anchored to the sea bottom), though during World War II the buoyant mines were replaced with ground mines (stationed on the sea floor). The mines were only deployed during times of war or for practice, otherwise they were stored disassembled ashore—the mines and their control cables became defective after extensive exposure in the water. The controlled mines were connected to shore by undersea cables and could be exploded by electrical switches from a control board on shore by the soldiers manning the mine defenses when a warship passing over these mines or by direct contact. Controlled mines were usually laid in rows across the key shipping channels to create a group of mines, usually 19, which would cover a space of about 2,000 feet long in water up to 250 feet deep. Several groups of mines were to be deployed to create a field of mines. The U.S. Army Coast Artillery Corps had dedicated units to man the mine planting vessels, fire control stations, mine and cable storage facilities, mine casemates and switchboards, and loading wharves.

The Endicott Program roughly covered a period from 1885 to 1905, and the coast artillery func-tion was a key mission of U.S. Army during this time (and made up a large percentage of total U.S. Army manpower). This also required a more technical trained soldiers to man them which led to the U.S. Army's Artillery branch to be reorganized in 1901 and 1907 to create the U.S. Coast Artillery Corps.

Planting a mine (Stillion Collection NPS, Gulf Shores Natl. Seashore)

Seacoast Defenses built after the Taft Board Report 1906-1916

In 1905, a new National Coast Defense Board headed by U.S. Secretary of War William Howard Taft was organized by President Theodore Roosevelt and charged with reviewing the progress of Endicott Program construction and update it. In the 20 years since the original Endicott Board report was presented, numerous technical and political developments had taken place. The Board, informally known as the "Taft Board" after its chairman, established new cost estimates and its recommendations were primarily concerned with modernization of existing coastal fortifications and adding coastal defenses to the overseas territories gained after the Spanish-American War including Hawaii and the Philippines and other locations.

The modernization of existing defenses included the electrification of lighting, communications, and ammunition handling equipment, both at the batteries and throughout the fort. The early emplacements had loading platforms widened and projectile hoists were installed to improve the rate of fire. The report recommended the use of searchlights for nighttime illumination of harbor entrances. During the Taft Program was the finalized development and implementation of a coordinated system of target information gathering and processing that greatly improved the target accuracy of the major caliber guns and mortars. Up to this time, the aiming of guns at a target had been generally done from each battery with basic sighting instruments and combination of luck and experience. The new system was based on triangulation using two observers with telescopic instruments at separate position finding stations or "base end" stations communicating with the newly developed telephones to a centralized battery plotting room that provided real-time tracking and firing coordinates on a moving target. The battery plotting room personnel would mathematically process this sighting information and other data into aiming instructions that would then be transmitted to each gun emplacement.

The 14-inch gun on a disappearing carriage the Taft-era Battary Osgood, Fort MacArthur, California (Fort MacArthur Museum)

While the Taft Board's recommendations on the construction of new fortifications was largely limited to existing defenses at Eastern Long Island Sound, San Diego, Puget Sound, Columbia River, and Chesapeake Bay in the continental United States, major new construction projects were planned for the Philippines, Panama, Hawaii, Cuba, Puerto Rico, Alaska, and Guam. Plans for Cuba, Puerto Rico, Alaska, and Guam were not carried out. New defenses were added for the port of Los Angeles in 1909. These "Taft Program" defenses varied little from the overall designs used during the Endicott Program. Variations from the Endicott Program were the product of advancing naval armaments and the U.S. Army's twenty years of experience of operating coast defenses. To match the increased caliber of naval guns, a new disappearing gun of 14-inch caliber was developed. Another characteristic of the Taft Program batteries was the increased dispersion of batteries. The reduce density of weapons can be seen in the construction of several one-gun, 14-inch batteries and the reduction of mortar batteries from 8 mortars to 4 mortars (4 per pit to 2 per pit).

During the Taft Program, several one of kind of projects were undertaken. The Endicott Board Report called for 16-inch guns but work on the development stalled after the construction of one gun tube in 1895. Two unique 16-inch disappearing batteries were finally built in Panama and the Long Island Sound. In the Philippines army-designed armored turrets were custom built for a very small island in Manila Bay. Four 14-inch guns were mounted in two turrets at Fort Drum, which also became known as the "concrete battleship."

World War I and the Interwar Period (1917-1939)

The march of technical improvements in naval weapons continued through improvements in naval fire control and the ability for naval turrets to elevate their guns. By 1915 the newer battleships had guns that could out range the effective range of the coast artillery emplaced during the Endicott and Taft Programs. The increased angle of fire of the newer battleships also threatened the disappearing carriage batteries which were not protected from the plunging fire of these new battleships. In 1915, a National Board of Review on the coast defenses of the U.S., Panama Canal, and the Insular Possessions recommended the construction of new batteries mounting 12-inch and 16-inch guns on higher elevation, longer range barbette carriages. While efforts to introduce these coast artillery weapons had begun, the demands of World War I placed

Firing one of the 12-inch guns of the post-WWI Battery Kingman, Fort Hancock, New Jersey (NARA)

the modernization of coast defenses on hold. Many Coast Artillery units were transformed into field and heavy artillery units for service in France. As the United States was short on long range field artillery,12-inch mortars, 10-inch, 8-inch, and 6-inch gun barrels were removed from several coast artillery batteries. These existing gun barrels, ranging from 6-inch to 14-inch in caliber were quickly mounted on railway and tractor-drawn carriages. While the United States involvement in World War I was brief, it resulted in the Coast Artillery Corps mission to be divided into three specialized areas as compared to its single mission before the war. These missions, based on armament type, were fixed coast defense weapons (including controlled mines), mobile seacoast artillery, and anti-aircraft artillery.

The development of the airplane as a ground attack weapon during the World War I added the task of defending both the mobile ground army and the shores of United States from attacks by aircraft to the Coast Artillery Corps' mission. The U.S. Army developed fixed and mobile anti-aircraft weapons, as well as accessory equipment such as aircraft sound locators, rangefinders, searchlights, specialized fuses, and mechanical fire direction calculators. The primary weapon for the defense against aircraft was the 3-inch gun on a fixed carriage (in batteries of three or four guns) located at existing coast defense posts. This weapon was later supplemented with .50 caliber machine guns and mobile 3-inch AA guns. By 1938 larger caliber anti-aircraft guns were introduced including the 90 millimeters (mm), the 105 mm, and 120 mm guns.

The mobile coastal defense mission came about because of the lack of U.S. heavy artillery for the troops in Europe. Existing gun barrels, ranging from 6-inch to 14-inch in caliber were quickly mounted on railway and tractor-drawn carriages. The construction of the new mobile carriages for guns, such as the railway mounts, took months and most of these weapons never reached European theater before the war ended. The availability of this ordnance material, especially considering the economics of using existing weapons and increased desirability of weapon mobility in the interwar period, made mobile coast artillery an attractive alternative to building new fixed coast defenses. The primary railway guns selected for coast artillery use from this large stock of World War I material were 8-inch guns and the 12-inch mortars mounted on new railway carriages. Added later was an improved version of the wartime 14-inch railway gun of which only four were constructed by 1920. The surplus mobile field artillery mounted on carriages designed for road movement included the 155-mm GPF gun (derived from a 1917 French design) of which almost a thousand were available. This powerful gun became the standard tractor drawn weapon for coast defense use against secondary targets.

While mobile coast artillery had the advantage of being able to respond to coastal areas most

a 155 mm G.P.F. mobile mount in a field position at Long Point, California (Ruhlen Collection)

threaten when enemy naval forces approached, both railway and tractor-drawn weapons lacked the accuracy and protection of fixed coast artillery. Without solid and steady firing platforms and the precision of pre-calibrated fire control networks, as well as the inability of the carriages of the mobile guns to quickly track horizontally moving targets made mobile artillery much less effective than weapons in fixed emplacements. Prepared locations with circular arcs of track were prepared at a few select locations. For the 155 mm GPF mobile artillery, simple circular concrete bases were designed. These circular bases improved stability during firing and provided for rapid azimuth adjustment for horizontal tracking. One of the most common base designs developed for the 155 mm GPF guns was a central pivot and a curved rail embedded in concrete, which the gun's split carriage would traverse. This design was first constructed in the Panama Canal Zone, so this design became known as the "Panama Mount". Given the limitations of mobile coast artillery, their use was primarily an augmentation of existing defenses or to provide protection during the construction of permanent fixed coast artillery. Due to the low level of military appropriations during the 1920s and 1930s, mobile coast artillery was the only available weapon to defend vital locations until new permanent defenses could be funded and constructed.

Given the low level of overall U.S. military funding during the 1920s and 1930s, the construction and development of new fixed coast defenses were limited. The need for economy and to allow for higher gun elevations led to the abandonment of the disappearing carriage and its complex two-level emplacements. Among the last disappearing carriages built were for two 16-inch single-gun batteries (one in Panama and the other in the Long Island Sound). A newly designed high angle barbette carriage for existing stocks of 12-inch Model 1895 guns allowed effectively doubled the range of the guns over the same 12-inch gun mounted on a disappearing carriage. Construction of fifteen long-range dual gun 12-inch batteries was started in 1917 and completed by the late 1920s. The emplacement design was a departure from those of the Endicott Program. The battery design had two guns located much further apart, each gun in the center of a large ground level concrete pad to allow for an all-around field of fire. Located between the two guns was an earth-covered reenforced concrete structure containing magazines for shells and powder, the power and plotting rooms, and storage rooms. Protection of the guns from naval fire was based on dispersion; the

Illustrated Directory with Maps   21

wide separation of the key elements of the battery. Other than camouflage and nearby anti-aircraft guns these batteries had no protection from air attack. The development of a new 16-inch gun and carriage with a range of nearly thirty miles which exceeded the range of all existing naval warships was completed in 1919. The 16-inch in emplacements that were very similar to those used for the long-range 12-inch barbette batteries, with an increased distance between the two guns of the battery and the dispersed location of magazines in simple storehouses connected by a rail system. Only a few of the U.S. Army designed barrels had been constructed when nearly sixty 16-inch barrels became available from the U.S. Navy. This windfall was due to the Washington Naval Treaty of 1922 that resulted in the cancellation of several U.S. battleships and battle cruisers then under construction. The naval 16-inch barrels were to be installed in modified U.S. Army barbette carriages after 1925. Six new twin-gun 16-inch batteries were built between 1922 and 1934. During the Interwar Period, the construction of new batteries including both long range 12-inch and 16-inch guns, amounted to little more than twenty new batteries. The coming of World War II would inject new life into building modern U.S. coast defenses.

The 1940 Modernization Program and World War II (1940 -1950)

During the 1930s the U.S. Army began discussing how to protect new coast artillery batteries from attack by aerial bombardment. The debate centered on the expense of designing and construction of turret mounts for 12-inch and 16-inch guns as compared to developing protective structures made of concrete and steel. It was practical economic and time frame requirements that resulted in the eventual selection of a concrete casemate structure design to protect the current type of barbette mounts.

The prototypes of this new type of major caliber battery were built at the San Francisco defenses, during 1937-1940. These emplacements were designed for two 16-inch guns located about six hundred feet apart with complete overhead cover. Located between the two guns along a service gallery were the ammunition magazines, power generators, and support areas. The 16-inch guns were enclosed in reinforced concrete casemates. The battery's structure was made up of eight to twelve feet of steel reinforced concrete which was topped by up to twenty feet of earth as additional protection. The entire battery structure was designed to withstand a direct hit from a naval projectile or an aerial bomb. When completed the southern San Francisco battery at Fort Funston emplacement looked like a small hill, especially when camouflage and natural ground cover was added to the structure. The only exposed portions of the battery were the casemates where the gun barrels projected out through armor shields and concrete canopies. A second casemated battery on a hilltop north of San Francisco was also undertaken. Four more casemated batteries were begun at Narragansett Bay, the Delaware River, and Chesapeake Bay in 1940-1941.

In 1940 the Harbor Defense Board was charged with developing a master plan to update the harbor defenses of the continental United States. Eighteen coastal areas in United States were selected for modernization due to their military and economic importance - Portland, Portsmouth, Boston, New Bedford, Narraganset Bay, Long Island Sound, New York, Delaware Bay, Chesapeake Bay, Charleston, Key West, Pensacola, Galveston, San Diego, Los Angeles, San Francisco, Columbia River, and Puget Sound. The Harbor Defense Board recommended the adoption existing stocks of 16-inch gun as the primary weapon and 6-inch gun as the secondary weapon for the modernization program. In all the board proposed building twenty-seven new 16-inch casemated batteries; the casemating of 23 existing primary batteries (both long-range 12-inch batteries and older 16-inch batteries; and building fifty new 6-inch two-gun barbette carriage batteries, which would provide long-range fire (15 miles maximum) against secondary warships. The new 6-inch batteries would be supported by 63 existing secondary batteries, mostly 6-inch and 3-inch barbette guns from the Endicott and Taft Programs, which would be retained. Upon completion of these new defenses 128 existing obsolete coastal batteries would be eliminated. The board estimated that the whole program would require three years to complete and cost about $82 million during 1941-3. Formal

One of the 16-inch guns of Battery Steele, Peaks Island M.R., Maine (Joel Eastman Collection)

A 6-inch gun of Battery Cravens, Peaks Island M.R., Maine, with a disguised SRC 296A radar behind
(Joel Eastman Collection)

approval of this modernization plan, which would become known as the "1940 Harbor Defense Modernization Program" or the "1940 Program," was approved in September 1940.

The 1940 Harbor Defense Modernization Program greatly simplified the task of Coast Artillery Corps by reducing the number of types of batteries as well as the overall number of batteries needed to carry out their coast defense mission. This allowed a reduction in personnel and the level of effort to maintain, training and supply the pre-1940 batteries. Some of coast artillery that was declared obsolete was shipped to Allied nations to supplement their defenses, but most were scrapped for the war effort. As the nation moved closer to war, additional coastal defense projects were added to the 1940 Program, especially at newly acquired overseas bases, such as Trinidad, Bermuda, Newfoundland, and in areas where the enemy threats seem greater, such as Alaska, Hawaii, Puerto Rico, and the Canal Zone. It also became apparent that planning, construction and emplacement of the many new batteries called for the 1940 Program was going to take a much longer time then original envisioned, especially as the program was competing with rapid expansion of the whole U.S. Army and U.S. Navy. By the middle of July 1941, only four 16-inch gun batteries were ready for action and construction work had been started on just five others. With pressure from the U.S. Army Air Corps, it was decided to limit active work to those batteries that could be completed by July 1944. As a result, all work on fourteen of the thirty-seven 16-inch batteries planned for the continental U.S. was discontinued. The expansion of overseas bases during 1941 impacted the construction of the new 6-inch gun batteries in the continental U.S. by priority assigned to the completion of twenty 6-inch batteries to guard these overseas bases.

The new batteries constructed under the 1940 Program were much more standardized that those of proceeding periods. The Army developed standardized designs for the 16-inch gun batteries and the 6-inch gun batteries which were used with only minor variations for local topography and soil conditions. Both the 16-inch and 12-inch guns, whether newly installed or retained from the Interwar Period, were emplaced within reinforced concrete casemates that limited their field of fire to about 180 degrees but gave them superior protection over the old open emplacements. The new 6-inch batteries were not casemated. A cast steel shield from four to six inches thick was placed around the gun and carriage. This shield would protect the gun and its crew from all but direct hits by heavy projectiles. Between the two 6-inch guns was an earth covered steel reinforced concrete structure contain the magazine, power generators, communications, air filtering equipment, storage, and plotting room. As these batteries were being built, they were assigned a "Battery Construction Number" for record keeping purposes. As many of the new batteries were never formally named, these construction numbers were the only designation they received. While the Army never referred to the 16-inch series of batteries as whole as the "100" series or the 6-inch series of batteries as whole as the "200" series, these terms are used by modern historians and are referred to as such in this work.

As the range of these new batteries was far greater than earlier batteries it was also necessary to update the fire control networks. The 16-inch batteries received new base end stations as far as twenty-five miles away from the gun's position to allow for gun's maximum range to be effectively used. These stations were built in wide variety of forms: houses, windmills, silos, water tanks, office buildings, or buried into hillsides. Radar was added as an early warning device and as a fire control instrument allowing the operation of coast artillery at maximum range during all weather conditions.

By the start of World War II, the Coast Artillery Corps' mobile coast artillery units had dwindled from the plans of the Interwar Period, especially the railway guns units. Several tractor-drawn 155mm GPF gun regiments were available in the continental U.S., but only part of one 8-inch railway regiment was on hand. The four 14-inch railway guns continued their role in Los Angeles and Canal Zone. The primary use of mobile coast artillery was to fill in for fixed coast artillery weapons until their completion or at secondary locations. The 155mm GPF gun units were reorganized into seventy-two 2-gun batteries along the Atlantic, Gulf and Pacific coasts. Using 12-inch railway mortars and 8-inch railway guns from storage, several CAC units were formed and sent to both domestic and overseas locations to provide temporary harbor defenses until permanent works could be constructed.

As with earlier periods, an integral part of harbor defenses was the use of controlled mines across key ship channels. These mine defenses were supplement by U.S. Navy contact mines and the use of submarine nets and booms. As the primary threat during World War II turned out to be enemy submarines at most of these ports, the U.S. Navy added detection devices in outer harbor approaches and conducted offshore patrols. Because of the need for both the U.S. Army and U.S. Navy to coordinate their coast defense activities, a centralized harbor entrance command was created in 1941. The Harbor Entrance Control Post (HECP) used both army and navy personnel to provide a link between higher command and all subordinate elements of a harbor defense. These centers were responsible for monitoring all movement of shipping in and out of the harbor. To support this effort a secondary gun battery was on duty as commercial shipping traffic was examined upon entering the harbor. One of the concerns at this time was an attack by fast moving torpedo boats combined with the lack of modern rapid-fire guns. To fill this void, the 90mm anti-aircraft gun was selected to replace the existing 3-inch pedestal guns of the Endicott and Taft Programs. In late 1942, special anti-motor torpedo boat (AMTB) batteries were installed along the Pacific and Atlantic coastlines. These batteries usually consisted of two fixed mounted 90mm guns and two mobile mounted 90mm guns, and two mobile 37mm or 40mm anti-aircraft guns. These guns would be protected by earthen revetments with protected magazines. The active harbor defenses received two, three, four, or more of these AMTB batteries beginning in 1943.

Outside the continental United States, where the threat of attack and invasion was greater, new coast defense construction proceeded with greater speed and with the use of armament on hand rather than waiting for weapons sto be provided by 1940 Program. The coast defenses of Hawaii are a good example, as the Japanese attack on Pearl Harbor made new defenses the highest priority due to concerns of an invasion attempt. A series of batteries were constructed, using excess naval guns, ranging from the 14-inch turrets from the battleship USS *Arizona* to 8-inch gun mounts from the aircraft carriers USS *Lexington* and USS *Saratoga*. Throughout the Pacific Islands and Alaska, surplus U.S. Navy guns (5-inch, 6-inch, 7-inch & 8-inch) were mounted on shore to defend U.S. Navy installations.

With the tide of the war shifting toward the Allies after 1942 and the demands to produce war material for the mobile army, the navy, and the army air corps, the 1940 Program was pared back. While the construction of structures could keep pace with the original plan, the manufacture of weapons and their accessories could not. In response to these pressures, the 1940 Program was scaled back even further. By the war's end, the modernization program resulted in the completion of nearly 200 new batteries in the continental United States at a cost of $220 million, or about one-half the number of installations proposed in the 1940 Program, but still the most powerful collection of coastal defenses in America's military history.

The development in military tactics and technology during World War II brought about numerous changes to the concept of coast defense. It was no longer thought necessary to defend one's seacoast using just coast artillery and controlled mines. Air power and naval forces were to replace breech loading rifles and reinforced concrete. Already at the end of World War II, all except a few 90mm AMTB batteries were placed on caretaking status. During the transition years of 1946 to 1948, some new batteries started during the war were completed while many other batteries were being disposed of and guns scrapped. By 1949, the process was completed as the last of guns were scrapped. In 1950, the remaining harbor defense commands were disbanded, and the Coast Artillery Corps was abolished as a separate U.S. Army branch with its remaining units, all anti-aircraft artillery, recombined into the Field Artillery. After 150 years of being one of America's military prime missions, the building and manning of permanent coastal fortifications was over.

The U.S. Coast Defense Objective in the Modern Era

The objective of seacoast defense is to provide protection of the coastline from invasion by an enemy, and specifically the defense of important harbors, which includes securing the anchorages and bases needed for naval operations. Coast defense is not only protective in its strength but protects the nation's ability to carry war beyond its own coastline.

It is impractical to fortify the entire extent of any nation's long coastline in such a way that an enemy in command of the sea could not land upon some portion of it. The cost of such an undertaking would be excessive, as maintenance of these defenses and number of men required would make it prohibitive expensive. An example of this type of defense was the "Atlantic Wall" built by Germany in World War II (which stretched from Norway to Spain), which failed to prevent the Allies from landing in Europe in 1944. It was essential, however, that certain selected points be permanently fortified to make invasion more difficult and to protect key naval shore installations and fleet anchorages and important commercial harbors that support the nation's economy.

The resources to defending the coastline during this era were divided into two kinds of troops. The first was the Coast Artillery troops, made up the regular U.S. Coast Artillery Corps and the U.S. Coast Artillery Reserves. These technical troops manned both the fixed and mobile seacoast artillery and controlled mines defenses. The second resources were the supporting troops of the mobile ground forces of the U.S. Army which protected the both the coast defenses and unfortified coastline from enemy landings. The second would been the local National Guard troops (formally militia), while the U.S. Navy's role in coast defense was through both offensive and defenses operations against enemy warships.

To carry out this mission, seacoast weapons were divided into classifications according to their capabilities against enemy warships. Primary armament were those weapons that could theoretically destroy the primary or capital warships of enemy naval force. Throughout most of the Modern Era primary weapons were defined as seacoast artillery of initially 8-inch and larger caliber. Controlled submarine mines were also considered part of the primary armament. The second group of seacoast weapons was the secondary armament, which were designed to counter secondary or non-capital warships, such as cruisers, destroyers, and torpedo boats.

The selection of the numbers and type of seacoast weapons was determined by such factors as the importance of the coastal area, the hydrograph profile of the approaches, the topography of area, and effectiveness of seacoast weapons in defending the coastal area. The positioning of seacoast artillery was based on the attainment of effective fire and protective factors, such as concealment, other weapons, and local defense against ground or air attacks. Attainment of effective fire refers to a position which offers the widest field of fire and greatest range over navigable water. Also considered was the need to provide coverage to all areas in which an enemy warship may operate and the placement of a suitable concentration of fire on critical areas such as harbor entrances, approaches to mine fields, and narrow portions of the channel. Consistent with these requirements, batteries were sited to provide mutual support and defense against all forms of attack. The considerations for the location of primary armament included the ability to protect friendly naval forces while entering, within or leaving the harbor, and preventing hostile naval forces from approaching within effective range of the defended coastal areas. Submarine mine fields would be placed in the seaward area of the harbor entrance and within effective range of searchlight and rapid-fire secondary armaments. Both controlled and uncontrolled submarine mines are located to prevent entry into or close approach to the harbor of enemy surface warships or submarines at all times, including during night or during conditions of heavy fog or smoke. The secondary armament would be located to provide protection for mine fields, nets, booms, and other obstacles; and the attack of hostile secondary warships engaged in raids, reconnaissance, laying of mines, and torpedo fire. Since targets of the secondary armaments were within range of visual observation and assumed to move at high speed on rapidly changing courses, these batteries were sited in direct fire positions. Protective factors in site selection included protection for the power plant, plotting room, magazines, communications, exposure of the gun crew and ammunition during the service of piece, gas protection for command post and plotting room, distances between emplacements, and concealment.

U.S. Coast Defense Armament and Equipment in the Modern Era

Few weapons of the Modern Era of coastal fortification remain today. This is the result of the advancing technology that quickly made weapons obsolete and given the economic value of high-grade steel the military sold these obsolete weapons to salvage companies. The scrapping of coast artillery material also holds true for most its supporting equipment, machinery, and instruments. As a result, today we mainly only have period images of these armaments or supporting equipment.

The development of new armament and equipment over this era usually went through cycles where the level of perceived external threats to the United States generated appropriations from Congress to allow the funding of new weapon systems. The development process for new weapons required several steps. First, was the design stage which led to the prototype and testing period and then to production and installation phase. Finally, while the weapon was in service it received modifications and improvements until it was declared obsolete. The life cycle of seacoast artillery varied from a few years to as long as fifty years.

The construction of the Endicott and Taft Programs defenses relied on the growth of heavy industry in the United States. Many of items used in coast artillery forts were invented specially for that purpose and represented the cutting edge of that technology. Early defense works relied on steam, coal, and manual energy to make things work. The use of oil and the advances in electricity brought motor driven equipment, telephones, radar, computers, and electric lighting to become key ingredients in U.S. coastal defenses.

It is also important to note that different U.S. Army branches had specialized functions that need to work together to complete a weapon system. A seacoast weapon would be designed and constructed by Ordnance Department while the emplacement was designed and constructed by the Corps of Engineers. These activities were all support by the Quartermaster Corps, Signal Corps, and so forth. The final product was then turned over to the Coast Artillery Corps for use. As you may imagine sometimes the priorities of these various organizations were not always in agreement, so delays or undesired weapons systems did occur.

A 10-inch rifle on a disappearing carriage
Battery Benson, Fort Worden, Washington (Puget Sound Coast Artillery Museum Collection)

6-inch rifles on dissapearing carriages
Possibly Battery Tolles, Fort Worden, Washington (Puget Sound Coast Artillery Museum)

A 6-inch gun on a pedestal mount, Battery Carpenter, Fort McKinley, Maine
(Joel Eastman Collection)

3-inch guns on pedestal mounts in Battery O'Rorke, Fort Barry, California (NPS, GGNRA)

For coast artillery material, the U.S. Army insured that all items were assigned a "type" and a "model". For seacoast artillery, the type for the gun or barrel refers to the size of bore (diameter) in inches while type for carriage or mount refers to the style of operation. Associated with the type is the model which refers to year of development and any subsequent modifications until 1930 when use of the year was dropped. This nomenclature extends to projectiles, fire control instruments, searchlights, submarine mines, ammunition hoists, power generators, radar, etc.

12-inch mortars in pit A, Battery Worth, Fort Pickens, Florida (Stillions Collection, NPS)

Carriage or mount types were either fixed or mobile, they allowed the guns to elevate and provide for some horizontal movement while taking up the recoil of the discharge and return the piece to the loading position. The major caliber fixed carriages were classified as Barbette (BC) carriage, which allowed the gun to remain above the parapet for loading and firing; Mortar (MC) carriage, which allowed a short-barrel gun to fire in a high arc; the Barbette long-range (BCLR) which allowed for greater firing elevations and ranges; and the Turret (TM) mount which was a barbette carriage protected by an armored housing with ammunition supplied from below. Guns of 7-inch or lesser caliber were mounted on the Pedestal (PM) mount, which had a fixed cylindrical base on which rotated a yoke that held the gun in a cradle equipped with recoil absorbing cylinders; the Anti-aircraft (AA) mount, a pedestal mount that allowed fire at high attitude. The Fixed retractable carriages included the Gun-Lift (GLC) carriage which was a BC on an elevator platform; the Disappearing (DC) carriage where the gun is raised above the parapet for firing and retracts behind the parapet for loading. The earlier smaller caliber guns had the Balanced Pillar (BPM) mounts and the Masking Parapet (MPM) mounts, which enabled the gun to be lowered below the parapet to protect it from view. Guns on mobile carriages were used in the Interwar Period as the Railway (RY) mount cars and Tractor-drawn (TD) mounts. Other temporary coast defenses made use of available weapons with a range of carriage types, primarily former naval models.

Controlled mines were anchored to the bottom of a harbor, either sitting on the bottom itself (ground mines) or floating (buoyant mines) at depths which could vary widely, from about 20 to 250 feet. These mines were fired electrically through a vast network of underwater electrical cables at each protected harbor. Mines could be set to explode on contact or be triggered by the operator, based on reports of the position of enemy ships. The networks of cables terminated on shore in concrete bunkers called mine casemates, that were usually partly buried beneath protective coverings of earth. The mine casemate housed electrical generators, batteries, control panels, and troops that were used to test the readiness of the mines and to fire them when needed. Each protected harbor also maintained a small fleet of mine planters and tenders that were used to plant the mines in precise patterns, haul them back up periodically to check their condition (or to remove them back to the shore for maintenance), and then plant them again. Each of these harbors also had onshore facilities to store the mines and the TNT used to fill them, rail systems to load

On the deck of mine planter (Stillions Collection, NPS)

and transport the mines (which often weighed over 750 lbs.) each when loaded), and to test and repair the electrical cables. Fire control structures were also built that were used first to observe the mine-planting process and fix location of each mine and second to track attacking ships, reporting when specific mines should be detonated. The preferred method of using the mines was to set them to detonate a set period of time after they had been touched or tipped, avoiding the need for observers to spot each target ship.

Key to the successful use of coast artillery was fire control and position finding as if the guns, mortars, and controlled mines failed to strike their intended targets their mission was incomplete. Early aiming efforts relied on the skill of the gunner to hit the target, but as the weapon's range increased so did the need for specialized fire control. Using geometry, optical instruments, telephones, timing interval bells, and mechanical devices a system was devised to point weapons successfully at their targets. Key equipment included the Depression Position Finder (DPF), the Azimuth Instrument (AI), Coincidence Range Finder (CRF), Plotting Board, Range Correction Board, Fire Adjustment Board, Deflection Board, Spotting Board, Range Percentage Corrector, Data Transmission Devices, Telephone Sets, and Timing Interval Bells. Many of the devices were replaced or supplemented by the development of radar (for both surveillance and fire control duties) and gun computers (combining many of plotting room devices) during the 1940 Program.

Until the advent of radar, the use of searchlights (plus star shells and airplane flares) was used to illuminate naval targets at night. Both mobile and fixed searchlights were used for both harbor defenses and anti-aircraft defense. At first 36-inch and 60-inch searchlight were used, but the 60-inch became the standard. Searchlights were located as close as possible to the water-edge to maximize their effective range of between 8,000 and 15,000 yards. Fixed searchlights were provided with a shelter to protect the searchlight from elements, to house the electrical generator, and provide concealment of the light when it was not in use. Some positions placed the searchlight on small rail cars that allowed the searchlight to move a short distance to a more exposed operating site. Searchlights were also housed in towers, pits, and even tower that "disappeared" by pivoting. After 1940 all new searchlights assigned to the defenses were the mobile type.

By 1943, two technological advances significantly changed coast artillery fire control. The most striking was the development of radar, which, as noted, could function in any weather or visibility. The use of radar greatly reduced the need for searchlights and for fire control stations as spotting enemy warships and aircraft could now be undertaken by radar units. In addition, after decades of experimentation and development, largely stymied by inadequate funding, the coast artillery adopted gun data computers, primarily for the last generation of batteries. These replaced the plotting boards and, coupled with direct-reading observation instruments, substantially automated the fire control process, reducing the human error that had always plagued the system.

Searchlight and shelter/powerhouse, Fort Flagler, Washington (Puget Sound C.A. Musuem)

A Depression Range Finder (left) and a azimuth scope (right) in a base end station
(Al Scroeder Collection)

U.S. Coast Defense Fire Control Structures in the Modern Era

The development and changes in the optical instrument fire control system from 1900 to 1945 was a long and complicated process that changed equipment, operating procedures, and designations frequently. The reader is encouraged to consult *American Seacoast Defenses: A Reference Guide* and articles in the *Coast Defense Journal* for more detail and references to U.S. Army manuals and reports. The maps included in this guide have an extensive set of symbols indicating the locations of the various fire control structures.

By 1909, each battery was under the immediate command of the officer stationed at the battery commander's station (BC). Each battery may have had one or more additional base end stations (B) with optical spotting instruments. Small caliber batteries usually had a coincidence range finder station (CRF) nearby. Mine commanders manned their posts at the mine primary (M') station. In the defended harbor areas, called the Coast Defense Command, batteries were grouped into Fire Commands, each under the overall command of the fire commander stationed at the fire command primary station (F'). The Fire Commands were then grouped together by geographical areas under the command of the officer in command of that entire sector of the coast defense. This command was initially called the Battle Command but later was changed to the Fort Command. This officer was stationed at the primary fort command station (C'). In 1925, this chain of command was changed slightly. All forts and/or groups were under the Harbor Defense command (H). Forts (F) were also used as tactical commands. Individual gun batteries were assigned to a gun group (G). Later an additional tactical organization, the groupment (C), was added below the Harbor Defense command composed of two or more groups.

In general, batteries in each harbor defense were assigned tactical number designations, generally in numerical sequence from the south (Tactical Battery #1) to the north (Tactical Battery #2, 3, 4, etc.) on the Atlantic coast; from the north to the south along the Pacific coast; and from the east to west on the Gulf coast and along the Puget Sound during the 1940s. Note that by 1940 base end stations (B) and spotting stations (S) were often combined. This is useful in deciphering the symbols for designating the fire control observation stations on the maps: $B^1_1S^1_1$, $B^2_1S^2_1$, etc. The lower number is the tactical battery number to which the station is assigned, the upper number (or "prime" mark) is the station designation number in the series of stations assigned to that tactical battery. The number of base end stations assigned to each battery ranges from a single station to as many as 14 stations. Each station had at least one azimuth scope and/or depression range finder (DPF) scope as well as connected telephone communication equipment. 3-inch small caliber batteries had one base station with a coincidence range finder (CRF) located close to the battery.

During the period 1905 to 1940 the fire control structures were generally located on existing military reservations. The location and identities of these stations can be found on the confidential blueprint maps; in the reports of completed batteries; in the reports of completed works; and in the harbor defense engineer notebooks that are part of the CDSG ePress harbor defense document collection. After 1940 the ranges for the new guns were longer and the fire control stations were more dispersed, which resulted in the acquisition of a number of new small reservations along the coastline of each active harbor defense.

Battery Plotting Room circa 1944 (NARA)

The U.S. Coast Defense Organization Before World War II

The following organization structure of the administration and tactical command of the U.S. Coast Artillery Corps is for the 1930s period. The organization prior to 1924 was on a company basis and after 1942 on a separate battalion basis and is not discussed here.

Earlier organizations had similar purposes but used different terminology. Earlier tactical structures had Artillery Districts that were divided into Battle Commands, Fire Commands, Mine Commands, and Battery Commands, each with their own commanders, while for administrative and training purposes the CAC was divided into companies which in turn were assigned to coast artillery forts or posts.

A Harbor Defense Command is a subdivision of a Defense Command, which would cover an entire region. All elements, including materiel and personnel of a Harbor Defense Command, were located at one or more coast artillery forts. These forts consisted of defined land areas within a harbor defense in which the harbor defense elements were assigned. The forts were organized primarily to provide a centralized control over administrative and technical components of the harbor defense. The materiel provided for a harbor defense may have included various types of seacoast artillery guns, anti-aircraft guns, searchlights, controlled submarine mines, underwater listening posts, radar, observation and fire control systems, and harbor patrol boats. Harbor defenses were designated by the name the harbor or coastal area which they were defending, or by the name of the largest city in their immediate area. Examples are "The Harbor Defenses of San Francisco" or the "Harbor Defenses of Chesapeake Bay."

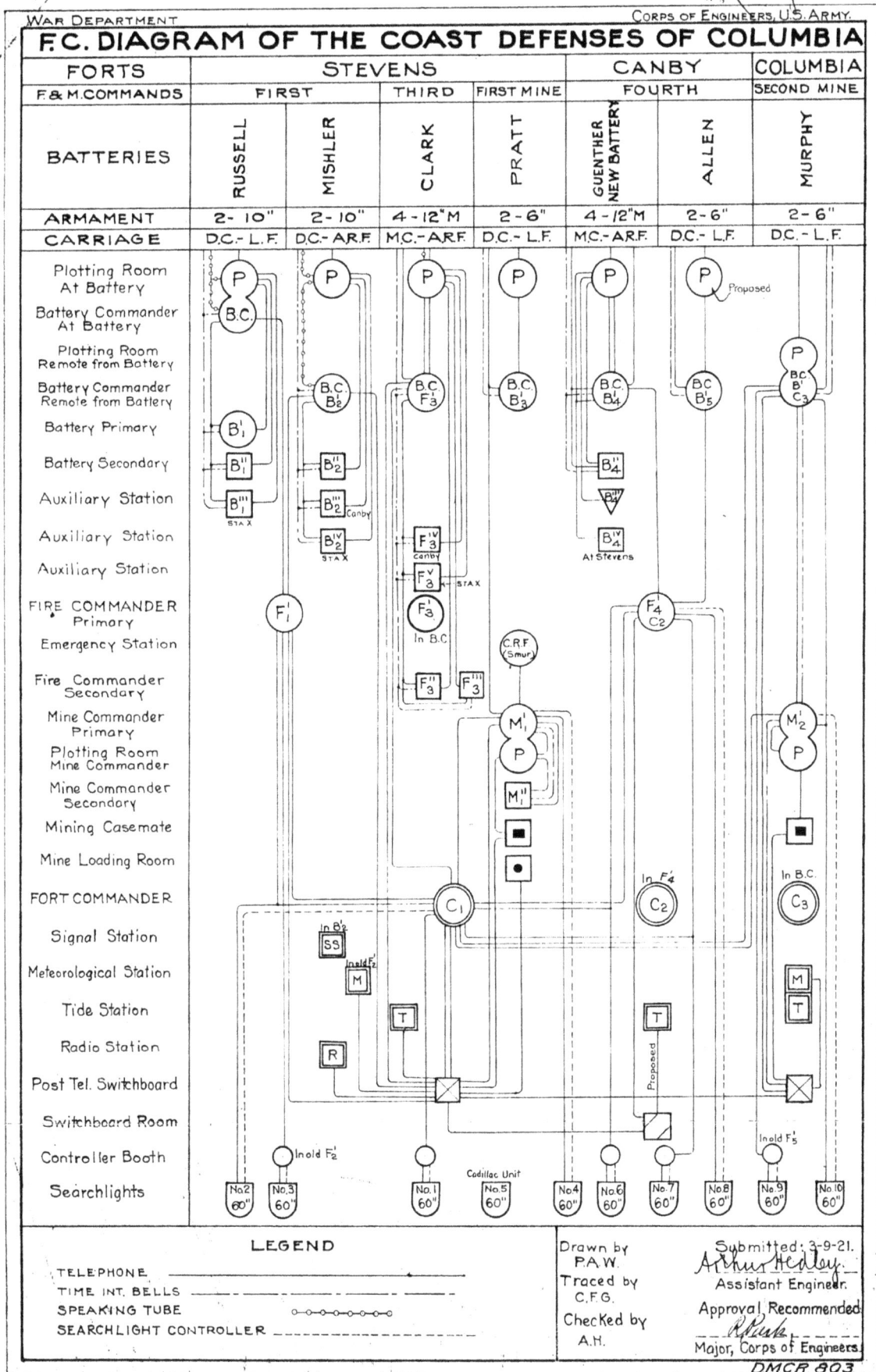

A fire control diagram showing the communications lines between the various stations.

A senior U.S. Coast Artillery Corps officer was usually designated the harbor defense commander responsible for both the administration and tactical commands. He was supported by a harbor defense headquarters staff and service units from the Quartermaster, Ordnance, Medical, Signal, Engineers, and Military Police organizations. The service units usually staff the administrative headquarters and the Coast Artillery Corps the tactical headquarters. Each fort was organized with its own headquarters and fort commander, who was responsible for the administration of the post. While the fort commander was not included in the tactical chain of command, he was responsible for the training and supervision of damage control to all the fort's structures and the activities of the service units.

The basic units of the coast defense tactical command were the battery, battalion, and group. The battery was the basic combat unit of the harbor defense and contained enough men required to man one primary battery. Batteries were classified by the type according to the material with which they were equipped. The gun battery consisted of one or more fixed or mobile guns of the same caliber and characteristics to be employed against a single target and of being commanded by a single individual. It included all structures, equipment, and personnel necessary for emplacement (or mobile weapons), the conduct of fire, and the performance of service. The strength and organization of a battery depended upon the type, number, and caliber of the guns of the battery. It was divided into a battery headquarters section, a range section (containing a battery commander's detail, an observing detail, and a range detail), a maintenance section, and a gun section for each gun or mortar. Special gun batteries were the anti-motor torpedo boat (AMTB) battery and the fixed anti-aircraft battery. The mine battery consisted of the personnel, structures, and equipment other than mine planters necessary for the installation, operation, and maintenance of all or part of the controlled mine fields. It was divided into a battery headquarters section, an operations section (containing a command post detail and range detail), a casemate section, a loading and property section (consisting of loading, cable, explosive, and maintenance details), a planting section (consisting of mine planter, distribution box boat, and small boat (yawls) details) and a maintenance section. The searchlight battery consisted of the personnel, material, equipment, and structures necessary for the operation and maintenance of seacoast and anti-aircraft searchlights.

These batteries were normally administerial combined into battalions with each battery commander reporting to the battalion commander. The battalion was organized to provide administrative, training, and tactical functions. Gun battalions were composed of from two to five-gun batteries, while a mine battalion consisted of the personnel, submarine mine material, structures and vessels necessary to plant, operate, and maintain part or all the controlled mine fields. The primary purpose of the coast defense battalion was providing effective fire direction through the coordination of various types of batteries. When a harbor defense command was large, battalions will be organized into groups. A group was a tactical command containing from two to five battalions or independent batteries. As with battalions, the primary mission for the group and the group commander was to provide effective fire direction. The use of groups occurred when the number of units is greater than can be controlled by the harbor defense commander. The basis for battalions or groups was to organize batteries that covered same field of fire or water area. When large number of batteries covered the same water area then the organization was based on target selection, such as primary and secondary armaments.

For administrative and training purposes, battalions were organized into regiments up to 1942. The garrison of a harbor defense consisted of part or all of one or more regiments, and the organization of different regiments varied to conform to the special requirements of the different harbor defenses. Generally, a coast artillery regiment assigned to fixed armament consisted of a headquarters battery, a searchlight battery, a band, and three battalions. The forts were assigned to Coast Artillery Districts. The district commander commands all coast artillery troops stationed within the territorial limits of the district, including the coast artillery units of the Organized Reserves and those of the National Guard when in the service of the U.S. At the start of World War II, the headquarters for Coast Artillery Districts were in Boston, MA (1st CAC

District), New York, NY (2nd CAC District), Fort Monroe, VA (3rd CAC District), Fort MacPherson, GA (4th CAC District), and Presidio, CA (9th CAC District). Overseas coast artillery units were assigned to local U.S. Army Departments, such as the Hawaii Department, etc.

View of Fort Flagler, Washington (D. Kirchner Collection)

A Typical U.S. Coast Artillery Fort in the Modern Era

While each coast artillery fort has its own unique design, it is possible to provide a general blueprint of the type and purpose of structures that you would find at a U.S. coast artillery fort built during the Modern Era. It is important to remember that each fort was like small self-contained city. All the services that were required to support the daily needs of its garrison and to operate the fort's weapon systems were included within the military reservation.

The reservation was typically surrounded by a fence. There was a main entrance gate with a guard house. While very few coast artillery forts had any land defenses, the use of security fencing was widespread. Recognizing this fencing is usually the first indication of a former U.S. military reservation. The main cantonment area contained a variety of buildings spread over a large area, not much different in appearance of a rural college campus. This support area was subdivided into functional sections surrounding by a large parade ground area. While the overall fort was under the Coast Artillery Corps, each of the support services (Quartermaster, Engineer, Medical, etc.) had their own buildings or reservations within the fort.

The main parade ground is the focal point of the post. The fort headquarters, officer's quarters, non-commission officer housing, service clubs, and enlisted barracks usually surround it. Most of the non-tactical structures at the forts constructed during the Endicott-Taft Programs were designed to be permanent structures. These wood-frame buildings were built on stone foundations with slate roofs, sided with local brick, clapboard, or stucco. The Quartermaster Corps architect's office created standard plans for all types of buildings. Those designed at the turn-of-the-century—when most Coast Artillery forts were constructed—were of Colonial Revival style with elements of Queen Anne style in the officers' quarters. As the century progressed, new styles were adopted, such as Italianate and Spanish Revival, and these styles were used when additional buildings were constructed. Store houses and pumping plants used more practical industrial or utilitarian styles.

Officer's quarters varied in size and elaborateness depending upon the rank of officer for whom the building was intended. The Commanding Officer's Quarters was usually the largest and most elaborate of the officer's quarters, and it was placed, if possible, on the highest and most prominent location on the parade ground. Other senior officers were assigned single quarters, while many of the quarters were double quarters for two families. Large forts had a Bachelor Officer's Quarters with its own mess. Non-Commissioned Officer's quarters were usually double sets.

Parade ground and officer's quarters, Fort Casewell (BW Smith Collection)

The interiors of buildings were finished with wood floors, plaster walls with wood trim, and pressed metal ceilings. All structures where officers and men lived or worked had electricity, running water and flush toilets. Each barracks was designed to house a company or battery of 100 men and was self-contained with its own kitchen, dining room, day room, barber shop, and tailor shop. Sleeping quarters were on the second floor, while the lavatory and latrine were located in the basement in northern climates. In the south, separate lavatory and latrine buildings were sometimes built. Large forts had double barracks–two 100-man barracks-built end-to-end–which functioned as two separate barracks. Forts which served as the headquarters post for a harbor defense usually had a band barracks.

Although the parade ground was used as a general athletic field, tennis and handball courts, and baseball fields were also built in open areas of the fort. A system of permanent roads served the entire fort, and the streets were usually named. Railroads and tramways were built during the construction of the forts, and these lines often continued to be used. These forts eventually had their own water, sewer, telephone, and electrical systems. If municipal water and commercial power services were available, the army used them, but at many sites the engineers built their own water and electrical plants and distribution systems. Sewer pipes ran into the ocean. Ice houses, and in northern areas, ice ponds, were also built to provide refrigeration for food in the years before electrical cooling became available. Systems for the disposal of garbage and rubbish were also created. Garbage and combustible waste were burned in crematoria, while non-combustible materials were disposed of in landfills or dumped into the ocean. The major fuel at forts was coal, and a system of unloading, transporting, and storing the fuel was developed, usually relying on mule-drawn wagons.

A large portion of fort's reservation would be devoted to the Quartermaster Corps. The Quartermaster was tasked with providing housing, supplies, and transportation for all the troops assigned to the fort. The Quartermaster oversaw the construction of most of fort's support buildings, as well as the installation of its own quartermaster wharf and tramway to transport supplies within the fort. Storehouses, commissary, workshops, and stables were usually centered near the quartermaster wharf.

Fort Terry buildings (BW Smith Collection)

The Corps of Engineers were responsible for construction the actual fortifications known as the tactical structures (emplacements, fire control stations, casemates, power houses, etc.); the Ordnance Department provided the weapons, machinery, and instruments that went into these structures; and the Signal Corps provided the technical equipment as new technology was developed. Near the shoreline were located the fire control stations along with protected telephone exchanges, command posts, meteorological stations, seacoast searchlights positions, and reserve magazines that support the fort's weapons systems.

The fort's two main coast defense weapon systems were controlled mine fields and seacoast artillery. Controlled mine fields required an extensive infrastructure within the fort. Principal structures for the mine defense included the mining casemates from which the mines were operated; the conduits connecting the casemates with the shore; the cable terminals on the shore; the cable tanks in which the mine cables were stored when not in use; the mine storehouses in which were kept the mine cases; the loading rooms in which the mines were loaded; the magazines in which the dynamite was stored; the range stations, plotting rooms, and dormitories, the mine wharves at which the mine planter used to land and receive the loaded mines; and the tramway connecting the wharves with the cable tanks, storehouses and loading rooms.

Closer to the shoreline are the emplacements of the fort's other main weapon system the large caliber gun batteries. These gun batteries consisted of both of large caliber breach loading rifles mounted on disappearing or barbette carriages and smaller rapid-fire guns on pedestal mounts during the Endicott and Taft Programs. The purpose of these gun emplacements was to provide a stable base for these guns and carriage and a convenient platform for the personnel serving the gun. The emplacement also designed to provide the armament and the personnel the maximum protection as possible, as well as providing a safe storage place of the ammunition. These thick concrete structures were covered with earthen fill on the seaward side, while they were partially buried, these batteries are easily accessible from the rear due their open back design.

The gun emplacements from the Interwar Period and 1940 Program are quite different from these earlier designs as American coast artillery responded to the progression of larger caliber naval weapons with longer firing ranges and the advent of military aviation with aerial bombs. These emplacements are usually one-story high but usually completely buried. The gun position consists of a gun well surrounded by a circular concrete pavement. Later many of these emplacements were completely rebuilt with thick reinforced

Fort Mott 1936 (NARA)

concrete casemates to protect their weapons from aerial attack and naval bombardment.

Another primary seacoast weapon of the Endicott and Taft Programs was the seacoast mortar, actually a short barrel breech loading rifle. These batteries by definition did not require direct fire, so they were often located away from the shoreline. They were located within or behind the fort's cantonment area. A typical mortar battery had a high reinforced concrete parapet with traverses that formed a series of pits. These pits were usually open to the rear, but early designs were completely surrounded with access through a tunnel. A battery had one to four pits with two or four mortars in each. Between the pits or around their sidewall were ammunition magazines, power generator rooms, shot truck areas, storerooms, and a plotting room.

The secondary gun batteries mounted rapid fire guns for the defense of the controlled mine fields from minesweepers and to repulse fast moving naval vessels and were installed after the first round of primary gun batteries. These are simple emplacements that basically provided a stable firing platform for the weapons and a protected magazine for their ammunition. Also located around some forts were groups of three or four concrete gun blocks for anti-aircraft guns that were added in later years.

A key feature of all coast artillery forts are the fire control stations which provided the target information for the mine and gun defenses. These stations come in all shapes and size. They range from a single below-grade room with observing slots to large multi-level, multi-room towers. Constructed of both wood and concrete, these stations have been disguised as non-military structures ranging from summer cottages to grain silos. Associated with World War II fire control stations were radar stations that by 1944 replaced their function. These radar stations had antennas which were mainly located on steel towers but could be mount on other structures. These antennas sent their signals to operating rooms where measurements provide location data to plotting rooms. Support these stations were power rooms and dorms for troops manning the stations.

Several to many of the structures at most of remaining U.S. coast artillery forts. However, you may only view piles of rubble and mounds of dirt as their status and condition are constantly changing. Nearly all the seacoast armament and equipment were scrapped after World War II which accounts for the lack of actual coast artillery at the forts.

Garrison Life at a U.S. Coast Artillery Fort in the Modern Era

The soldiers assigned to the defenses experienced a great change in quality of life during the years from 1890 to 1950. The early years were certainly the roughest. In general military service in the U.S. armed forces was not well compensated or widely respected in some quarters. As the permanent posts were being established, physical living conditions were sometimes poor, and relationship with the local, civilian community at times strained. Officers could afford higher standards of living for themselves and their families as well as greater social involvement with the local community.

63rd Coast Artillery Company on parade gound at Fort Worden, Washington in 1908
(Puget Sound C.A. Museum)

Soldiers in barracks (Stillions Collection, NPS GSNS)

By the end of the early modernization programs in the 1910s, the living and work conditions had greatly improved. In particular the Coast Artillery was an elite assignment, with considerable prestige. The Coast Artillery Corps was relatively well funded and equipped, had a strong technical and professional dedicated career officer contingent, and was based on teamwork activity that encouraged close camaraderie. Opportunities for duty at oversea bases in exotic tropical locations like Hawaii, the Philippines, and the Panama Canal had its advantages, especially as many of the tropical diseases had been conquered. Training was emphasized, but in all the workload was reasonable. Pay was not extravagant for the enlisted man– but decent food, recreation, and athletic events were provided on post. Soldiers tended to stay in this branch of service, often re-enlisting, and were quite good at what they were taught and with what equipment they practiced on.

The daily schedule of the Coast Artillery troops focused on drill, inspections, maintenance, meals, and recreation. The center of activity for enlisted men was their barracks which was designed to house one or two companies or batteries of 100 men each with its own kitchen, lavatories, dining room, day (recreation) room, barber shop, and tailor shop. The barrack along usually arrayed around a parade ground. The officer quarters were usually located on the opposite side of the parade ground. The day would begin with meal, roll call, and assignment of duties. This usually was training/drill in the mornings with maintenance tasks or recreation events in the afternoons. Recreation was considered important by the U.S. Army after 1900 as it was believed that it not only maintained physical fitness but promoted competitiveness which made the men more effective in combat. Most large forts were provided with a gymnasium and bowling alley, as well as athletic fields, handball and tennis courts. Other recreation activities included visiting the post theater, service clubs, libraries, chapel, and the Post Exchange, as well as leave to visit the local cities and towns. Another aspect of garrison life were weekly inspections and parades, and soldiers who failed these inspections would end up spending their weekend cleaning barracks and latrines, rather than having a weekend pass to visit the local communities.

Mess hall set for Christmas Dinner 1911, 126th Coast Artillery Company, Fort Worden Washington
(Puget Sound CA Museum)

MODERN ERA SEACOAST FORTS TODAY

In 1950, the remaining harbor defense commands were disbanded, and the U.S. Coast Artillery Corps was abolished as a separate branch with its remaining units, all anti-aircraft artillery, moved into the Field Artillery. Meanwhile, the responsibility for limited harbor defense, primarily underwater defenses, was transferred to the U.S. Navy. The U.S. Army retained several of the old coast artillery forts for other missions, while the Navy acquired several reservations for thier use including for its new role in harbor defense. Other federal agencies had an opportunity to claim all or portions of the former coast artillery sites. Those not transferred were turned over for disposition to the U.S. General Services Administration (GSA), who offered them to state, county, and local governments, and finally to private citizens. Many of the smaller, independent plots of land which had been leased or purchased for fire control and searchlight positions were returned to original owners or sold to private owners, before selling or transferring these former forts, the U.S. Army either returned to its depots all usable equipment or auctioned items in lots to the public.

Several coast defense sites had been abandoned by the U.S. Army as active defenses by 1928 including those at the Mississippi River, Mobile Bay, Tampa Bay, Savannah, Port Royal Sound, Cape Fear River, Baltimore, and the Potomac River, and the smaller inner harbor defenses at East New York, San Francisco, and Puget Sound. Many of these reservations were reclaimed for use during World War II. The next large-scale transfer of harbor defense properties from the U.S. Army began in 1947 and continued through the mid-1950s. In the early 1970s a general series of military base closures occurred throughout the U.S. Department of Defense to reduce basing costs. Several large former harbor defense sites, including military reservations around San Francisco, New York, and Pensacola, were included. Given the large size and value of these properties, Congress passed several laws that directed the ownership of these former forts to be transferred to the U.S. National Park Service (NPS). Base closure commissions in the 1990s, 2000s, and 2010s recommended the closure and transfer of other former harbor defense sites, which included the Presidio of San Francisco, Fort Wadsworth on Staten Island, Fort Monroe in Hampton, and Fort Trumbull in New London. Only a handful of old coast defense reservations remain in military hands in 2025 — Fort Story, VA, Fort Hamilton, NY, a large part of Fort Rosecrans, CA, Fort Kamehameha, HI, Fort Hase, HI and a few other sites. Other national agencies, state agencies, and local governments acquired numerous coast artillery sites for parks and recreation areas, since they inevitably had scenic river or ocean views. Depending on how diligently the GSA protected the sites, and the length of time it took to dispose of them, some sites and structures survived in excellent condition, while others suffered at the hands of salvagers and vandals.

While many of the Modern Era forts and batteries are now located within parks, they have not been accorded the same level of protection or care as the remaining brick and stone forts. Most of the old coast defenses structures are considered to be, at the worst, a legal liability or at best, an eyesore to the park. Remaining structures have been built on, fenced in, buried, or destroyed. They have been removed as interfering with the park's primary mission of providing recreation space. Vandalism has caused considerable damage over the years. Abandoned and neglected coast defense structures have suffered from freeze-thaw cycles cracking and spalling the concrete and brick, rusting metal rebar and materials has hastened deterioration. Unchecked vegetation growth has caused some structures to collapse. And rising sea levels and increasingly violent storm surges are eroding away shoreline and destroying major structures. While most gun emplacements have been constructed in such a way to resist these attacks, many other tactical structures have collapsed, and even brick structures have been damaged or destroyed by vandals and neglect. Non-tactical structures, particularly officers' quarters, have survived at many parks and government-owned sites through adaptive reuse, but at some former posts such structures have been completely removed.

However, public interest in the history of American coast defenses has grown since the publication of *Seacoast Fortifications of the United States: An Introductory History*, by E.R. Lewis in 1970. The book publication was a pivotal event, giving the public and park personnel a well-documented interpretive history of American coast defenses. A group of coast defense history enthusiasts gathered at a meeting in 1978 and organized the Coast Defense Study Group (CDSG) in 1985. The CDSG's annual conferences, Journal, Newsletter, web site, and reprints of key coast defense books have played important roles in fostering interest in the history of American coast defense and assisting both the public and park staffs in understanding the fascinating history of these defenses and to interpret their surviving elements. These massive seacoast batteries have been able to withstand both the natural climate and economic development longer than other military features from the same periods. These structures incorporated the leading edge of technology of their time and that draws interest in studying them and interpreting purpose and history. Hopefully this will translate into efforts to preserve and restore these sites for current and future generations.

Battery Winchester, Fort Armistead, Baltimore, Maryland (Terry McGovern)

PERIOD MILITARY MAPS

This book contains a compilation of maps for all the harbor defense reservations utilized during the period 1900 to 1950. The harbor defense projects show a general map of the location of elements (sites) for each harbor and the individual site maps showing the fire control elements. A series of map symbol and abbreviation keys from 1921 and 1945 are included.

The directory is organized by Harbor Defense around the United States clockwise from Portland Maine, down the Atlantic coast to Key West Florida, across to Gulf coast to Galveston Texas, then up the Pacific Coast from San Diego California to the Puget Sound in Washington, then up to the Alaskan defenses of WWII, and followed by the defenses in Hawaii, the Philippines, Panama, the Caribbean, Bermuda, and Newfoundland, Canada. While the status information is fairly comprehensive of the larger fort and military reservations, the status of many of smaller WWII-era fire control stations is not. The authors would appreciate receiving any updated information to correct or add to what has been presented here.

Notes on Coast Defense Maps

Site maps; site plans; exhibits from project plans, supplements and annexes; confidential blueprints; D-series maps—these are all terms that have been used to describe various maps which depict sites used by the U.S. Army, at one time or another, in connection with harbor defense fortifications and fire control. These maps have been keys to ferreting out the identification of the various remaining structures during site visits, yet there is some confusion over where these maps come from, what their cryptic symbols mean, and even what they are called.

Most maps of harbor defense installations are located in the Cartographic Branch of the National Archives. Many of the more frequently seen maps have come from a variety of National Archives holdings. The two concrete-era (1890-1950) map formats most frequently seen are the Confidential Blueprint map series (1900-1935 and 1940-1948) and the exhibits from the annexes/supplements to the harbor defense projects (1940s), which cover the 1940 Modernization Program (WWII-era) construction.

Confidential Blueprint Series Maps (1915-37)

As new construction finished, maps were created, revised, and updated by the Corps of Engineers. A series of maps was reproduced as negatives from a master positive in blueprint style, which meant maps were composed of white lines on a blue or dark background. As they were classified "confidential" by the War Department, they became known as "confidential blueprints."

A number of these confidential blueprints have been found in various cartographic and textual Corps of Engineers records in the National Archives. The confidential blueprint series of maps have general maps of each defended harbor, and general maps of each of the forts and military reservations in the harbor defense. If it was warranted, larger scale maps of parts of some forts were also included. These were labeled "D" for "detail" and followed in series, D-1, D-2, D-3, etc., as required. These maps show the location of batteries, various components of the fire control and communication system, mine facilities, and all the post buildings. Identification of each structure was shown by name, symbol, abbreviation, or number.

After 1900 an optical system for fire control based on trigonometric principles was developed for more precisely aiming coast artillery guns. The structures that were built to house the optical and communication elements of this system were often numerous and small in relation to the other major buildings on a military reservation, and many required a detailed description making it complicated to label them on a map, so a set of map symbols was developed to indicate the fire control structures. As these fire control structures were built in the years following 1905, they were incorporated into the maps on which the

Corps of Engineers recorded the location of all the structural elements of the fortifications in the seacoast defenses.

Keys to the fire control map symbols began appearing in coast artillery manuals, such as drill regulations, training regulations, and later field manuals. A complete update of these maps was performed during the years 1920-1922, just after the major construction projects of the Endicott and Taft programs were completed and before some of the smaller harbor defense areas were eliminated. These maps were kept as part of the records of the various Corps of Engineer district offices around the country. Copies were turned over to qualified parties in the army, such as the Coast Artillery Corps, the Quartermaster Corps, etc. On July 12, 1922, the Coast Artillery Board at Fort Monroe requested a complete list and set of these maps for their records, which were provided in August 1922. The 1922 collection contains about 290 maps of 29 harbor areas. Other versions of these maps were found in the notebooks kept by the engineer assigned to each harbor defense. In due course, the records of the Corps of Engineers and the other branches of the army have been turned over to National Archives. The map collections have been scanned and digitally "cleaned up" to remove extraneous lines and smirches from the scanning process.

WW II-Era Harbor Defense Project Maps

The 1940 "Modernization Program" brought a new set of harbor defenses, some on existing reservations, some on entirely new reservations. The fire control system was much more widespread and frequently located on newly obtained smaller reservations located around the harbor defense shoreline. Maps for these works in this guide come from the 1944-46 supplements to the various harbor defense projects published by the army.

A Harbor Defense Project was a written document which described all existing and projected harbor defense elements, including structures, first prepared in 1932-33. Supplements to the Harbor Defense Projects were prepared 1943-44 and updated during 1945-46. The supplements detailed the progress on the construction of the new 1940s modernization program defenses with descriptions and a set of maps that showed where these new structures were located, the field of fire of the guns, radar coverage, etc. The supplements provide extensive detailed information on all tactical and physical aspects of the harbor defenses on the date of the annex, both existing and proposed, and a number of exhibits detailing the locations of elements. The supplements are generally composed of 7 annexes:

A- Armament
B- Fire Control (including optical instruments and radar installations)
C- Seacoast Searchlights
D- Underwater Defenses (mines)
E- Antiaircraft Artillery
F- Gas Defense
G- Equipment (usually detailing what was on hand and what was needed)
H- Real Estate Requirements (usually detailing sites not yet obtained

These supplements and other forms of the Harbor Defense Projects have been scanned from the National Archives and are available from the CDSG ePress as electronic PDFs. These supplements contain a very comprehensive listings and exhibits of everything that was to be in place at the completion of new rearmament program and are the key references to consult for information on the final state of the American seacoast fortifications in 1945.

A few comments on the items that appear on the confidential blueprint maps and the Harbor Defense Project maps—

A **Harbor Defense** (called a "Coast Defense" before 1925) consisted of a series of land reserves (named as "Forts" and in some cases "Camps" and "Military Reservations") on which the various components of the seacoast defense fortifications were built to guard a major commercial and naval seaport. When the harbor defenses of the United States were modernized in 1890-1910, a new system of defensive works were created. The modern forts consisted of tactical and non-tactical structures spread over hundreds of acres of land. The U.S. Army Corps of Engineers selected the locations, purchased additional land, sited, designed, and constructed the tactical structures—gun batteries, mine facilities, observation stations, plotting rooms, power plants, switchboard rooms, and searchlight shelters.

Gun Batteries: The modern seacoast artillery consisted of guns, mortars and antiaircraft weapons mounted in concrete support structures varying from the simple to the quite complex. Guns were mounted on barbette, pedestal and "disappearing" carriages. Mortars were emplaced in protected pits. Antiaircraft weapons, usually the 3-inch guns, were mounted in simple concrete platforms. The term "battery" was used to describe a set of guns under a single commander together with the entire structure erected for the emplacement, protection, and service of those guns.

Fire Control Structures: The target range and azimuth for seacoast artillery guns were determined using command and equipment systems collectively referred to as fire control and position finding. The standard systems of position finding used by seacoast artillery were based on trigonometry. Components of the system included widely spaced base end stations, command stations, plotting rooms, tide stations, meteorological stations, and cable linked telephone communication systems with protected switchboards. Radar installations were deployed for the major gun batteries and as general surveillance after 1942. The radar installations included power/control buildings and antenna towers.

Searchlights: Most searchlights installed during the period 1901-1920 were fixed, located in a structure for concealment and protection during the day, with their electrical power generator. Over the years after WWI, the mobile searchlights became more reliable, durable, and rugged. By the late 1930s, the Coast Artillery switched to using mobile searchlights and replaced fixed searchlights where at all possible so after 1940 the US seacoast defenses used mostly mobile searchlights.

Controlled Mine Facilities: Throughout the modern or "concrete" era of American harbor defenses (1890-1950), mines were considered to be one of the primary harbor defense weapons. Mines were only deployed during times of war or during limited training expertises. The mines and cables were stored ashore between use. The mine shore facilities included torpedo storerooms, loading rooms, mine wharfs, explosives storage, tramway systems, cable tanks, mine casemates, and cable vaults.

Electrical Generator Power Plants: By the turn of the century, electricity had become a vital necessity for the Coast Artillery. It was used to traverse and elevate some of the large guns, to light emplacements, to operate ammunition hoists, to power searchlights, to control submarine mines, and for communications, in addition to standard garrison uses. Most large forts had a central power plant with electrical generators. The requirement that coast defenses be self-contained resulted in power rooms being included in most batteries and mining casemates, and separate searchlight powerhouses were constructed.

Protected Switchboard Rooms: As seacoast defense artillery covered increasing distances, a need for remote accurate and instant communications was required. Telephones connected by phone lines were integrated into the fire control system utilizing protected switchboard rooms after 1906. As radio communication developed in the 1930s, fixed radio sets were often integrated with the telephone communication system in their protected switchboard rooms or housed in separate protected structures.

Garrison Buildings: These are shown in the Confidential Blueprint series maps but not on Supplement series maps.

The system of numbering for buildings was the same for all Confidential Blueprint maps in the period 1915 to 1937. All buildings of the same "type" were given the same number on all the maps. For example all barracks buildings were numbered "7."

1.	Administration Building
2.	Commanding Officer's Quarters
3.	Officer's Quarters
4.	Hospital
5.	Hospital Steward's Quarters
6.	Non-commissioned Officer's Quarters
7.	Barracks
8.	Guard House
9.	Post Exchange

10 to 19 and 100 to 199	Post Buildings
20 to 29 and 200 to 299	Quartermaster Buildings
30 to 39 and 300 to 399	Ordnance Buildings
40 to 49 and 400 to 499	Engineer Department Buildings
50 to 59 and 500 to 599	Signal Corps Buildings
60 to 69 and 600 to 699	Reserved for future requirements
70 to 79 and 700 to 799	Religious and Social Buildings
80 to 89 and 800 to 899	Government Buildings not under War Dept. Control
90 to 99 and 900 to 999	All Private Buildings (Private dwellings, stores, contractor's buildings and buildings purchased with the land but not assigned to public use.)

Fort Columbia, Washington 1913 (NARA)
From left to right is the Post Exchange, a Company Barracks, the Administration Building,
a Double Officer's Quarters and the Commanding Officer's Quarters.
Just visible behind the front row of buildings is the Quartermaster's Storehouse and the Post Hospital

Symbols and Abbreviations—1921 Confidential Blueprints

Name	Abbr.	symbol	Sta. w/o roof
Fort Commander's Station	C	ⓒ	(C)
Primary Station, Fire Command	F'	(F')	(F')
Secondary Station, Fire Command	F''	[F'']	(F'')
Supplementary Station, Fire Command	F'''	[F''']	(F''')
Primary Station of a Battery	B'	(B')	(B')
Secondary Station of Battery	B''	[B'']	(B'')
Supplementary Station of a Battery	B'''	[B''']	(B''')
Battery Commander's Station	BC	(B.C.)	(B.C.)
Primary Station, Mine Command	M'	(M')	(M')
Secondary Station, Mine Command	M''	[M'']	(M'')
Supplementary Station, Mine Command	M'''	[M''']	(M''')
Double Primary Station, Mine Command	M'-M'	(M'+M')	(M'+M')
Double Secondary Station Station, Mine Command	M''-M''	[M''+M'']	(M''+M'')
Separate Plotting Room	P	(P.)	
Separate Observing Room	O	(O.)	(O.)
Self-contained Horizontal Base	C.R.F.	(C.R.F.)	(C.R.F.)
Emergency Station	E	(E.)	(E.)
Spotting Station	Sp	(S.)	(S.)
Meteorological Station	Met	[M.]	
Tide Station	T	[T.]	
Searchlight (30, 60, etc., relates to the size of the lights)	S	(N° 60)	(N° 36)
Controller Booth	C.B.	●	
Watchers Booth	W	⊕	
Signal Station	S.S.	[SS]	
Radio Station	R	[R]	
Cable Terminal	C.Ter.	⊟	
Post Telephone Switchboard	P.S.B.	⊠	
Mining Casemate	M.C.	▣	

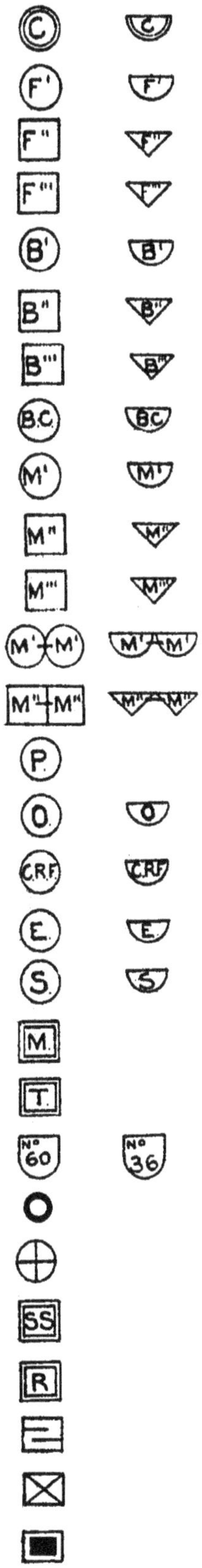

Name	Abbr.	symbol
Loading Room	L.R.	
Switchboard Room	S.W.B.	
Central Powerhouse	C.P.H.	
Powerhouse (and Searchlight Powerhouse)	P.H.	
Combined Stations, in same room		
Combined Stations, in communicating rooms		
Combined C and F' Station in same room		
Differentiation of auxiliary plants		

Abbreviations used on maps

Cable Gallery	C.Gal.
Cable Tank	C.T.
Cable Hut (commercial cable)	C.H.
Coast Guard Station	C.G.S.
Engineer Wharf	Engr. Whf.
Gasoline Tank	G.Tk.
Guard House	G.H.
Latrine	L.
Lighthouse	L.H.
Lighthouse Wharf	L.H.Whf.
Magazine	Mg.
Mining Boathouse	M.B.H.
Mining Derrick	M.D.
Mining Tramway	M.T.
Ordnance Machine Shop	O.M.S.
Mine Wharf	M.Whf.
Private Wharf	Pvt.Whf.
Radio (commercial station)	Rad.
Railway Wharf	Ry.Whf.
Saluting Battery	Sl.B.
Searchlight Shelter	S.Sh.
Service Dynamite Room	S.D.R.
Steamship Wharf	S.S.Whf.
Sunset Gun	S.G.
Tide Gauge (not a Tide Station)	T.G.
Torpedo Storehouse	T.S.
Tower	Tw.
Water Tank	W.Tk.
Weather Bureau	W.B.

Additional Symbols and Abbreviations
Name   Abbr.   symbol

Pumping Plant	P.P.	
Radio Powerhouse	R.P.H.	
Searchlight Powerhouse	S.P.H.	
60 inch Searchlight No. 7	S.$^{60}_{7}$	
Coincidence Rangefinder	C.R.F.	
Quartermaster Wharf	Q.M.Whf.	

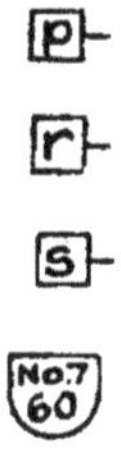

Subscripts for use in both Legend and on Face of Plat are—
  Imp.   Improvised.   B" imp.
          (for temporary fire control structures only.)
  p.   Portable.   S$^{36}_{p2}$
          (Principally used for portable searchlights etc.)
  s.   Superseded.   24s.
          (for abandoned buildings, etc.)
  t.   Temporary.   19t.
          (For all uses except fire control structures.)

Datum Point—location indicated by intersection of lines
or by dot at end of arrow.

Triangulation Station.

Intersection Point.

Benchmark.

Lighthouse.

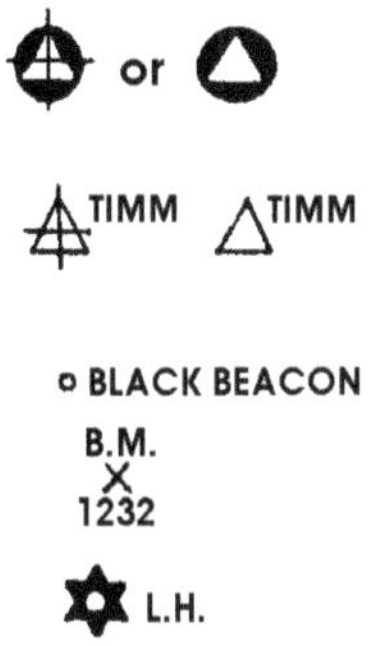

Such other topographic signs as were necessary were taken from the *Engineer Field Manual* (Professional Papers, Corps of Engineers, No. 29) pages 74 to 97.

Note: Maneuver buildings were classed as post buildings.

SYMBOLS and ABBREVIATIONS 1940
FM 4-155, Reference Data (Seacoast Artillery and Antiaircraft Artillery) 1940
TABLE C.-Symbols for seacoast artillery fire-control maps, diagrams, and structures

Part 1.—Basic symbols

Name	Abbreviation	Symbol
Harbor defense command post	H D C P	 (H)
Groupment command post	Gpmt C P	 (C)
Fort command post	Ft C P	 (F)
Gun group command post	G C P	 (G)
Mine group command post	M C P	 (M)
Seacoast battery command post	B C P	 (BC)
Harbor defense observation station	H D O P	 △(H)
Groupment observation station	Gpmt O P	 △(C)
Fort observation station	Ft O P	 △(F)
Gun group observation station	G O P	 △(G)
Mine group observation station	M O P	 △(M)
Battery observation station	B O P	 △(B)
Emergency observation station	E O P	 △(E)
Antiaircraft observation post	A A O P	 AA △
Battery spotting station	S O P	 △(S)
Separate observation station	O P	 △(O)

Name	Abbreviation	Symbol
Operations and plotting room	O P R	
Plotting room	P	
Self-contained base range-finder station	R F	
Magazine	Mg	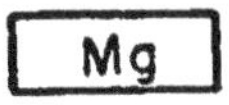
Shellroom	S Rm	
Temporary or improvised fire-control structures	Imp	
Mine casemate	M C	
Mine loading room	L R	
Searchlight, 60-inch seacoast	S L	
Searchlight, seacoast, other than 60-inch	S L	
Antiaircraft searchlight	A A S L	
Searchlight shelter	S Sh	
Searchlight powerhouse	S P H	
Searchlight controller booth	C B	
Data booth	Data B	
Watchers booth	W Bth	
Meteorological station	M E T	

Name	Abbreviation	Symbol
Tide station	Td	 T
Signal station	S S	 SS
Fire Control switchboard room	F S B	
Post telephone switchboard room	P S B	
Combined fire-control & post telephone S B room	F S B P S B	
Cable terminal	C Ter	
Powerhouse	P H	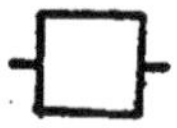
Radio powerhouse	R P H	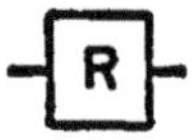 R
Central powerhouse	C P H	 O
Pumping plant	P P	 P
Datum point		 ● OR
Triangulation station		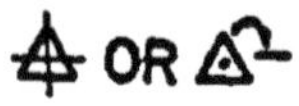
Intersection point		 O Black Beacon
Benchmark	B M	 BM ✕ 1232
Lighthouse	L H	 ★

Other abbreviations used in this guide

BS - base end station & spotting station

HECP - harbor entrance command post

HDOP - harbor defense command observation post

HDCP- harbor defense command post

SBR -telephone system switchboard or radio room

AMTB- Anti-motor torpedo boat BC station

BC - battery commander's station

C - fort commanders station

G- group command station

M- mine station

SCR - signal corps radar

SL - searchlight

Part 2.-Numbers for harbor defense installations.—a. In harbor defense, seacoast artillery installations of each type are numbered consecutively from right to left, facing the center of the field of fire of the harbor defense. Antiaircraft installations pertaining to the harbor defense may be numbered in any convenient sequence.

b. Groupments, gun groups, mine groups, batteries, and all installations functioning directly under the harbor defense commander, such as harbor defense observation stations, searchlights, and underwater listening posts, are numbered consecutively, each type in a separate series, beginning with number 1. These numbers normally are shown as subscripts to the letter included in the appropriate symbol. Exceptions are included among the examples that follow.

Name	Abbreviation	Symbol
Harbor defense observation station	$H D O P_3$	
Fort observation station	$Ft O P_3$	
Antiaircraft observation post	$A A O P 2$	
Magazine or shell room	$Mg 2$ or $S Rm 2$	

c.Groupment, group, and battery observation and spotting stations assigned to a unit are numbered consecutively within the unit, each type in a separate series, beginning with number 1. These numbers are shown as superscripts to the letter included in the appropriate symbol, the unit number remaining as the subscript.

Name	Abbreviation	Symbol
Groupment observation station	$Gpmt_2 O P_2$	
Gun group observation station	$G_2 O P_1$	
Mine group observation station	$M_2 O P_1$	
Battery observation station	$B^1_1 O P$	
Spotting station	$S^1_3 O P$	
Emergency observation station	$E_2{}^1 O P$	
Temporary or improvised fire control structures	$B_3{}^2 Imp.$	

d. In certain cases it is desirable to show additional information regarding an installation, such as its size and whether fixed, portable, or mobile. Such information is placed either in the symbol or to the right thereof.

Name	Abbreviation	Symbol
60-inch seacoast searchlight; fixed, portable or mobile.	SL 2F (P or M)	2F(P or M)
Seacoast searchlight other than 60-inch	SL$^{36}_{3P}$	$\frac{3}{P}$ 36'
Antiaircraft gun battery or composite battery, fixed or mobile.	A A No. 2 (F or M)	AA 2 (F or M)

e. Where two stations are combined in one room, the symbols are superimposed one upon the other, and the letters representing each station are inclosed in the combined symbol.

Name	Abbreviation	Symbol
Combined groupment command post and fort command post.	Gpmt Ft Cp	CF
Combined battery observation and spotting station.	$B^2_1 S^2_1$ O P	$B^2_1 S^2_1$
Combined group command post and battery command post.	$G_1 B_2 C$ P	G_1 BC_2
Combined battery command post and battery observation station.	$B_2 C$ P $B^2_2 O$ P	B^2_2 BC_2

f. Where stations are adjacent in the same structure, the symbols are tangent to each other and are arranged to show the relative location, as:

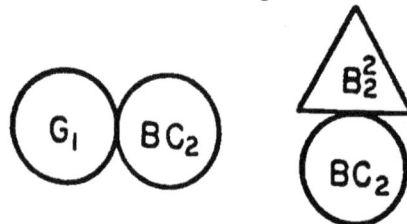

g. Where communication may be had by voice through a passage, door, window, or voice tube, the symbols are left open at the point of contact, as:

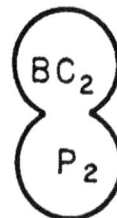

Part 3.—Communications symbols for use on harbor defense fire-control charts and diagrams.

Telephone cable (numerals indicate number of pairs and gage)	26-19
Speaking tube	
Mechanical data transmission line	
Electrical data transmission line	x x x
Searchlight controller line	
Zone signal and magazine telephone line	
Firing signal line	
Time interval bell line	
Submarine cable (numerals indicate number of pairs and gage)	50-19

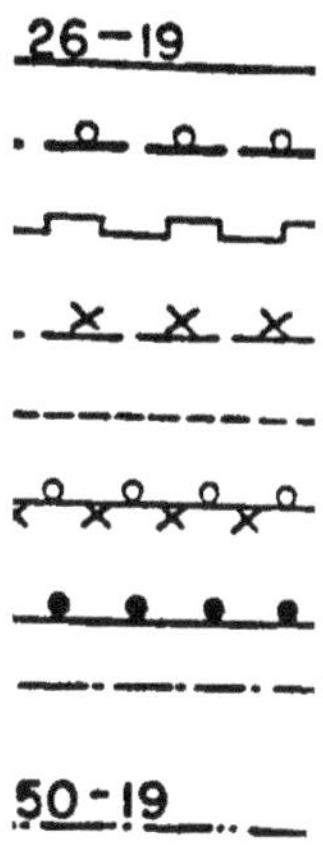

Part 4.-Abbreviations

Cable gallery	C Gal
Cable tank	C T
Cable hut (commercial cable)	C H
Coast Guard station	C G S
Engineer wharf	Engr Whf
Gasoline tank	G Tk
Guardhouse	G H
Latrine	L
Lighthouse wharf	L H Whf
Mine boathouse	M B H
Mine derrick	M Drk
Mine tramway	M Tmy
Mine wharf	M Whf
Ordnance machine shop	O M S
Private wharf	Pvt Whf
Radio (commercial station)	Rad
Railway' wharf	Ry Whf
Saluting battery	Sl B
Service dynamite room	S D R
Steamship wharf	S S Whf
Quartermaster wharf	Q M Whf
Superseded (for abandoned buildings, etc.)	24 s
Temporary (for all uses except fire-control structures)	19 t
Sunset gun	S G
Tide gage	T G
Torpedo storehouse	T S
Tower	Tw
Water tank	W Tk
Weather bureau	W B

A DIRECTORY OF AMERICAN SEACOAST DEFENSES 1890-1950

This directory is a comprehensive guide to all the major locations and sites used for harbor defense, with maps showing what was at each site and comments on the current status of each site (extant, in ruins, destroyed, privately owned, current U.S. military use, federal, state, county, city parks, etc.) as far as information is known to the authors. While the status information is fairly comprehensive for the larger forts and military reservations, the status of many of smaller World War II-era fire control sites is not. The authors would appreciate receiving any updated information to correct or add to what has been presented here. Terms used in this reference work to describe the various periods of construction such as "Endicott-Era," "Taft-Era," "Post-World War I-era," "World War II-Era," the "100-Series" and "200-Series" batteries, etc. are terms used by modern historians and were not used by the Army to describe these programs in progress. Note that several of the planned batteries in the 1940 program were cancelled before any work was done as denoted by their *battery # in italics* and as (planned).

This directory does not cover the following artillery used for seacoast defense at various times between 1898 and 1945—the Rodman guns emplaced or re-emplaced during the Spanish American War; the Navy guns and mounts installed during the World War II years, mostly in the Pacific theater; Hawaii's World War II temporary and provisional defenses; the fixed antiaircraft gun batteries emplaced in the defenses from 1920; mobile artillery which had prepared positions including those for 12-inch railway mounted mortars and 8-inch railway guns; the Panama mount positions for the tractor drawn 155 GPF guns; and the positions on Oahu for the 240 mm howitzers.

The directory is organized by Harbor Defense around the United States clockwise from Portland Maine to the Puget Sound in Washington State, followed by the Alaskan defenses of World War II, and the defenses in Hawaii, the Philippines, Panama, the Caribbean, Newfoundland, and Bermuda.

This directory includes detailed brief histories of modern-era American coast artillery concrete gun batteries. Glen Williford created this as a personal reference guide over many years of research and study of U.S. Coast Artillery history. This battery listing includes the histories of all modern (post-1886) "fixed" or permanent concrete seacoast gun batteries emplaced by the U.S. in the country and outlying territories. The emplacements were mostly built by the Corps of Engineers, and manned by the Coast Artillery. Each battery description includes the following information where possible. The battery name in capital letters if an officially conferred name, in lower case if just an informal, local, or construction designation. The description then briefly covers the purpose for construction and the general location on the reservation, particularly in relation to other elements. In most cases the act or source of original funding (which does not include the cost of coast artillery) and date of plan submission follows. Major design features or significant variations from Mimeograph Type plans are discussed. The general dates of construction, transfer date to the Coast Artillery and engineering costs may also be included. This is followed by a description of the armament, including gun and carriage models and specific serial numbers and date of mounting if known. In general, the manufacturer of the guns and carriages are only designated only when there are multiple producers and thus duplicate serial number runs. The general order and date, that names each battery is included with a brief description of person honored. Subsequent service events, including any major alterations, accidents, armament, or name changes follows. The date of gun dismounting or at least the date of authorization for deletion is covered. A brief statement on whether the battery still exists, or when destroyed, and park or status if on public property concludes the description.

The major sources consulted were: Reports of Completed Batteries, and Reports of Completed Works, Engineer Letters of Submission, surviving Fort Record Books, Seacoast Gun Record cards and earlier Ordnance Department Seacoast Gun Ledger Books, Annexes to Seacoast Projects, General Orders naming citations, supplemented by various records in archive primary engineer correspondence files, annual reports of the Chief of Engineers, private Williford studies on emplacement accidents, temporary defenses, defenses of the Spanish American War.

THE HARBOR DEFENSES OF SAN DIEGO – CALIFORNIA

San Diego Bay is a large natural estuary that was developed as a mission, town, and port by the Spanish beginning in 1769. The U.S. Navy developed it as a port and station from the 1850s and it remains a key Navy facility today. The US fortified the harbor during the Endicott and Taft Programs and again during the 1940s. Point Loma, a large natural ridge along the western side of the harbor entrance received the major seacoast fortifications to guard the bay and its harbor. The U.S. Army reservations were declared surplus in 1949 and transferred to the U.S. Navy by 1959. All but a small portion of the lands remain under U.S. Navy control today and generally off limits to the public.

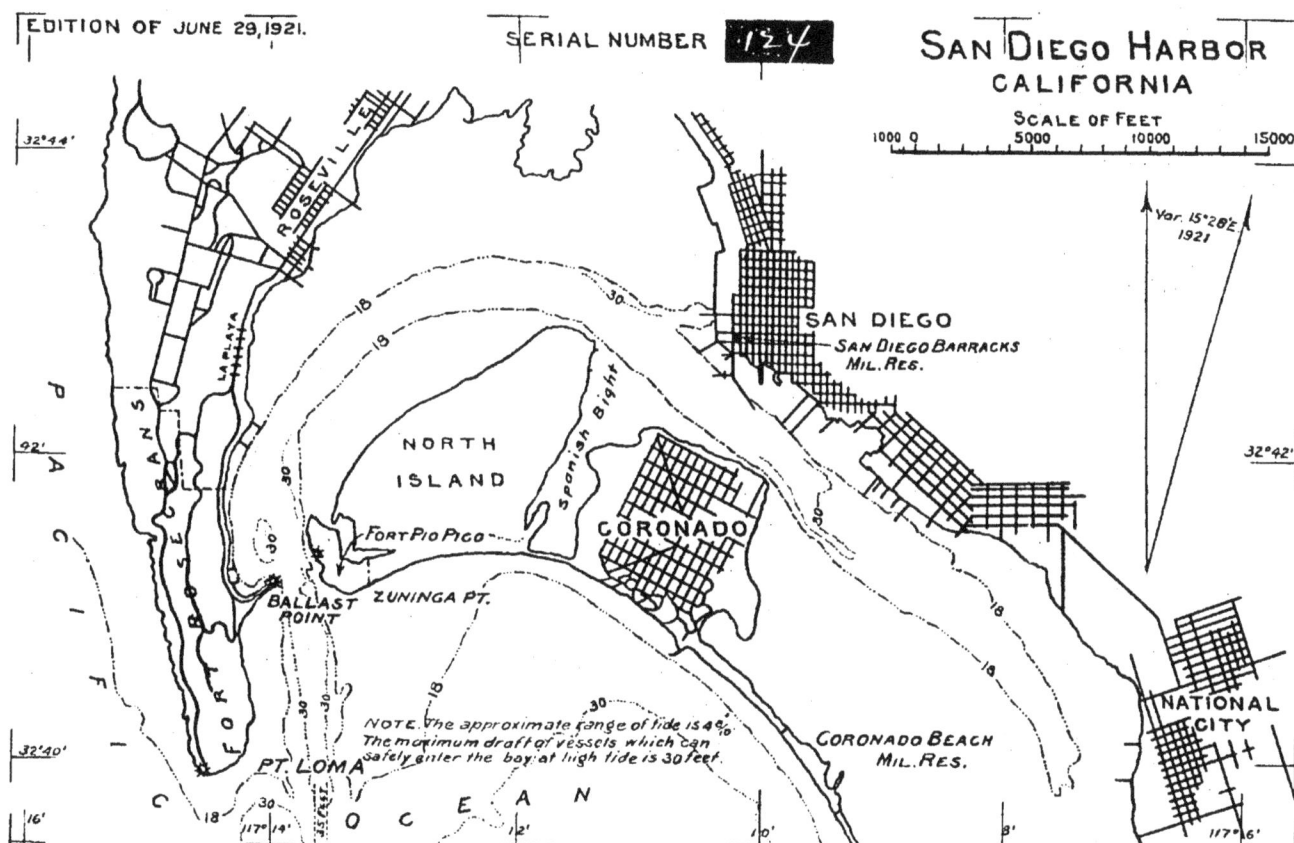

Fort Rosecrans (1899-1958) was located on southern end of the Point Loma headland that borders the west side of San Diego Bay. The first fort on Point Loma was Fort Guijarros at Ballast Point (on the bay side of Point Loma), which was built by the Spanish in 1797. Point Loma was to be armed in 1870 with a series of batteries that were started but not completed. It was not until 1899 when Fort Rosecrans came into existence. The Endicott Program saw the construction of four concrete batteries and a controlled mine facility near Ballast Point and one 3-inch battery across the channel on North Island (the short-lived Fort Pio Pico). It was named in General Orders 134 of 1899 for Maj. Gen. William S. Rosecrans, USV, and Fort Pio Pico Was named in General Orders 20 of 1906 for Pio Pico the last Mexican governor of Alta California. The site received new set of garrison buildings as well. The rather meager harbor defenses were upgraded in 1915 by the construction of two 12-inch mortar batteries and a new submarine mine casemate. In 1937, 7-inch and 5-inch naval guns were installed on the Pacific Ocean side of Point Loma. With the approach of World War II, a new 8-inch gun battery was built in 1938, followed by the 1940 Program adding a 16-inch casemated Battery Ashburn, two 6-inch shielded batteries (Batteries Humphreys and Woodward). Also added were two 90mm AMTB batteries and several Panama mounts for 155mm GPF mobile artillery. The end of World War II and the phase-out of the Coast Artillery Corps resulted in Fort Rosecrans becoming surplus in 1949. After 1958 the property was transferred to the U.S. Navy for use as a submarine base and a naval weapons testing and development facility. Most of the fort and all of the major batteries remain under in U.S. Naval control, so permission to visit must be sought. A section of the old fort is now the Fort Rosecrans National Cemetery. The area around the Cabrillo Point Lighthouse is the National Park Service's Cabrillo National Monument. The site features a large visitors center, and a number of trails leading to fire control and searchlight structures.

Fort Rosecrans Gun Batteries

- **CALEF – WILKESON:** An Endicott battery for four 10-inch disappearing guns emplaced on the lower, Ballast Point reservation, of Fort Rosecrans. Plans for the first two emplacements were submitted on October 29, 1896, followed by the third (No. 2 emplacement) on June 17, 1897, and the fourth (No. 1 emplacement) on July 5, 1898. They were in a line, though the eastern two were flank emplacements with increased field of fire to the east (channel) than the western two internal emplacements. This was a low site (trunnion height of 35-feet), following the trace of the older 1870s earthen work on the small peninsula known at Ballast Point. The battery fired to the southeast on an azimuth direction of 205-degrees. General layout of platforms was according to type plans. Gun spacing was 124-feet, magazines were on the lower left. Ammunition service was by balanced platform lift, but chain hoists were installed and transferred in 1905. Very large traverses were built between guns, with a sharp angle back on the left flank. Tunnel corridors were built through the traverses to allow communication between guns. The traverse between guns two and three held the reserve power plant for the emplacement. Work was done from 1896-1899. Transfer of the first three emplacements was made together on March 22, 1898. The fourth emplacement was transferred on February 15, 1900. All four were calculated at a cost of $217,262. As completed the battery had three 10-inch guns Model 1888 on Model 1896 disappearing carriages (Bethlehem #10/carriage #7, Watervliet #10/carriage #5, and Watervliet #4/carriage #6. Emplacement No. 4 had a 10-inch Model 1895 Watervliet gun #8 on Model 1896 carriage #53). All four were originally named on General Orders No. 16 of February 14, 1902 for Lt. Colonel Bayard Wilkeson, killed in action at Gettysburg in 1863. Under General Orders No. 36 of June 9, 1915, the battery was tCTICALLY split by name to form two units. Emplacements No. 1 and No. 2 being named for Colonel Haskell Calef, an artillery officer. Shortly thereafter the names were switched in order to conform with standard Coast Artillery naming procedures. The battery served for many years without armament modification. But the emplacement did receive new hoists in 1905 and slightly later enlargement of loading platforms and new battery commander stations. The guns were authorized for removal in August 1942, that being done in early 1943. The emplacement was later used by the U.S. Navy for storage and still exists at the Naval Base Point Loma. The battery is not open to the public.

- **McGRATH:** A battery for two 5-inch rapid-fire guns emplaced at the juncture of the Ballast Point peninsula with the Point Loma headlands. Plans were submitted on June 20, 1899 using funding from the Act of July 7, 1898. Plans adhered to the current mimeograph recommendations for rapid-fire guns. It had two platforms with gun centers of 45-feet. Magazines were on the lower left of each platform; ammunition service was entirely by hand. The location was on the rise of the ridge to the east of the 10-inch battery, which was at a 285-foot distance. Interior crest was 73-feet, it pointed and fired to the southeast. Work was started on excavation on August 21, 1899, concrete work was done from October to December 1899. It was done, awaiting armament on June 30,

FORT ROSECRANS, CAL.

SERIAL NUMBER

GENERAL MAP.

EDITION OF JUNE 29, 1921.
REVISIONS: APR. 16, 1925.
OCT. 20, 1928. OCT. 30, 1934.

True Meridian

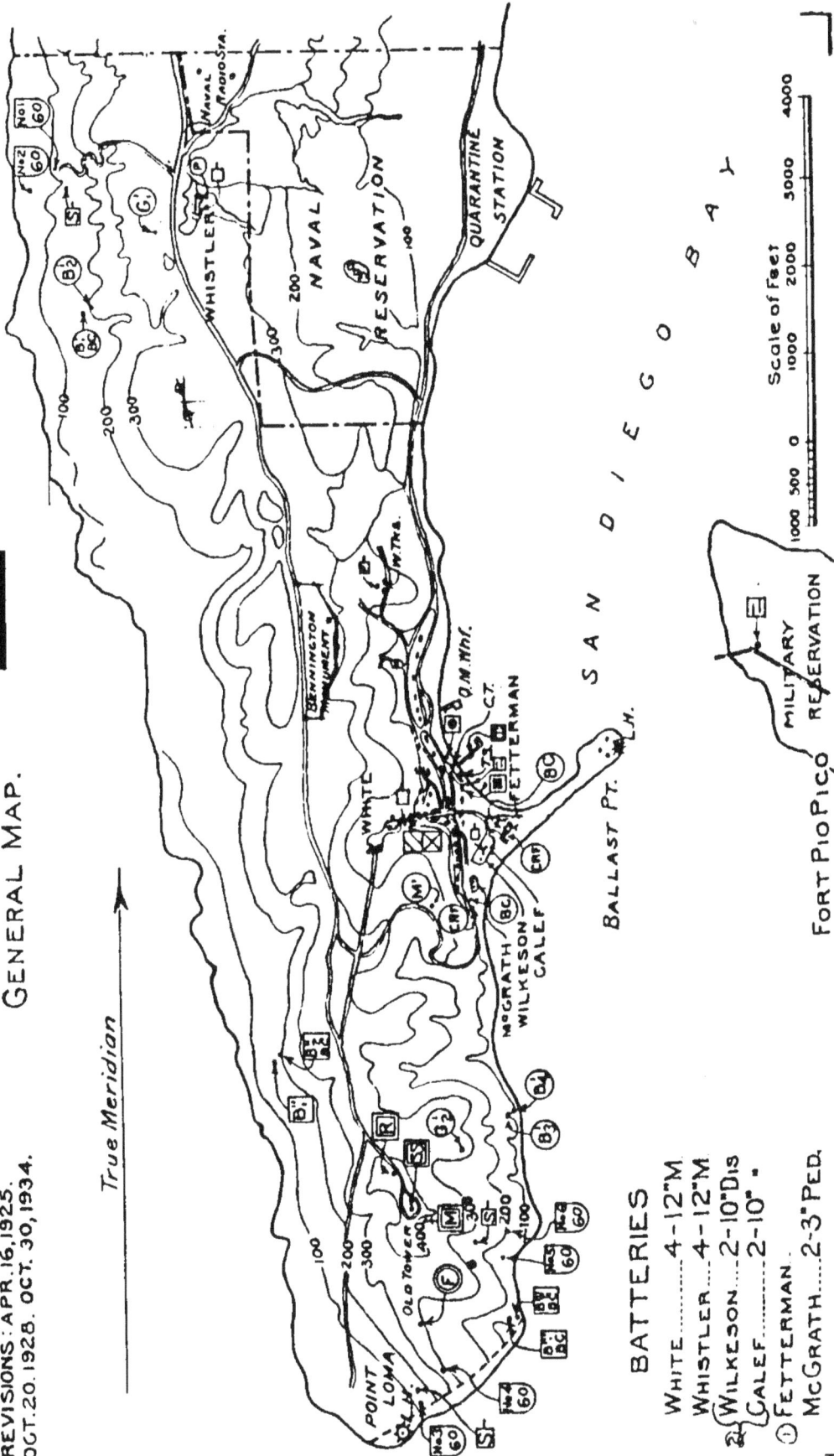

NAVAL RADIO STA.

WHISTLER

NAVAL RESERVATION

QUARANTINE STATION

BENNINGTON MONUMENT

Q.M. WHF.

FETTERMAN

WHITE

POINT LOMA

OLD TOWER

BALLAST PT.

L.H.

S A N   D I E G O   B A Y

McGRATH
WILKESON
CALEF

SCALE OF FEET

1000 500 0   1000   2000   3000   4000

FORT PIO PICO
MILITARY RESERVATION

On Caretaking Status.

BATTERIES

WHITE	4-12"M
WHISTLER	4-12"M
WILKESON	2-10"Dis
CALEF	2-10" "
FETTERMAN	
McGRATH	2-3"PED.

SAN DIEGO HARBOR, CAL.

FORT ROSECRANS-DI

SERIAL NUMBER 1.2.4

SCALE OF FEET

EDITION OF JUNE 29, 1921.

TRUE MERIDIAN

SAN DIEGO BAY

Q.M. WHARF

WHITE

WILKESON

CALEF

McGRATH

FETTERMAN

LEGEND

1. ADMINISTRATION BLDG.
2. COMMANDING OFF. QRS.
3. OFFICERS' QUARTERS.
4. HOSPITAL.
4a. HOSPITAL WARD.
5. HOSPITAL SERGEANT.
6. N.C. OFFICERS QRS.
7. BARRACKS.
8. GUARD HOUSE.
9. POST EXCHANGE.
10. FIRE STATION.
11. BAKERY.
12. BOAT HOUSE.
13. OIL HOUSE.
14. WAGON SHED.
15. BLACKSMITH SHOP AND COAL SHED.
16. CIV. EMPLS. QRS.
17. TRANSFORMER AND SWITCHBOARD BLDG.
18. BATH HOUSE.
19. CHICKEN HOUSE.
21. Q.M. WORKSHOP.
22. Q.M. & COM. BLDG.
23. Q.M. STABLES.
24. Q.M. & COM. STO. HO.
31. ORD. REPAIR SHOP.
32. ORD. STOREHOUSE.
40. E.D. OFFICE.
41. E.D. SHOP.
42. E.D. STOREHOUSE.
43. E.D. REPAIR SHOP.
71. OFFICERS' CLUB.
72. SCHOOL, E.&R.
73. SERVICE CLUB.
101. MESS HALL.
102. LAVATORY.
103. M.T. GARAGE.
104. ART. ENGR. QRS.
105. DENTIST.
106. STOREROOM.
107. STOREHOUSE.
108. ART. ENGR. OFF. & ST. HO.
109. PLUMBER & ELECTRICIAN.

BATTERIES

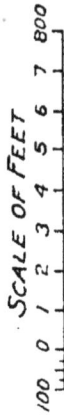

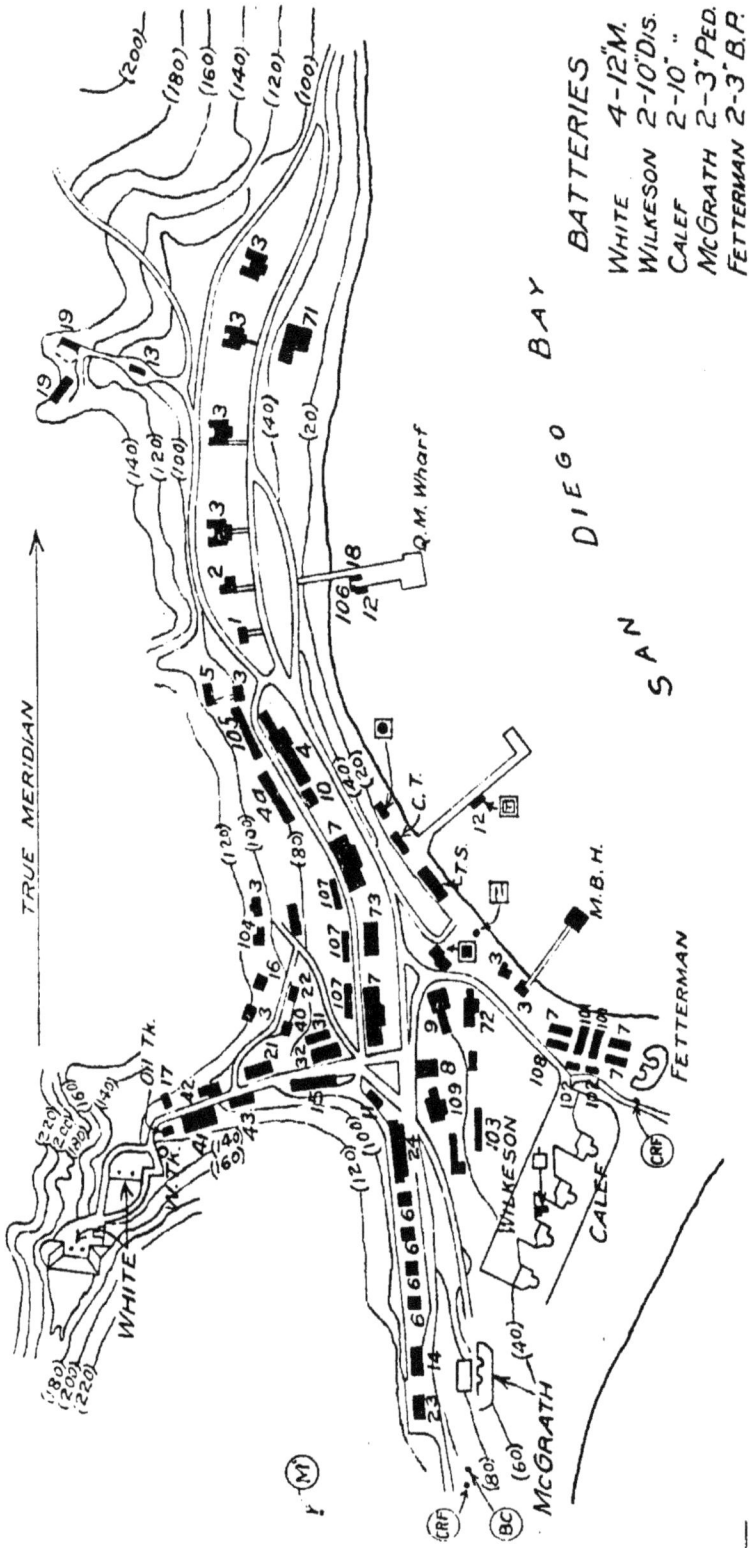

WHITE	4-12" M.
WILKESON	2-10" DIS.
CALEF	2-10"
McGRATH	2-3" PED.
FETTERMAN	2-3" B.R.

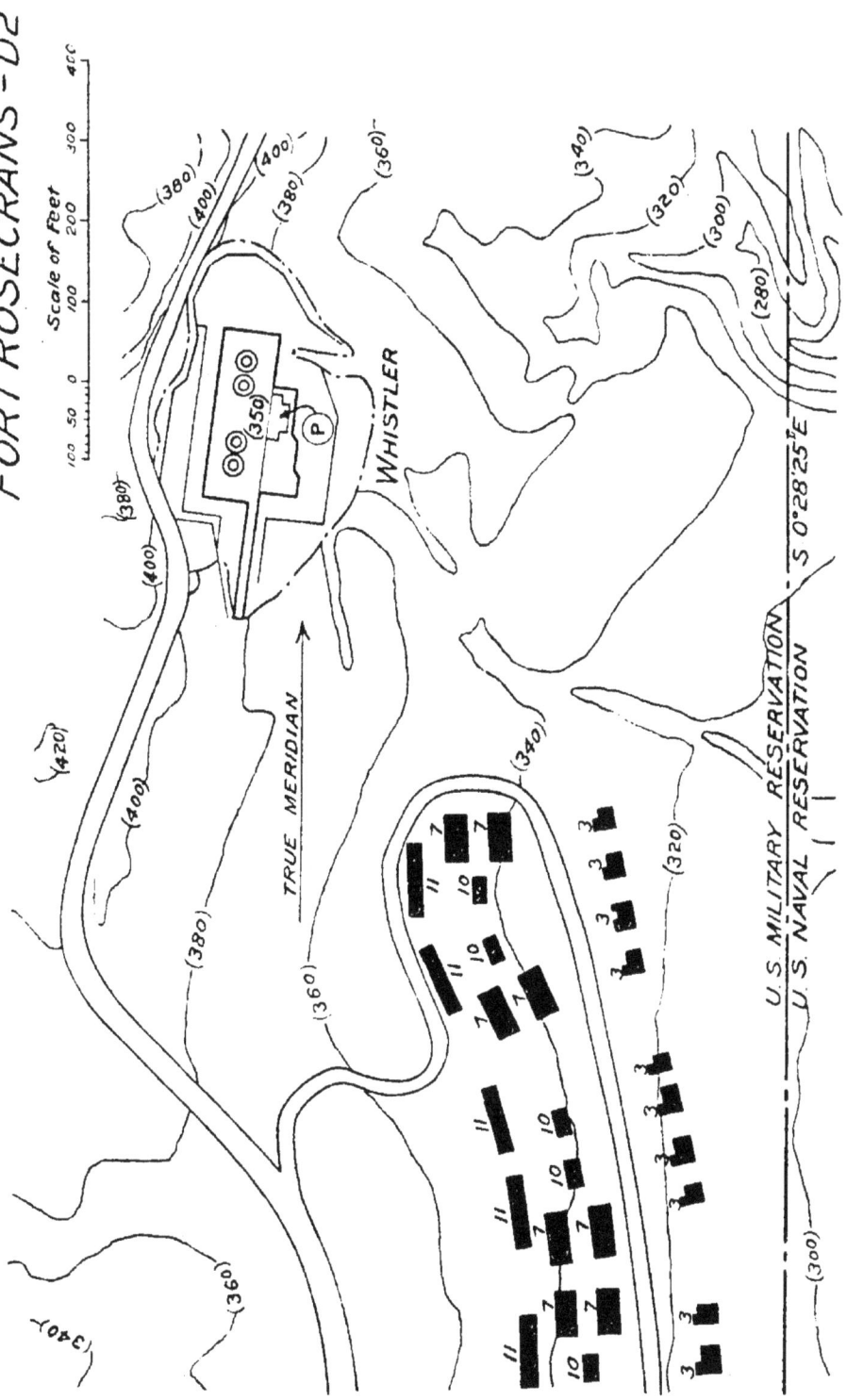

SAN DIEGO HARBOR, CAL.

FORT ROSECRANS - D2

Scale of Feet

SERIAL NUMBER 112

EDITION OF JUNE 29, 1921.

LEGEND.

1.
2.
3. OFFICERS' QUARTERS
4.
5.
6.
7. BARRACKS.
8.
9.
10. LAVATORY.
11. MESS HALL & KCHN.

TRUE MERIDIAN

WHISTLER

U.S. MILITARY RESERVATION
U.S. NAVAL RESERVATION

S 0°28'25" E

BATTERIES.
WHISTLER.....4-12"M.

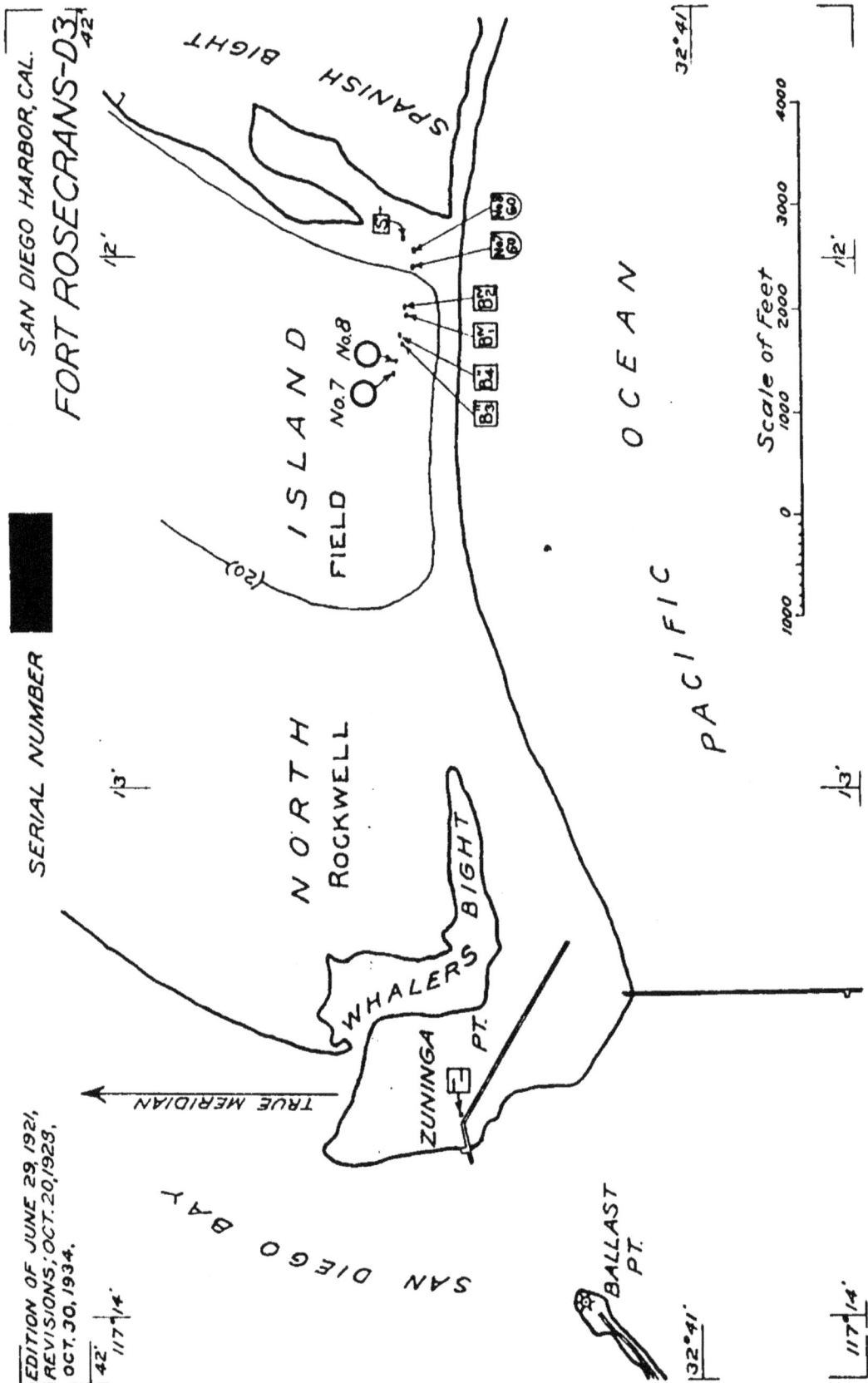

SAN DIEGO HARBOR, CAL.

FORT ROSECRANS-D3

SERIAL NUMBER

EDITION OF JUNE 29,1921,
REVISIONS; OCT.20,1923,
OCT.30,1934.

SPANISH BIGHT

ISLAND
FIELD
No.7   No.8

NORTH
ROCKWELL

WHALERS BIGHT

ZUNINGA   PT.

TRUE MERIDIAN

SAN DIEGO BAY

BALLAST
PT.

PACIFIC   OCEAN

Scale of Feet

On Caretaking Status.

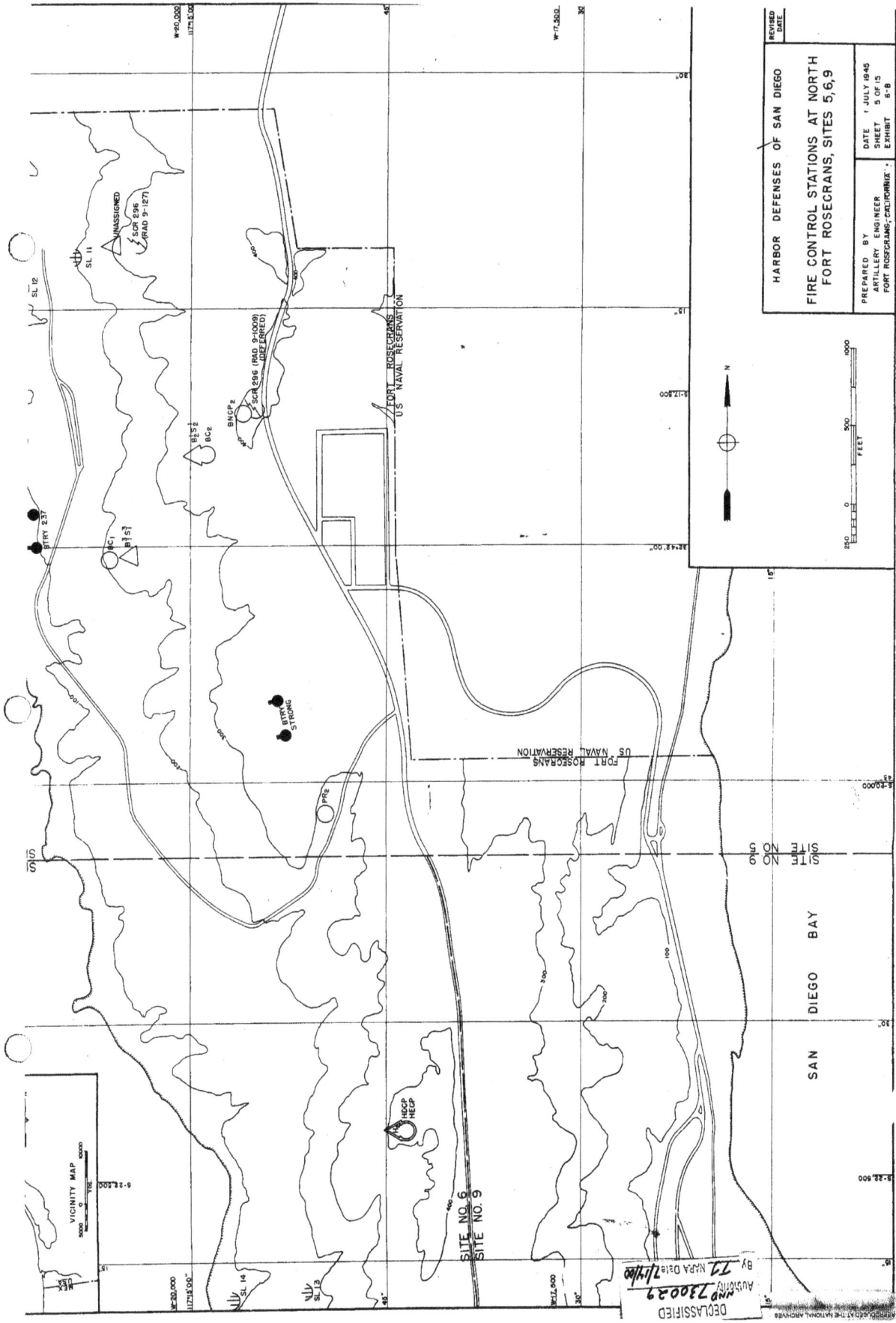

HARBOR DEFENSES OF SAN DIEGO

FIRE CONTROL STATIONS AT NORTH
FORT ROSECRANS, SITES 5, 6, 9

	DATE	1 JULY 1945
PREPARED BY	SHEET	5 OF 15
ARTILLERY ENGINEER	EXHIBIT	6-B
FORT ROSECRANS, CALIFORNIA		

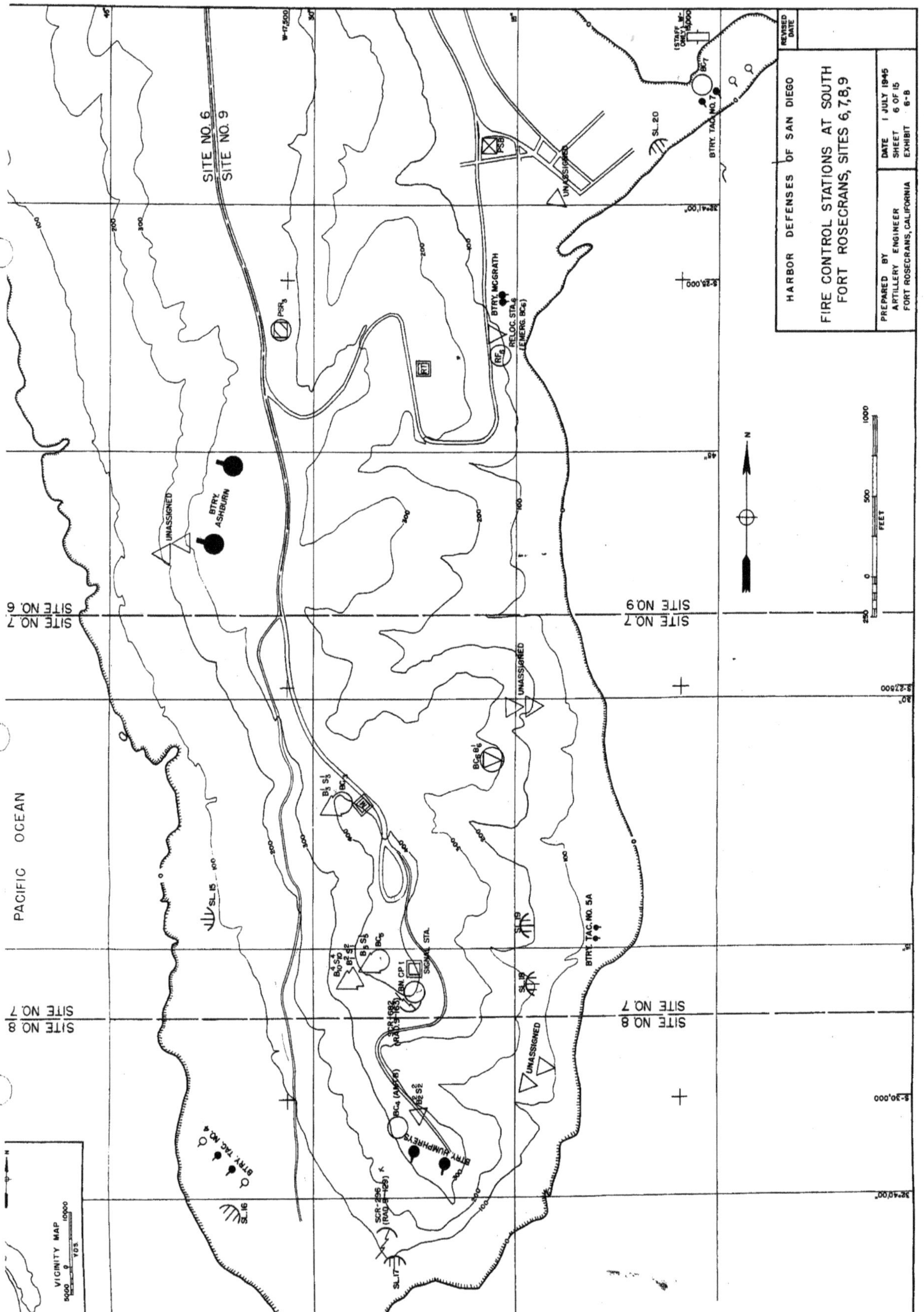

HARBOR DEFENSES OF SAN DIEGO
FIRE CONTROL STATIONS AT SOUTH FORT ROSECRANS, SITES 6,7,8,9
PREPARED BY ARTILLERY ENGINEER FORT ROSECRANS, CALIFORNIA
DATE 1 JULY 1945
SHEET 6 OF 15
EXHIBIT 6-B
REVISED DATE
PACIFIC OCEAN
SITE NO. 6
SITE NO. 9
SITE NO. 7
SITE NO. 8
BTRY. ASHBURN
BTRY. MCGRATH
BTRY. HUMPHREYS
BTRY. TAC. NO. 5A
BTRY. TAC. NO. 4
BTRY. TAC. NO. 7
SIGNAL STA.
UNASSIGNED
VICINITY MAP
FEET

(0·7-110N-4AB)(11-27-37-11A)(12-4000) FORT ROSECRANS, SAN DIEGO, CALIF. (SECRET)

Fort Rosecrans 1937 (NARA)

1900. Transfer was made on November 17, 1900 for a cost of $18,203.17 It was named in General Orders No. 16 of February 14, 1902 for Major Hugh J. McGrath mortally wounded during the Philippine Insurrection in 1899. It was armed with two 5-inch guns Model 1897 on Model 1896 balanced pillar mounts (Bethlehem guns #7/#29 and #14/#30). The 5-inch guns were removed on July 26, 1918. Two 3-inch pedestal guns, formerly at Battery Meed at Fort Pio Pico were moved to be mounted in this emplacement. The platforms were raised in January 1919 and new base rings laid, and shortly after it received the two 3-inch Model 1903 guns and pedestal mounts (#47/#67 and #69/#66). These then served until deleted under authority granted on March 7, 1946. The emplacement was used for a while for explosive storage by the Navy. It still exists on the Naval Base Point Loma. The battery is not open to the public.

- **FETTERMAN**: A battery for two 3-inch masking parapet guns emplaced on fill land along the spit extending from Ballast Point. Plans were submitted on March 27, 1899. It was placed on the then recently widened spit; enlarged with the spoil removed by excavation for the 10-inch battery. It followed type plans, with the right platform a straight, interior one, and the left one a curved flank emplacement. Magazines were on the lower left flank of each platform; a latrine was also part of the design. Gun centers were 29-feet, trunnion height was 22-feet. It fired to the southeast covering the mine field. Construction was done from 1899-1900. It was transferred on April 30, 1903 at a cost of $8,865. It was named in General Orders No. 16 of February 14, 1902 for Colonel William Fetterman killed in action with Sioux Indians near Fort Phil Kearny, WY in 1866. It was armed with two 3-inch Model 1898 Driggs-Seabury guns on balanced pillar mounts (#69/#95 and #95/#96). These were converted to the M1898M1 fixed pedestal mounts in about 1916. The armament was removed in June of 1920 with the declared obsolescence of this type of mount. The emplacement itself was destroyed under authority of August 1939 to make room for other buildings on the site. No traces remain today of the emplacement.

- **WHITE**: A battery for four 12-inch mortars emplaced in 1915 in a ravine on the eastern slopes of the Point Loma ridge of Fort Rosecrans. Proposals for new mortar emplacements at San Diego began in May 1914. A special board recommended two new emplacements with a total of 16 mortars—though these were to be relocated from other sites rather than from new production. It was soon noted that the number of mortars could be safely reduced to a two-per-pit arrangement similar to recent Panama and Los Angeles emplacements. This two-pit battery, with two mortars in each pit, was located in a ravine known as Powerhouse Canyon. Plans were finalized for submission on March 25, 1915. Due to the narrow dimensions of the canyon, it was actually two totally separate pits arranged in echelon (one in front of the other, and at a 20-foot elevation difference). They faced and fired just 5-degrees west of south. Even though thoroughly protected by the nature of the ravine, they were given considerable concrete protection. Work was done from November 8, 1915 to mid-1917. Transfer was made on August 19, 1919 for a cost of $144,200. The battery was named in General Orders No. 15 of April 25, 1916 for Colonel John V. White, CAC who had recently died in 1915. The armament for this battery had been dismounted from Delaware Battery Best and Battery Rodney in 1915. As received and mounted, Battery White was armed with four 12-inch Model 1890M1 mortars on Model 1896MII carriages (Builders tube #3/#76, Bethlehem tube #27/#83, Builders tube #9/#56, and Builders tube #4/#87). The battery served up until listed for disposal on November 3, 1942. It was subsequently used for a variety of storage purposes by both the army and navy. It still exists on property of the Naval Base Point Loma. The battery is not open to the public.

- **WHISTLER**: A battery for four 12-inch mortars emplaced on the northern extension of the army reservation of Fort Rosecrans at Point Loma. It was part of the 1914 proposal for a new mortar batteries for relocated mortars as was Battery White. This battery was located further to the north, at about the top of the central Point Loma ridge on part of the navy reservation (which had to be transferred to the army at Fort Rosecrans). Plans were submitted on October 26, 1914. It was a more conventional design than White: two adjacent pits containing two mortars each, with a traverse between them, and magazines in the forward, western parapet. Because of the relative simplicity of the design, it cost almost $50,000 less that Battery White. It fired to the west and had a parados in the reverse (eastern) face to protect the pits from the possibility of fire from enemy ships in the bay. The guns had a trunnion height of 354-feet. For a brief period in 1916 consideration was given to manufacturing new Model 1912 mortars for these batteries, but lack of funds meant that the original plan for re-locating mortars was retained. Work was done from March 15, 1916 to early 1917, for transfer on August 19, 1919 for a cost of $118,000. It was named in General Orders No. 15 on April 25, 1916 for Colonel Garland Nelson Whistler, a Coast Artillery officer. Four guns and carriages were transferred from Battery Laidley at Fort DeSoto in April of 1917. They were mounted at this battery by the end of 1917. As finally mounted, they were Model 1890M1 Watervliet mortars on Model 1896MII carriages (#76/#186, #85/#188, #107/#189, and #127/#190). This was carried well into World War II, being listed for disposal on December 3, 1942. In 1959 the abandoned emplacement itself was heavily modified to serve as the Arctic Submarine Laboratory for the U.S. Navy (which used several large caliber barrels as pressure chambers). In 1999 the ASL vacated Battery Whistler and the emplacement was cleared of all the ASL equipment. The old emplacement still exists on property of the Naval Base Point Loma. The battery is not open to the public.

- **STRONG**: An interwar period battery for two 8-inch barbette guns emplaced on the western slope of Point Loma at the northern limit of the government reservation. It was the prototype design for the new, open, barbette emplacements for 8-inch guns. While never exactly duplicated, it served as design proof that evolved into the standard overseas 8-inch emplacements used in Hawaii, Alaska, and Roosevelt Roads. It featured a central traverse magazine and power room with ammunition carts riding on rail tracks leading to the two open gun platforms. It was armed with two older navy 8-inch, 45-caliber guns transferred to the army emplaced on a newly designed Army barbette carriages. The battery also had an entirely separate plotting and radio room. The basic design emerged during 1932-1934, and the construction at Fort Rosecrans was accomplished from February to July 1937. Transfer was made on October 12, 1942 at a cost of $100,725.59. The plotting room was transferred at the same time for another $19,146. The battery was named for Major General Frederick V. Strong, 4th U.S. Artillery. It was armed with two 8-inch navy guns MkVIM3A2 on Model M1 barbettes (#195L2/#1 and #193L2/#2). These were received in January 1941 and proofed in April 1941. The battery served until disarmed in 1946. The emplacement was subsequently modified for use in testing navy electronics. The emplacement still exists with its magazines used as lab space at the Navy Base Point Loma annex. The battery is not open to the public.

- **ASHBURN**: A 1940 Program dual 16-inch casemated, barbette battery emplaced on the high ridge of Point Loma. This was intended to be the most important single battery for the modernization efforts for San Diego. During planning and construction, it was known as Battery Construction No. 126. It was located on the ridge height of Point Loma, in the north central sector, positioned to have its main field of fire to the west. In the September 11, 1940 national priority list, it was assigned #12, but that was advanced to #9 by August 11, 1941. Work was begun under the FY-1942 Budget, actual work taking place between June 12, 1942 and March 16, 1944. Transfer was made on October 19, 1944 at a cost of $1,323,912.38. Emplacement work was accompanied by a separate PSR room which was transferred on June 16, 1944 at $168,141.72. It was of standard design for the 100-series, with gun houses separated by 500-feet and a standard magazine layout in the traverse. Proofing of the weapons occurred in July 1944. The battery was armed with two 16-inch guns MkIIM1 on carriages M1919M4 (#71/#31 and #97/#39). Carriages were equipped with 4-inch-thick shields. It was named in General Orders No. 69 of December 22, 1942 for Major General Thomas Quinn Ashburn, U.S. Army. The battery was an active part of the late-war defenses, was retained under the 1946 Program but finally deleted on May 10, 1948. The emplacement still exists, though actively utilized by the U.S. Navy at the Space and Naval Warfare Systems Center Pacific. The battery is not open to the public.

- **HUMPHREYS**: A 1940 Program dual 6-inch battery for San Diego at Fort Rosecrans. It was located on the cliff slope at the southern end of the Point Loma peninsula, firing to the south. Originally it was planned as Battery Construction No. 238. It is of fairly conventional 200-series designs, though the No. 1 magazine entry is twisted with the ready ammunition boxes arranged to fit the local topography. Work was started on February 2, 1942 and completed on October 14, 1943. It was transferred on June 20, 1944 for a total cost of $199,661.23. The battery had begun with national priority #14, and started with the FY-1942 Budget appropriations, but apparently was pressed for early completion during work. It was named in General Orders No. 28 of June 5, 1942 for Captain Charles Humphreys, an artillery officer. The battery was armed with two 6-inch Model 1903A2 guns on Model M1 carriages. On January 29, 1944 during a firing drill, one tube exploded due to premature detonation of the charge, resulting in five deaths and seven injured in the gun crew. The tube was replaced. After this incident the armament consisted of guns and carriages #44/#100 and #9/#101. It then served until finally deleted in 1948. The emplacement still exists on U.S. Navy property and used for testing and storage. The battery is not open to the public.

- **BATTERY #237**: Another 1940 Program dual battery for Fort Rosecrans. This was a dual 6-inch battery, begun as Battery Construction No. 237. It was located on the beach bench (trunnion height of 105-feet) on the northwest side of the Rosecrans reservation, firing to the west. It was assigned national priority #9 under the original 1940 project list and funded under the FY-1942 appropriations. It was of conventional design for the 200-series batteries, with no major variations. Work was done from March 4, 1943 and completed to October 29, 1943. It was transferred on August 31, 1944 at a cost of $255,912.24. It was un officially named for Colonel Charles G. Woodward, the commander at Fort Rosecrans from 1906-1907. The armament was mounted in October 1944 and consisted of two 6-inch guns Model 1903A2 on Model M1 barbette carriages (#40/#103 and #55/#109). The battery served through the war, being disarmed in 1946. The emplacement still exists on U.S. Navy property and used for testing and storage with its power generaing equiment still in place.. The battery is not open to the public.

- **AMTB #4**: A 1943 Program AMTB battery of two 90mm fixed and two 90mm mobile guns. It was emplaced on the western tip of the Point Loma peninsula, firing to the southwest. Informally known as Battery Cabrillo. Work was done from June 29 to August 26, 1943. Transfer was made on December 14, 1943. The guns were mounted in September 1943 and dismounted in 1946. No remains of the gun blocks or other parts of the battery appear to still remain. Site is now part of the Cabrillo National Monument.

- **AMTB #7**: A 1943 Program AMTB battery of two 90mm fixed and two 90mm mobile guns emplaced at Ballast Point. Informally the emplacement was also known as Battery Fetterman, taking the name of the old 3-inch battery which used to be in the immediate vicinity. The AMTB emplacement consisted of simple gun blocks for the two fixed guns and adjacent ammunition magazine. Work was done from June 29 to August 26, 1943 for transfer on December 14, 1943. The guns were ordered removed after the war, but parts of the emplacement still exist at the Naval Base Point Loma. The battery site is not open to the public.

Fort Pio Pico (1901-1915) is located on North Coronado Island across the channel from Point Loma. Located here was Battery Meed (1902-1914). It was damaged in a storm in 1914, and the guns were later transferred to Battery McGrath at Fort Rosecrans in 1919. The fort was demolished in 1922. Site is now part of the North Island Naval Air Station. The site is not open to the public.

Fort Pio Pico Gun Battery

- **MEED**: An Endicott battery for two 3-inch, rapid-fire guns emplaced at Zuniga Point on North Island, across the main channel from Ballast Point and Fort Rosecrans. Plans were submitted on August 24, 1900. The emplacement generally followed type plans with a few modifications. It had two platforms with 29-foot gun centers. Emplacement No. 1 was an internal type, and No. 2 a rounded, flank emplacement. The left (west) flank of the battery ended in a long extension of a retaining wall to help avoid potential erosion. Also, as the adjoining land at the time of construction was an active hunt club, the army specified steel doors instead of wooden on the magazines and barbed wire along the property boundary. Work was done in 1900-1901. However, on August 17, 1901 local engineers were notified that the gun type was going to be changed from masking parapet pillars to pedestals, and to delay setting the base rings. It took until February 1910 to re-

Cabrillo National Monument (Terry McGovern)

Battery Ashburn (Terry McGovern)

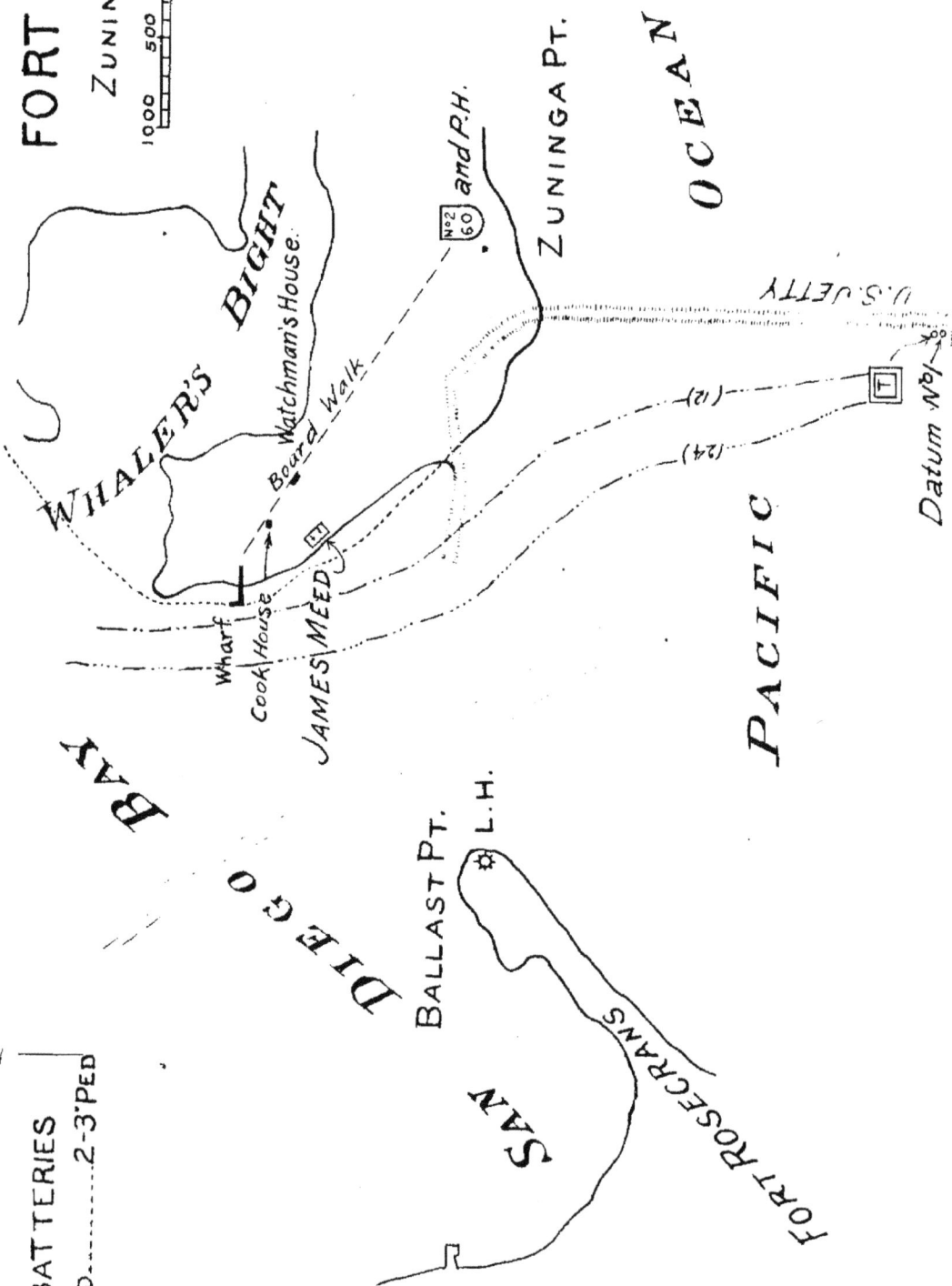

SERIAL NUMBER **128**

SAN DIEGO HARBOR CAL.

FORT PIO PICO.

ZUNINGA POINT

True Meridian.

WHALER'S BIGHT

Watchman's House.

Board Walk

No.2
60   and P.H.

ZUNINGA PT.

Wharf

Cook House

JAMES MEED

OCEAN

U.S. JETTY

Datum No.1

PACIFIC

SAN DIEGO BAY

BALLAST PT.

L.H.

FORT ROSECRANS

EDITION OF MARCH, 4, 1914.
REVISIONS: DEC. 7 1915, JUNE 9, 1916.

BATTERIES
MEED...........2-3"PED

ceive and mount the guns. The battery was transferred on May 10. 1902 for $10,000. The battery was named in General Orders No. 20 of January 25, 1906 for Captain James Meed who was killed in action at Frenchtown, MI in 1813. It was armed with two 3-inch Model 1903 guns on Model 1903 pedestal mounts (#47/#67 and #59/#66). The exposed emplacement position near the shoal point led to extensive storm and erosion damage. The storm of February 23, 1913 was particularly severe. It was determined that expenses to repair and further protect the battery were too excessive and the location was abandoned by the Coast Artillery. The guns and carriages were soon removed and eventually used to rearm Battery McGrath at Fort Rosecrans. The emplacement was not subsequently used, and eventually totally destroyed by erosion. Nothing remains today.

Coronado Beach Military Reservation (1897-1946) is located 4.5 miles south of the city of Coronado on Highway 75. The reservation was located on the sand-spit that forms the outer edge of San Diego Bay between Coronado and Imperial Beach. The planned location of an Endicott-era mortar battery that was never built. AMTB Battery Cortez (1943-1946, proposed name Breitung) was located here. Two World War II fire-control towers were once located here. The sand spit that connected the beach to the mainland was frequently breached during high tide. The reservation is now the site of Silver Strand State Beach.

Silver Strand M.R. Gun Battery

- **AMTB #8**: A 1943 Program AMTB battery consisting of two 90mm fixed and two 90mm mobile guns emplaced on the sand beach along the Silver Stand between Coronado and Imperial Beach. It was one of three such batteries built for the San Diego defenses from June 29 to August 26, 1943. Informally known as Battery Cortez. They were transferred on December 14, 1943 at an individual battery cost of $19,600. It consisted of simple gun blocks and wooden, earth-covered magazines. The battery was disarmed in 1946. No remains of the emplacement exist today

Fort Emory (1942-1948) is located at Coronado Heights at the southern end of San Diego Bay, near the city of Imperial Beach, California. It was named in General Orders No. 67, December 14, 1942 for Brigadier General William Hemsley Emory, United States Army. Built as part of the 1940 Program, Fort Emory was a sub-post of Fort Rosecrans. The 1940 Program construction resulted in one 16-inch battery, #134, and one 6-inch battery, Battery Grant (#238). Also added was one 90mm AMTB battery and several Panama mounts for 155mm GPF mobile artillery. The end of World War II and the phase-out of the Coast Artillery Corps resulted in Fort Emory becoming surplus in 1948. After the armament was removed, the military reservation was turned over to U.S. Navy. It was transformed into the Imperial Beach Naval Radio Station, which was phased out of operation after 1990. Currently the U.S. Navy uses the site as a training facility for its SEALS as the Silver Strand Training Complex. Battery #134 was destroyed in 2016 clearing the way for a new training facility. The site is closed to the general public.

Fort Emory Gun Batteries

- **Battery #134**: A 1940 Program dual 16-inch casemated battery intended for the southern defenses of San Diego Harbor. This unit was not originally on the 1940 projects list but added later in 1941 (hence being out of sequence for numbering). A number of possible location sites were proposed, before being approved for the Coronado Heights reservation by August 11, 1941—where it was assigned national priority #24. Initial startup construction was delated until authority granted November 13, 1942. Physical construction was begun on March 27, 1943. Work was suspended on February 21, 1944. At that time most of the concrete work was completed, but much equipment

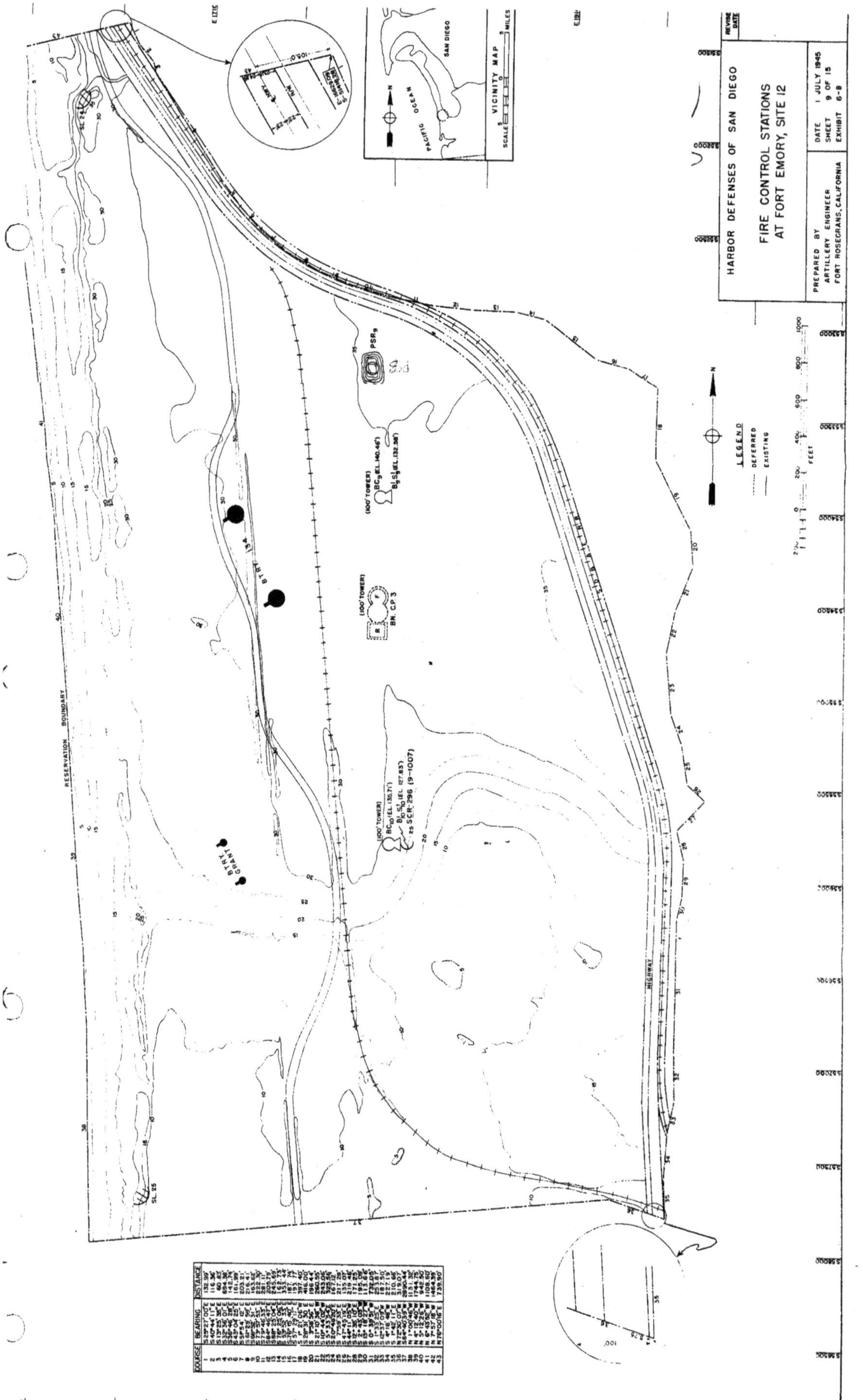

HARBOR DEFENSES OF SAN DIEGO
FIRE CONTROL STATIONS
AT FORT EMORY, SITE 12
PREPARED BY
ARTILLERY ENGINEER
FORT ROSECRANS, CALIFORNIA
DATE 1 JULY 1945
SHEET 9 OF 15
EXHIBIT 6-8
VICINITY MAP
PACIFIC OCEAN
SAN DIEGO
LEGEND
DEFERRED
EXISTING
FEET
RESERVATION BOUNDARY
HIGHWAY
SL 24
SL 25
PSR9
BTRY 134
BTRY GRANT
(100' TOWER)
BN. C.P. 3
SCR-296 (9-1007)

outfitting was still required. The battery was of standard late-war design, with gun rooms separated by 500-feet and the usual traverse magazine arrangement. It was transferred provisionally on November 11, 1944 at a cost of $1,044,907.29. The separate PSR room to the northeast was transferred on April 25, 1944 for $121,348.15. There is some indication that additional work was taken up for a short while postwar, during 1946, still never completed. No armament was ever mounted in the battery. While never officially named, the emplacement was sometimes referred to locally as Battery Cortez or Battery Gatchell. The emplacement was present for many years at the Imperial Beach Radio Station (and later as the Silver Strand Training Complex) but was finally destroyed by the U.S. Navy in 2015 to build a new SEAL training facility. The battery's PSR structure remains and is used by the U.S. Navy.

- **GRANT**: A 1940 Program dual 6-inch barbette battery also planned for emplacement at Fort Emory. Built as Battery Construction No. 239, it was assigned initially a rather low priority and not funded until the FY-1943 Budget. It was placed on the Emory reservation just to the south of the site for the 16-inch battery, firing to the east. Work was begun on June 5, 1942 and completed by December 8, 1943. It was transferred on April 25, 1944 for $218,851.95. It was of standard 200-series design type. It was armed with two 6-inch guns Model 1905A2 on Model M1 barbette carriages (#324/#56 and #20/#57). The battery was named in General Orders No. 69 of December 22, 1942 for Colonel Homer Grant, Coast Artillery Corps. The battery was retained under the 1946 Program but soon abandoned and the guns eventually removed. The U.S. Navy took over the emplacement for several functions since that time. The U.S. Navy continues to use the emplacement as part of the Silver Strand Training Complex (SSTC). The battery is not open to the public.

The demolition of Battery #134 Fort Emory (Marvin Henzie)

PACIFIC OCEAN

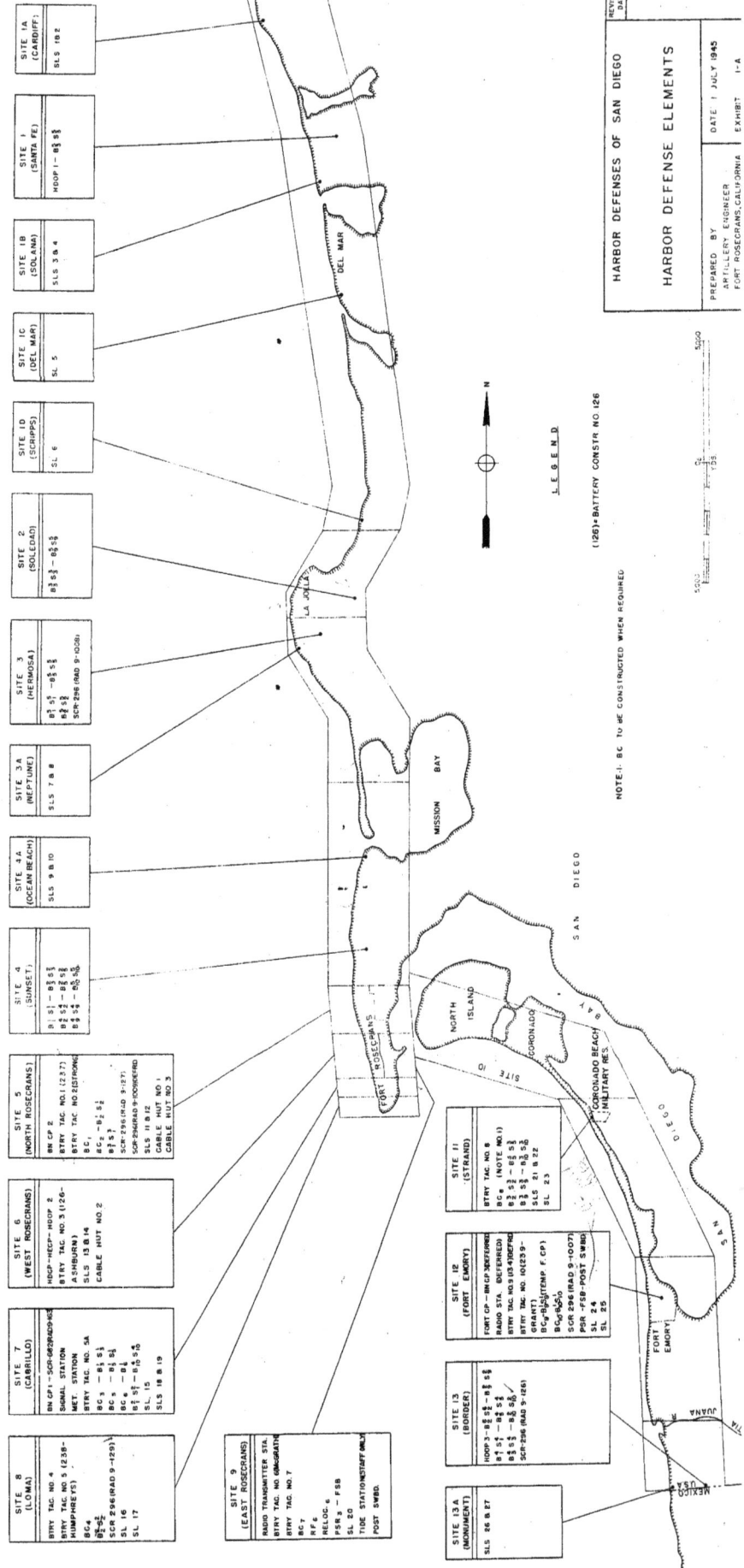

SITE 1A (CARDIFF)
SLS 1 & 2

SITE 1 (SANTA FE)
HDGP 1 — B½ S½

SITE 1B (SOLANA)
SLS 3 & 4

SITE 1C (DEL MAR)
SL 5

SITE 1D (SCRIPPS)
SL 6

SITE 2 (SOLEDAD)
B½ S½ — B½ S½

SITE 3 (HERMOSA)
B½ S½ — B½ S½
B½ S½
SCR 296 (RAD 9-102B)

SITE 3A (NEPTUNE)
SLS 7 & 8

SITE 4A (OCEAN BEACH)
SLS 9 & 10

SITE 4 (SUNSET)
B½ S½ — B½ S½
B½ S½ — B½ S½
B½ S½ — B½ S½

SITE 5 (NORTH ROSECRANS)
BN CP 2
BTRY TAC NO. 1 (2,3,7)
BTRY TAC NO. 2 (STRONG)
BC,
BC₁ — B½ S½
B½ S½
SCR 296 (RAD 9-127)
SCR 296 (RAD 9-100) & CRD
SLS 11 & 12
CABLE HUT NO 1
CABLE HUT NO 3

SITE 6 (WEST ROSECRANS)
HDGP — HDGP HDGP 2
BTRY TAC NO. 3 (126-ASHBURN)
SLS 13 & 14
CABLE HUT NO. 2

SITE 7 (CABRILLO)
BN CP 1 — SCR-682 & RADIO STA
SIGNAL STATION
MET. STATION
BTRY TAC. NO. 5A
BC₃ — B½ S½
BC₆ — B½ S½
B½ S½ — B.4,5,6
SL 15
SLS 18 & 19

SITE 8 (LOMA)
BTRY TAC NO. 4
BTRY TAC NO. 5 (2,38-HUMPHREYS)
BC₄
B½ S½
SCR 296 (RAD 9-129)
SL 16
SL 17

SITE 9 (EAST ROSECRANS)
RADIO TRANSMITTER STA.
BTRY TAC NO. 6
BTRY TAC. NO. 7
BC₇
RF₆
RELOC 6
PSR₃ — FSB
TIDE STATION & STAFF & MET.
POST SWBD.

SITE 11 (STRAND)
BTRY TAC. NO. 8 (NOTE NO.1)
BC₈ (NOTE NO.1)
B½ S½ — B½ S½
B½ S½ — B½ S½
SLS 21 & 22
SL 23

SITE 12 (FORT EMORY)
FORT CP — BN CP 3 (DEFERRED)
RADIO STA. (DEFERRED)
BTRY TAC. NO.9 (3-KEEFE)
BTRY TAC. NO. 10 (239-GRANT)
BC₉ — B½ S½ (TEMP F.CP)
BC₁₀ — B½ S½
SCR 296 (RAD 9-100?)
PSR — FSB-POST SWBD
SL 24
SL 25

SITE 13 (BORDER)
HDGP 3 — B½ S½ — B½ S½
B½ S½ — B½ S½
B½ S½ — B½ S½
SCR 296 (RAD 9-126)

SITE 13A (MONUMENT)
SLS 26 & 27

DEL MAR

LA JOLLA

MISSION BAY

SAN DIEGO

FORT ROSECRANS

NORTH ISLAND

CORONADO

SAN DIEGO BAY

CORONADO BEACH MILITARY RES.

SITE 10

FORT EMORY

TIA JUANA R.

MEXICO

LEGEND

N

(126) = BATTERY CONSTR NO. 126

NOTE I- BC to BE CONSTRUCTED WHEN REQUIRED

5000

5000

HARBOR DEFENSES OF SAN DIEGO

HARBOR DEFENSE ELEMENTS

PREPARED BY
ARTILLERY ENGINEER
FORT ROSECRANS, CALIFORNIA

DATE 1 JULY 1945

EXHIBIT I-A

San Diego World War II-era Site Locations. Stations housed in a single structure are connected by dashes (-)

location	Loc#	Purpose
Sante Fe	1	BS5/126, HDOP 1
Cardiff	1A	SL 1,2
Solana	1B	SL 3,4
Del Mar	1C	SL 5
Scripps	1D	SL 7,8
Soledad	2	BS3/126-BS5/134
Hermosa	3	BS5/237-BS5/238, BS5/Strong, SCR 296
Neptune	3A	SL 7,8
Sunset	4	BS1/237-BS2/126, BS4/Strong-BS2/238, BS5/239-BS4/134
Ocean Beach	4A	SL 9,10
Fort Rosecrans North	5	Batt. Tact. #1 BCN 237, Batt. Tact. #2 Strong, BnOP2, BC 237-BS2/237, BC Strong-BS1/Strong, BS3/126, SL 11,12, SCR296, SCR 296
Fort Rosecrans West	6	Batt. Tact. #3 Ashburn, HECP-HDEP-HDOP2, SL 13,14
Cabrillo	7	BnOP1, BC 126-B1/126, Met, SCR 682
Loma	8	Batt. Tact. #4 Humphries, BC/Humphries, BS2/Strong, SCR 296, SL 16,17
East Fort Rosecrans	9	Batt Tact. #5 McGrath, BC McGrath, BC7 RF6, PSR 126, SL 20, Tide, SBR
North Island	10	FC unassigned
Coronado Beach/ Strand	11	BC8, BS3/Strong-BS4/238 , SL 21,22,23
Fort Emory	12	Batt. Tact. #6 BCN 134, Batt Tact. #7 Grant, BnCP, BC 134-B1/134, BC 239-B1/239, SCR296, PSR 134, SL 24,25
Border	13	HDOP3-BS6/Strong-BS2/134, BS4/237-BS4/238, BS4/126-BS2/239, SCR296
Monument	13A	SL 26,27

Thompson, Erwin N., and Howard B. Overton. *Historic Resource Study, the Guns of San Diego*, Cabrillo National Monument, NPS. San Diego, CA, 1991.

Joyce, Barry A. *A Harbor Worth Defending, a Military History of Point Loma.* The Cabrillo Historical Association. San Diego, CA, 1995.

Everett, H.R. *WWII Harbor Defenses of San Diego*, CDSG Press, McLean, VA 2021

Shell hoist Battery Calef-Wilkeson Fort Rosecrans USN (Mark Berhow)

THE HARBOR DEFENSES OF LOS ANGELES – CALIFORNIA

Los Angeles did not have a natural protected harbor until a breakwater was completed in 1909. By then Los Angeles had become a bustling transportation center and port. A special act of Congress provided funding for the construction of the defenses in 1910 which were completed in 1919. The 1940 Modernization Program construction provided four new modern batteries, and Fort MacArthur was the headquarters for the Nike defenses of Los Angeles from 1954 to 1972. After 1974 the U.S. Army phased out its use of the facilities transferring the properties to the City of Los Angeles and the US Air Force.

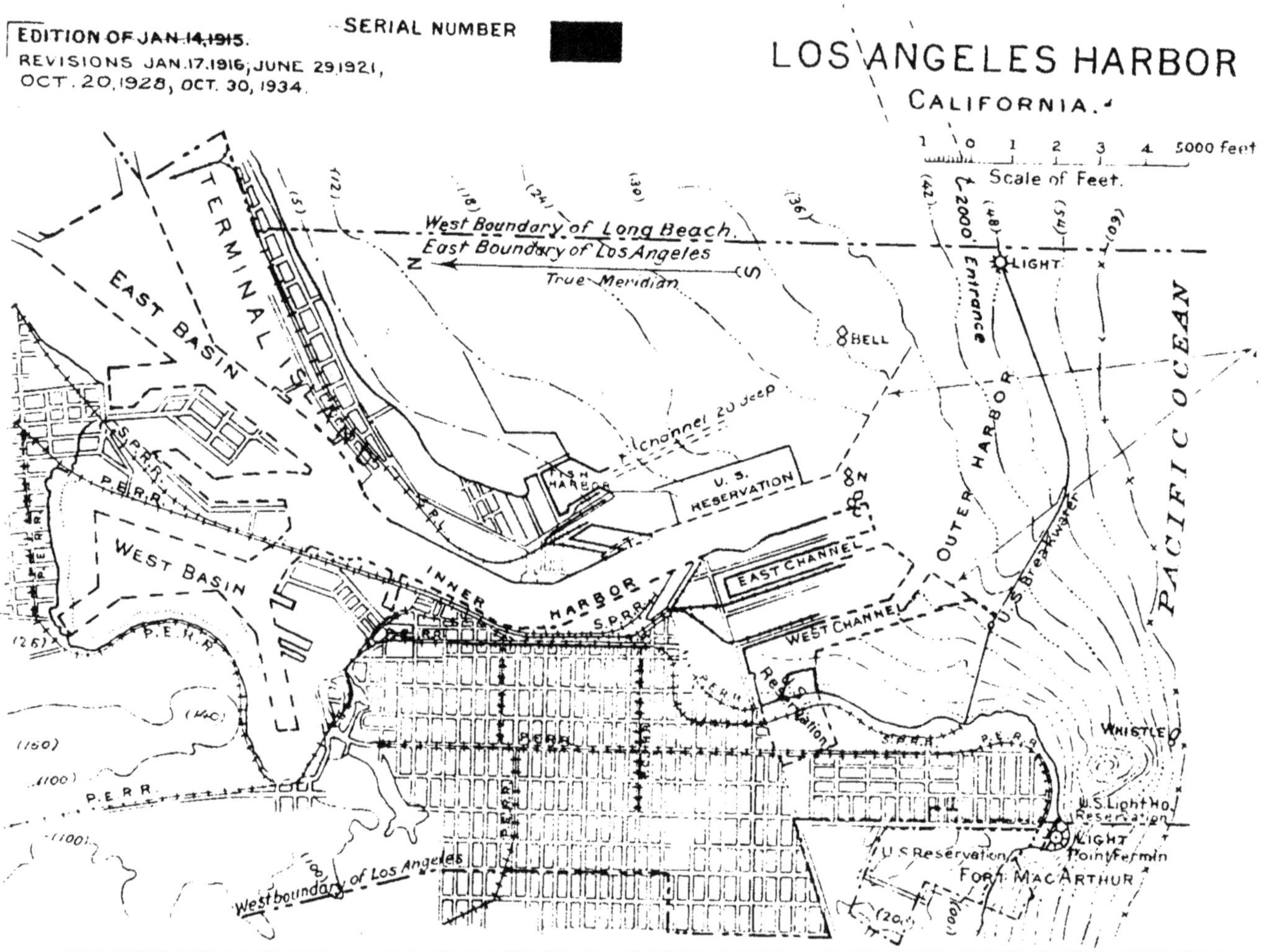

Angels Gate Park San Pedro (Terry McGovern)

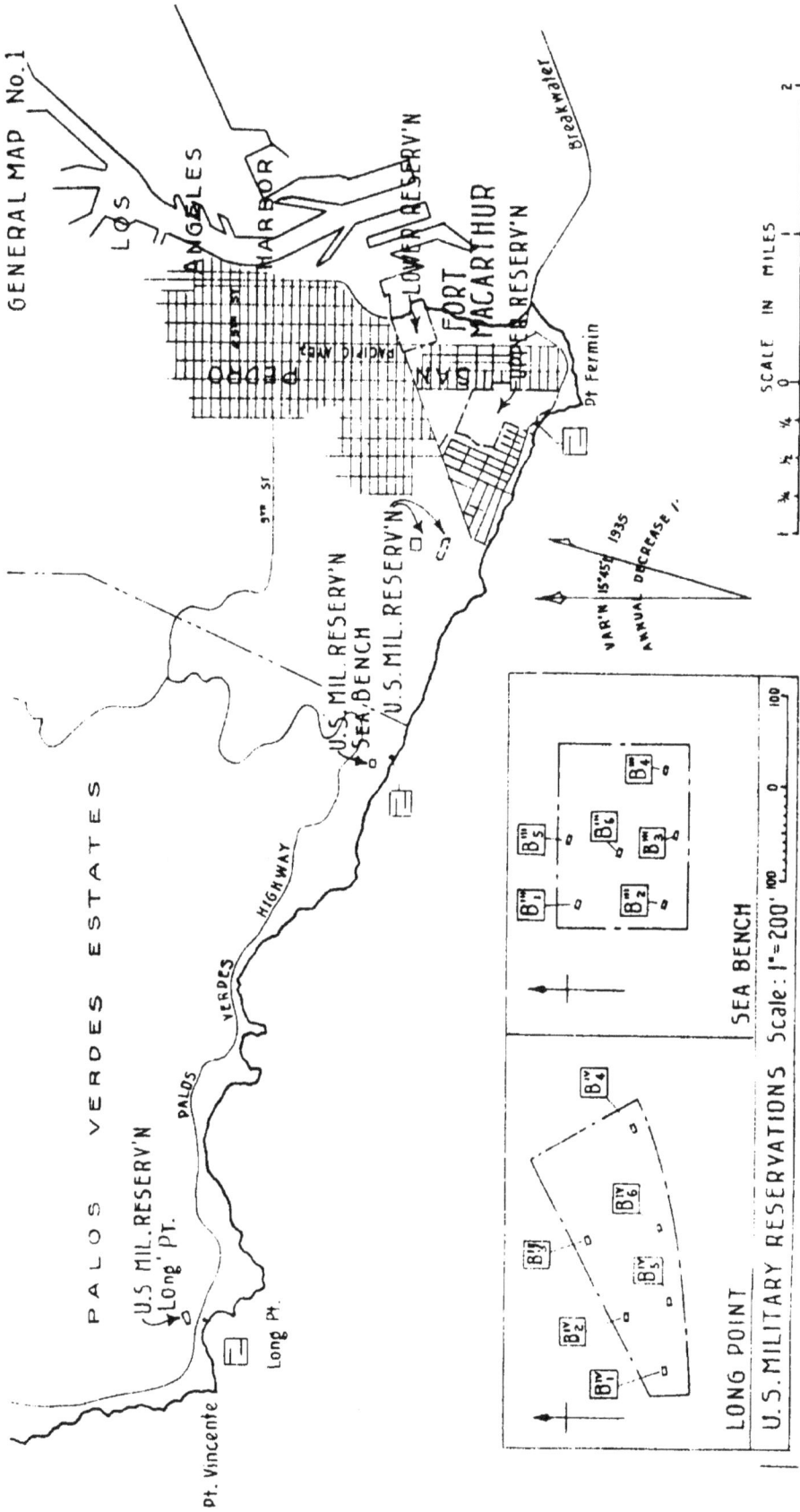

LOS ANGELES HARBOR
FORT MACARTHUR
GENERAL MAP No.1

SERIAL NUMBER

EDITION OF OCT. 20, 1928.
REVISIONS: OCT. 30, 1934.

PALOS VERDES ESTATES

U.S. MIL. RESERV'N
Long' Pt.

Pt. Vincente

Long Pt.

PALOS

VERDES

HIGHWAY

U.S. MIL. RESERV'N
SEA BENCH
U.S. MIL. RESERV'N

9TH ST

SAN PEDRO

LOS ANGELES

HARBOR

4TH ST

Pacific Ave.

LOWER RESERV'N

FORT MACARTHUR
UPPER RESERV'N

SAN

Pt. Fermin

Breakwater

VAR'N 15°45' 1935
ANNUAL DECREASE 1'.

SCALE IN MILES

Active Status

U.S. MILITARY RESERVATIONS   Scale: 1"=200'

LONG POINT

B_1^{IV} B_2^{IV} B_3^{IV} B_5^{IV} B_6^{IV} B_4^{IV}

SEA BENCH

B_1^{III} B_5^{III} B_6^{III} B_2^{III} B_3^{III} B_4^{III}

Long Point/Point Vicente Military Reservation (1942-1975) is about ten miles west of Fort MacArthur, near Point Vicente. Initially a small reservation with six fire control stations built in the late 1920s for the guns at Fort MacArthur, more property was acquired for the construction of the 1940 Program shielded 6-inch Battery Barnes (#240). A two-gun 155mm battery (1942) with Panama mounts was located on Long Point (unofficially named Battery Tucker). It has been destroyed. Two 1942 fire-control stations have been destroyed. Six older 1930s fire-control stations still exist in good condition. Nike Missile Site LA-55L was built there in 1954 which remained until 1972. The site was turned over the City of Ranchos Palos Verdes which uses the Nike Administration buildings as a civic center and the Nike launch area as a maintenance yard, while the Battery Barnes is locked up but abandoned.

Long Point Gun Battery

- **BARNES:** A standard 1940 program dual 6-inch barbette battery emplaced at a new military reservation at the southwestern extreme of the Palo Verde headlands. The emplacement was located on the ridge that extended to the south just inland at Point Vicente. It fired to the southwest. During planning and construction, it was known as Battery Construction No. 240. It was not funded until FY-1943 but pressed to relatively early completion. It was named postwar on General Orders No. 1 of January 9, 1948 for Colonel Henry C. Barnes, Coast Artillery Corps. The battery was completed for proof firing on October 25, 1943. It was turned over at a cost of $219,000. It was armed with two 6-inch Model 1903A2 guns on Model M1 barbette carriages (#63/#105 and #64/#106). The emplacement was retained in the 1946 Review but was deleted and disarmed in 1948. The emplacement still exists on Coast Guard property at the site of a radio beacon. The battery site is open, but the interior is not open to the public.

Battery H.C. Barnes Ranchos Palos Verdes City Administration Center (Terry McGovern)

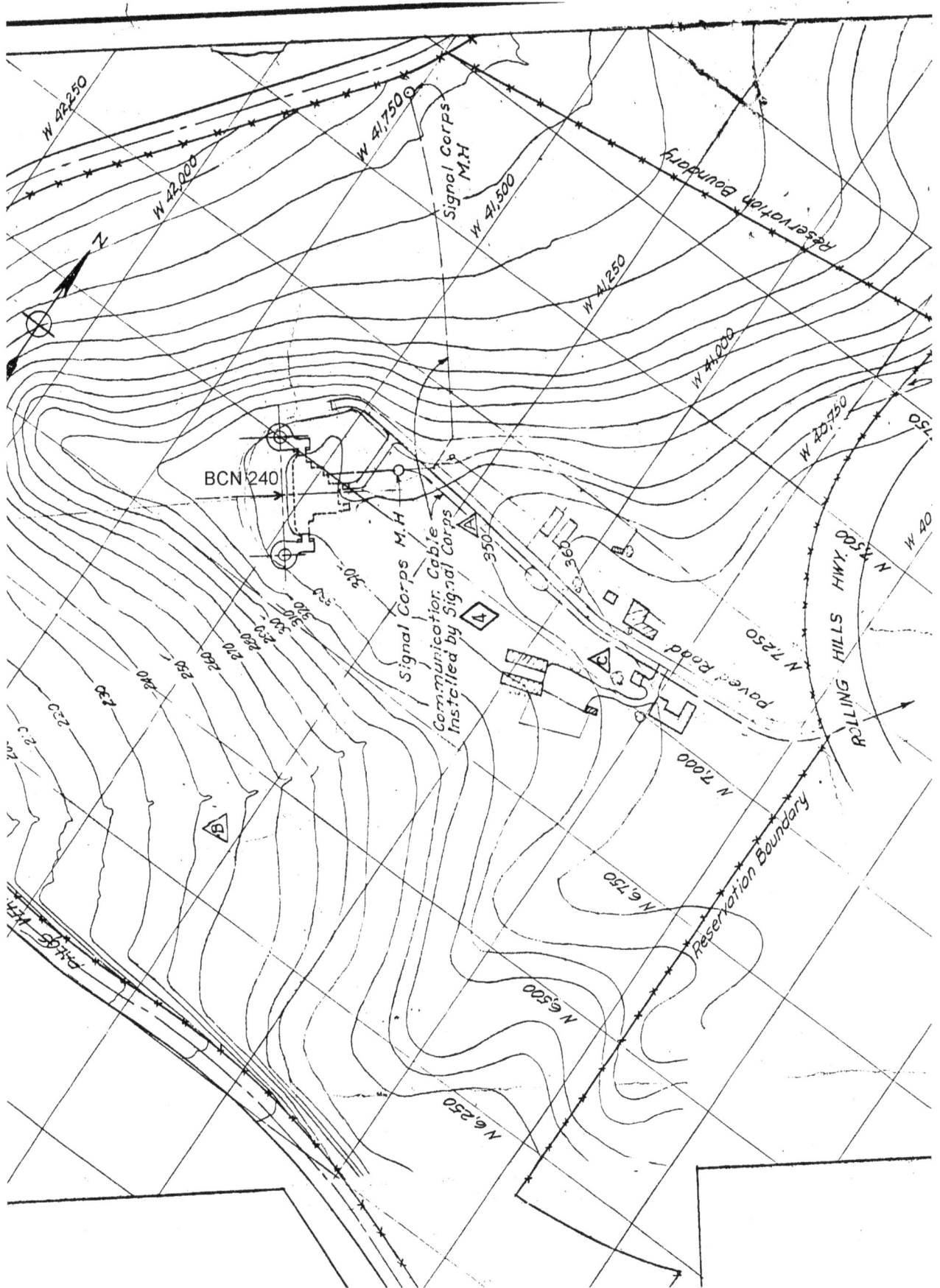

LOS ANGELES HARBOR
FORT MACARTHUR
GENERAL MAP No.2

SERIAL NUMBER

EDITION OF JUNE 29,1921
REVISIONS: APR 16,1925
OCT 20,1928, OCT. 30, 1934.

BATTERIES
Saxton .... 4-12"
Barlow .... 4-12"
Osgood .... J-14"L
Farley .... I-14"D
Leary .... I-14"D.
Merriam .... I-14"D.
*Lodor .... 4-3"Pz
A-Anti. aircraft gun+
J-14" Pt. GUN M-1920
Battery dismounted
Gun-housed o/pres
on Upper Reservation

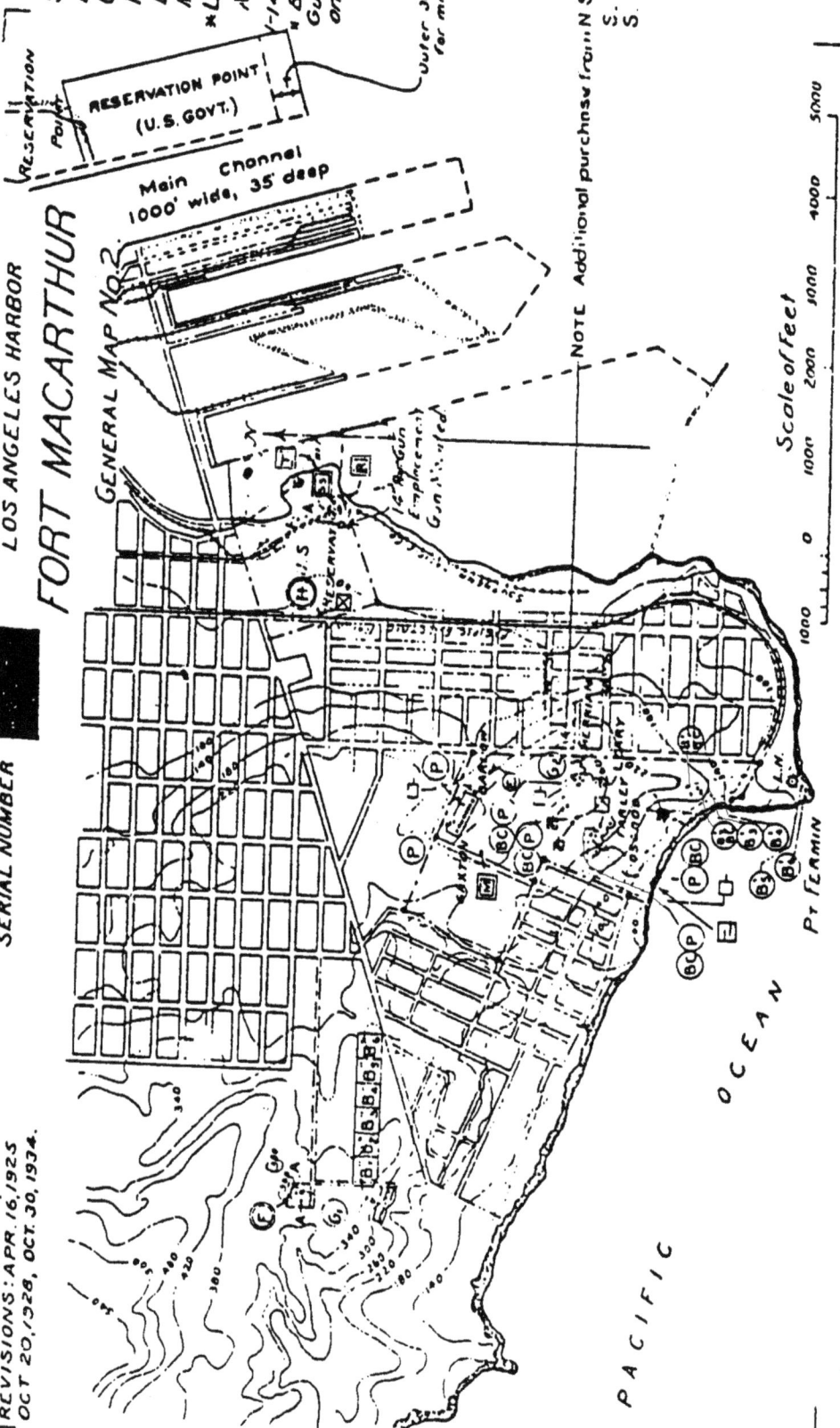

RESERVATION POINT
(U.S. GOVT.)

Main Channel
1000' wide, 35' deep

Note: Additional purchase from N Side 36"st to
S - 34th . not to
S - 33rd

Scale of Feet

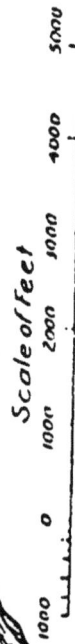

Active Status.

PACIFIC OCEAN

Pt FERMIN

White Point Military Reservation (1942-1975) is about three miles west of the Fort MacArthur Upper Reservation. This reservation was home to several fire control stations in the 1920s, and the location of 16-inch Battery Bunker which was completed as part of the 1940 Program and armed by 1945. The reservation was used as the launch area for Nike missile Battery LA-45. Currently the site is now home to the White Point Nature Preserve and Education Center, which is housed in the slightly modified missile assembly and warhead assembly buildings. Battery Bunker is accessible to the public, but its interior is closed.

White Point Gun Battery

- **BUNKER**: A modern 1940 Program battery for two 16-inch casemated guns. The site selected was on the ridge that ran east-west roughly parallel to the shore inland, and to the west of the Fort MacArthur upper reservation. During construction it was known as Battery Construction No. 127. It was allocated national priority #11 on the September 11, 1940 list, advancing to #8 on the list of August 11, 1941. Funds were allocated in the FY-1942 Budget. Actual construction was done from April of 1942 until December 1943. It was transferred on September 11, 1944 for a cost of $1,256,409.80. It was a nearly standard 100-series design for casemated guns, with a 500-foot spacing between mounts. The western, No. 1 emplacement gun house was turned 24-degrees from the straight axis in order to provide for a better field of fire to the west. An accompanying PSR building was constructed during the same dates and transferred also on September 11, 1944 for $132,189.37. The battery construction necessitated the removal of several base end stations previously on this ridge. The battery was armed with two 16-inch navy guns MkIIM1 on M1919M4 barbette carriages (#54/#30 and #72/#45). The battery was one of the very few named postwar, that coming in General Orders No. 51 of June 10, 1946 being named for Colonel Paul D. Bunker, the Coast Artillery commander from Corregidor who died in captivity during World War II. The battery was retained in the 1946 Program and was not ultimately disarmed until 1948 or 1949. The emplacement still exists as part of the White Point Nature Preserve. The interior of the battery is sealed, but the exterior of the battery can be visited.

Battery Bunker White Point Nature Preserve (Terry McGovern)

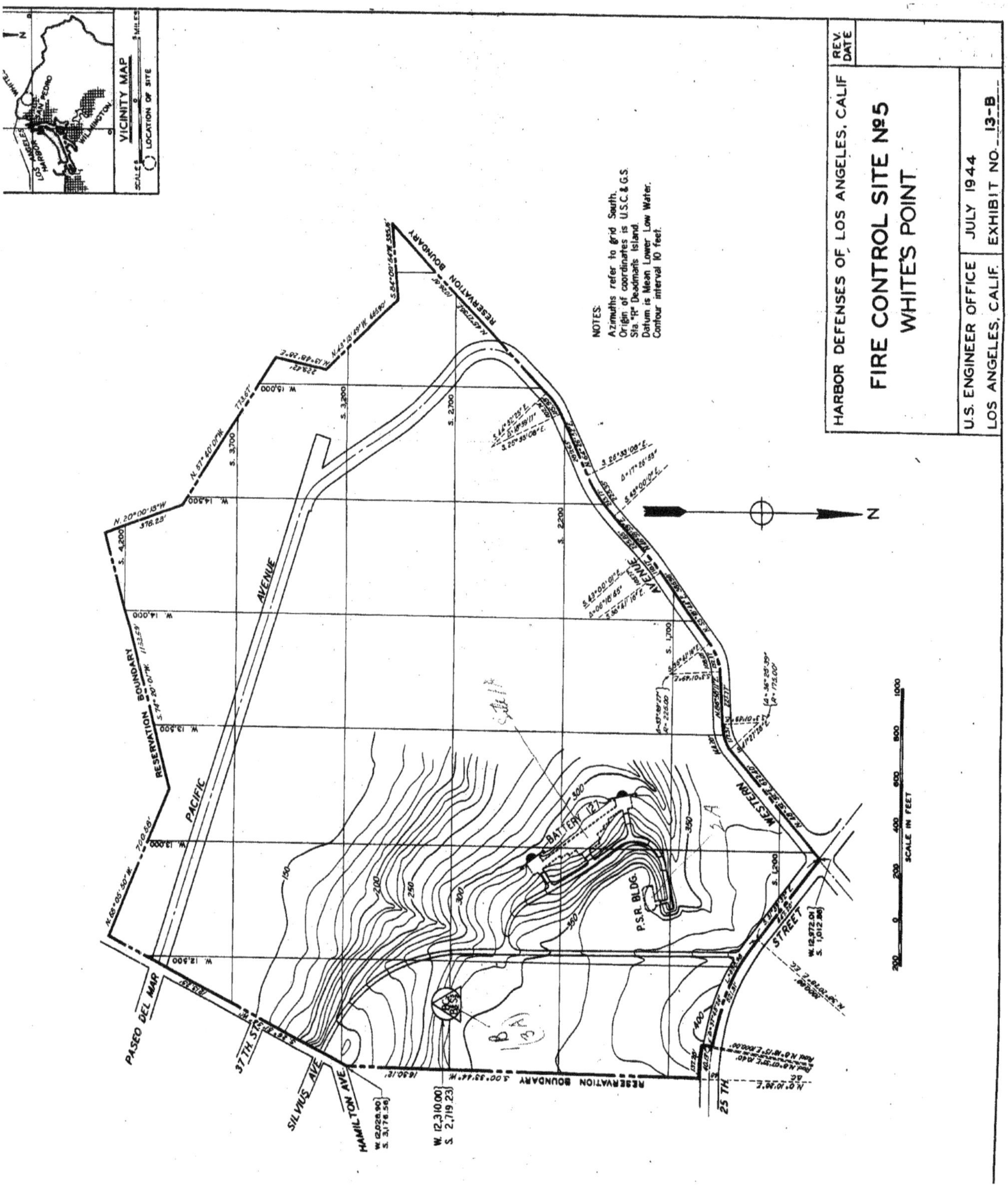

HARBOR DEFENSES OF LOS ANGELES, CALIF
FIRE CONTROL SITE Nº5
WHITE'S POINT
U.S. ENGINEER OFFICE
LOS ANGELES, CALIF.
JULY 1944
EXHIBIT NO. 13-B
REV. DATE
VICINITY MAP
LOCATION OF SITE
NOTES:
Azimuths refer to grid South.
Origin of coordinates is U.S.C.& G.S. Sta. "R" Deadman's Island.
Datum is Mean Lower Low Water.
Contour interval 10 feet.
RESERVATION BOUNDARY
PACIFIC AVENUE
WESTERN AVENUE
STREET
PASEO DEL MAR
37TH ST.
SILVIUS AVE
HAMILTON AVE
25 TH
BATTERY 127
P.S.R. BLDG.
SCALE IN FEET
N

Fort MacArthur (1914-1975) is located in San Pedro near Point Fermin on the Palos Verde Peninsula. The harbor defenses were built during the years after the Taft Program. It was named in General Orders 1 of 1914 for Lt. Gen. Arthur MacArthur, U.S. Army. The post was divided between a lower reservation with the garrison buildings off of Pacific Avenue in San Pedro, and the Upper Reservation off of Gaffey Street which held the major coast artillery armament. In 1917, three large concrete emplacements (two 14-inch DC batteries and one 12-inch mortar battery) were constructed on the Upper Reservation and one 3-inch battery was built across the channel on Deadman's Island. The harbor defenses were upgraded by the addition of two 14-inch railway guns which were stationed at Fort MacArthur's Lower Reservation after 1925. The 1940 Program construction resulted in a 6-inch battery added to the defenses of Fort MacArthur. Also added at the fort and around the harbor were three 90mm AMTB batteries and several Panama mounts for 155mm GPF mobile artillery. After the war Fort MacArthur served as a reserve training facility and a command center for Nike missile systems, the upper reservation was used as a IFC site for a Nike missile battery. In 1975, the fort's Upper Reservation was transferred to the City of Los Angeles as a park, while the U.S. Air Force took over the Lower Reservation, which retained a significant portion of the garrison area including the barracks, HQ buildings, post exchange, guard house and officer's quarters. It is not open to the general public. Angels Gate Park is now home to the Angeles Gate Cultural Center, a marine mammal rescue center, a high school and other school district activities, the Korean Friendship Bell. Battery Osgood-Farley is home to the Fort MacArthur Museum. The battery retains much of its service hardware, and has a restored generator room, plotting room, switchboard room, and battery commander's station. The museum's interior is currently closed pending some remediation and organizational issues.

Fort MacArthur Gun Batteries

- **OSGOOD – FARLEY**: A two-gun 14-inch disappearing battery emplaced on the bluff above Point Fermin at the Fort MacArthur upper reservation. This was the western battery of two guns, situated to fire to the south and provide the major coverage to the roadstead south of the highlands. Osgood was the western, No. 1 emplacement, and Farley the eastern, No. 2 mount. Site plans were submitted on July 8, 1914, construction plans on November 10, 1914. The battery was on the western side of the heights that were used for sister battery Leary-Merriam, at a crest of 240-feet. It utilized a single-level type plan with well-spaced gun platforms and adjacent, shared magazine in the central traverse between the guns. It needed no hoists. The rear portion of the traverse was used for the emplacement power plant, other support rooms were in a separate building to the rear of the traverse. The battery was approached by an access road from the right side. Work was done from September 1916 to November 1917. Transfer came on October 10, 1919 for a cost of $211,426. The battery names were issued on General Orders No. 15 of April 25, 1916 for Brigadier General Henry B. Osgood and Brigadier General Joseph P. Farley. Battery Farley was armed with one 14-inch gun Model M1910M1 on Model M1907M1 disappearing carriage (#17/#21). Battery Osgood was armed with one 14-inch gun Model M1910M1 on Model M1907M1 disappearing carriage (#13/#8). The battery served until late into World War II, the armament not being removed until January 1944. The emplacement has not been materially altered and several rooms, including the power room, were restored and interpreted by the Fort MacArthur Museum. However, the museum interior is currently closed to the public.

- **LEARY – MERRIAM**: A two-gun 14-inch disappearing battery emplaced on the bluff above Point Fermin at the Fort MacArthur upper reservation. These were the eastern pair of guns, situated to fire to the southeast and provide the major coverage to San Pedro Harbor. It occupied the summit of the hill above Pt. Fermin on the reservation, the emplacement being at a crest of 285-feet.

LOS ANGELES HARBOR, CAL.
FORT MACARTHUR-DI

SERIAL NUMBER 124

EDITION OF JUNE 29, 1921.

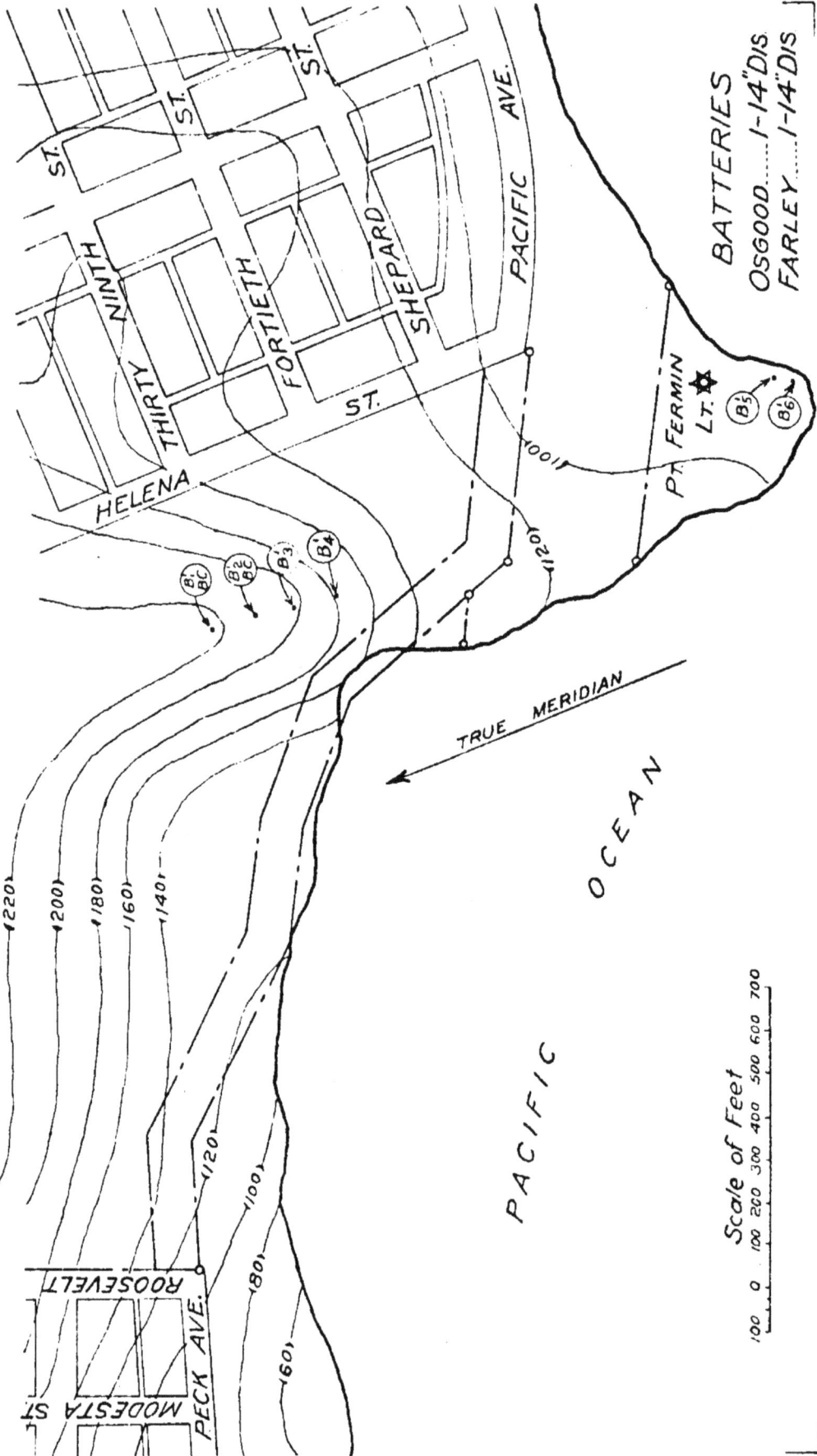

BATTERIES
OSGOOD......I-14"DIS.
FARLEY....I-14"DIS.

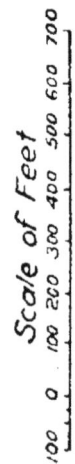

TRUE MERIDIAN

PACIFIC OCEAN

Scale of Feet
100 0 100 200 300 400 500 600 700

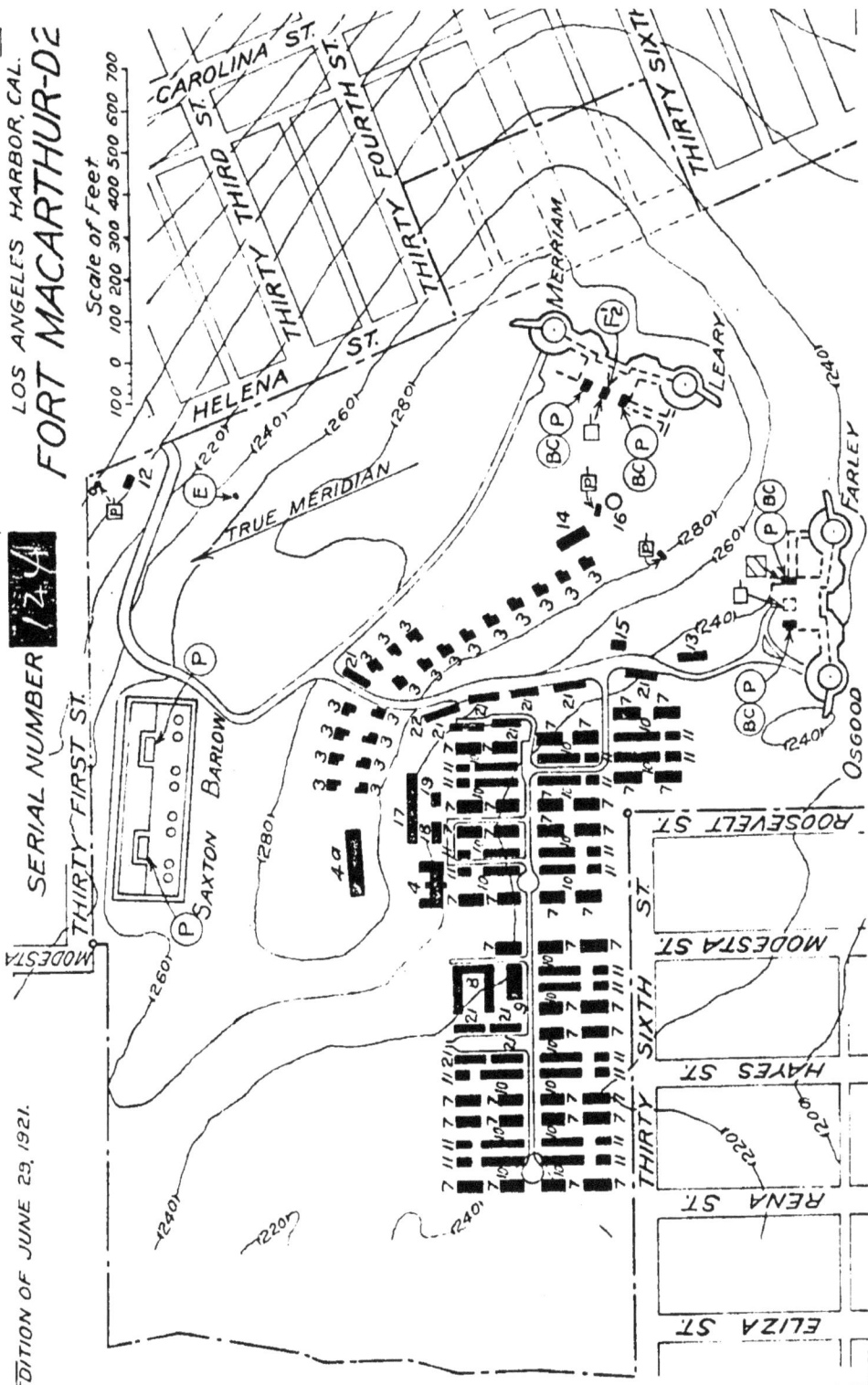

LOS ANGELES HARBOR, CAL.

FORT MACARTHUR-DE

Scale of Feet
100 0 100 200 300 400 500 600 700

SERIAL NUMBER 1244

EDITION OF JUNE 29, 1921.

TRUE MERIDIAN

LEGEND

1.
2.
3. OFFICERS' QUARTERS.
4. INFIRMARY.
4a HOSPITAL WARD.
5
6.
7. BARRACKS.
8. GUARD HOUSE.
9. POST EXCHANGE
10 MESS HALL.
11. LAVATORY.
12. GAS INSTR. BLDG.
13 AUTO REPAIR SHOP.
14. BLACKSMITH SHOP.
15 FIRE STATION.
16 RESERVOIR.
17 HOSPITAL BARRACKS.
18 HOSPITAL MESS.
19. HOSPITAL LAVATORY.
21 Q.M. STOREHOUSE.
22 Q.M. WAREHOUSE.

BATTERIES
SAXTON....4-12"M.
BARLOW....4-12"M.
OSGOOD.....1-14"DIS
FARLEY......1-14"DIS
LEARY.......1-14"DIS
MERRIAM..1-14"DIS

CAROLINA ST.
THIRTY THIRD ST.
THIRTY FOURTH ST.
THIRTY SIXTH
HELENA ST.
THIRTY FIRST ST.
MODESTA
SAXTON
BARLOW
MERRIAM
LEARY
FARLEY
OSGOOD
ROOSEVELT ST.
MODESTA ST.
HAYES ST.
THIRTY SIXTH ST.
RENA ST.
ELIZA ST.

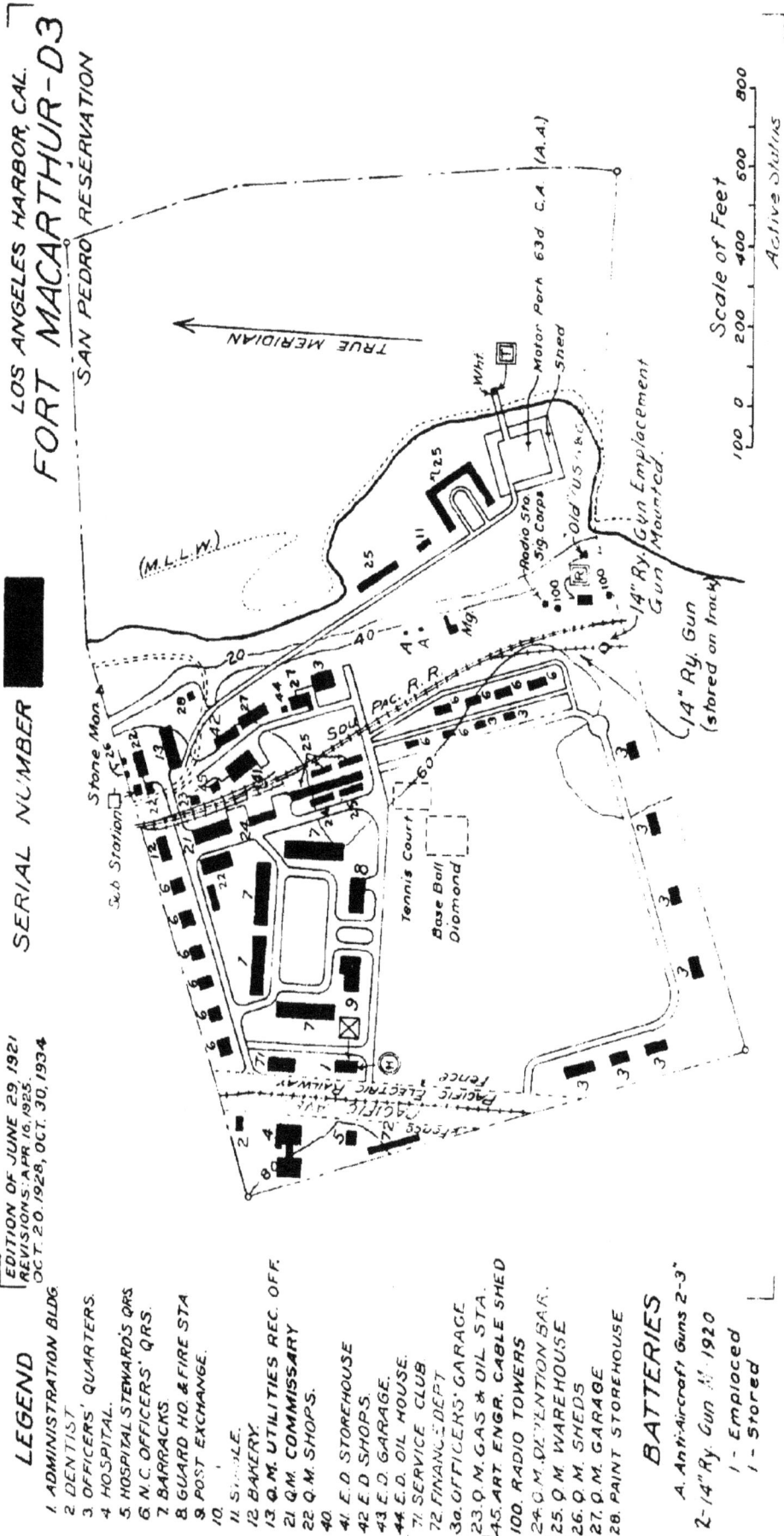

LOS ANGELES HARBOR, CAL.
FORT MACARTHUR-D3
SAN PEDRO RESERVATION

SERIAL NUMBER

EDITION OF JUNE 29, 1921
REVISIONS: APR. 16, 1925,
OCT. 20, 1928, OCT. 30, 1934.

LEGEND
1. ADMINISTRATION BLDG.
2. DENTIST
3. OFFICERS' QUARTERS.
4. HOSPITAL.
5. HOSPITAL STEWARD'S QRS
6. N.C. OFFICERS' QRS.
7. BARRACKS
8. GUARD HO. & FIRE STA
9. POST EXCHANGE.
10.
11. STABLE.
12. BAKERY.
13. Q.M. UTILITIES REC. OFF.
21. Q.M. COMMISSARY
22. Q.M. SHOPS.
40.
41. E.D STOREHOUSE
42. E.D SHOPS.
43. E.D. GARAGE.
44. E.D. OIL HOUSE.
71. SERVICE CLUB
72. FINANCE DEPT
30. OFFICERS' GARAGE.
23. Q.M. GAS & OIL STA.
45. ART. ENGR. CABLE SHED
100. RADIO TOWERS
24. Q.M. DETENTION BAR.
25. Q.M. WAREHOUSE
26. Q.M. SHEDS
27. Q.M. GARAGE
28. PAINT STOREHOUSE

BATTERIES
A. Anti-Aircraft Guns 2-3"
2-14" Ry. Gun 1920
1- Emplaced
1- Stored

TRUE MERIDIAN

Scale of Feet
100 0 200 400 600 800
Active Status

Motor Park 63d C.A. (A.A.)
14" Ry. Gun Emplacement Gun Mounted
14" Ry. Gun (stored on track)
Old U.S. R.R.
Radio Sta Sig Corps
Shed
Wht
Sub Station
Stone Mon
Tennis Court
Base Ball Diamond
Sou. Pac. R.R.
PACIFIC ELECTRIC RAILWAY
PACIFIC AVE
defense fence
(M.L.L.W.)

Angels Gate Park (Terry McGovern)

Battery Barlow-Saxton (Terry McGovern)

LOS ANGELES HARBOR, CAL.
DEADMAN'S ISLAND
RESERVATION POINT.

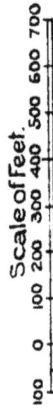

Scale of Feet.
100   0   100 200 300 400 500 600 700

SERIAL NUMBER 124

Line 300 Yards out beyond Low Water Mark (Ceded to U.S. by Act of Legislature of the State of Cal. Mar. 9, 1897.)

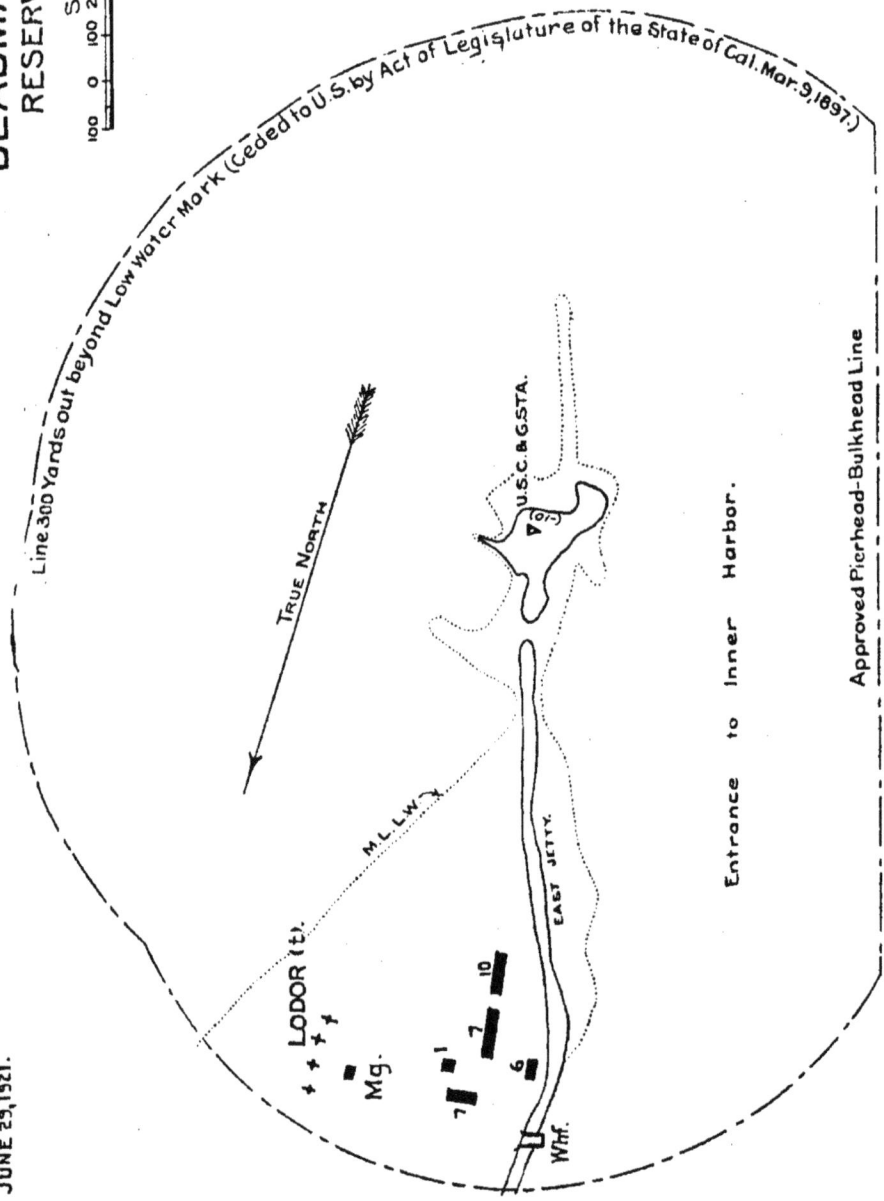

TRUE NORTH

U.S.C. & G. STA.

M.L.L.W.

LODOR (t).

Mg.

7

1

7

10

6

Whf.

EAST JETTY.

Entrance to Inner Harbor.

Approved Pierhead-Bulkhead Line

LEGEND.
EDITION OF JAN. 17, 1916.
REVISED SEPT. 12, 1918;
JUNE 29, 1921.

1. ADMINISTRATION BUILDING.
6. QUARTERS.
7. BARRACKS.
10. MESS HALL.

BATTERIES.
LODOR   4-3" Ped.

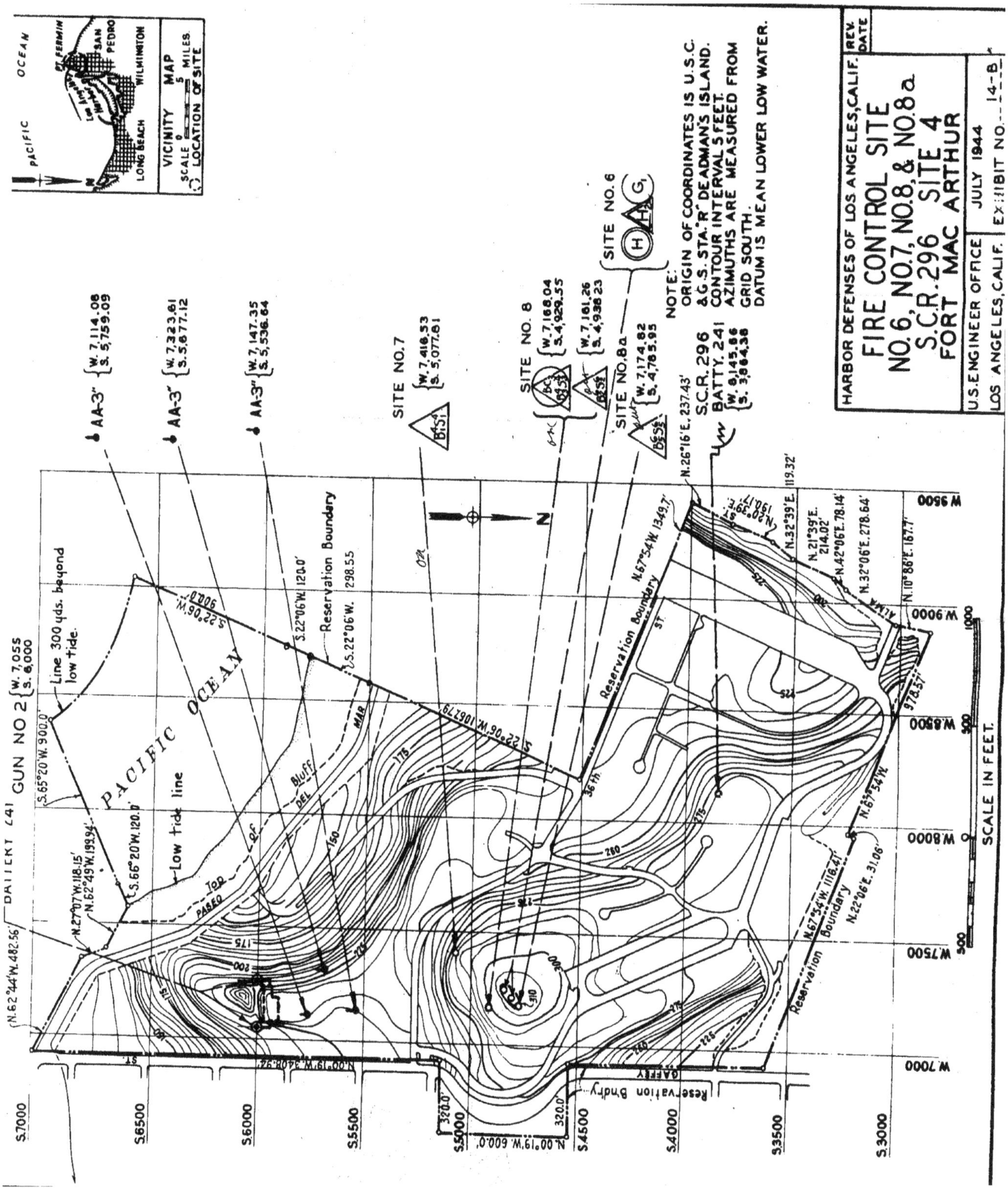
HARBOR DEFENSES OF LOS ANGELES, CALIF.
FIRE CONTROL SITE
NO. 6, NO. 7, NO. 8, & NO. 8a
S.C.R. 296 SITE 4
FORT MAC ARTHUR
U.S. ENGINEER OFFICE
LOS ANGELES, CALIF.
JULY 1944
EXHIBIT NO. 14-B
REV. DATE
NOTE:
ORIGIN OF COORDINATES IS U.S.C. & G.S. STA. "R" DEADMAN'S ISLAND.
CONTOUR INTERVAL 5 FEET.
AZIMUTHS ARE MEASURED FROM GRID SOUTH.
DATUM IS MEAN LOWER LOW WATER.
VICINITY MAP
SCALE 0 5 MILES
LOCATION OF SITE
PACIFIC OCEAN
LONG BEACH
WILMINGTON
SAN PEDRO
PT. FERMIN
BATTERY 241 GUN NO 2 W. 7,055 S. 6,000
Line 300 yds. beyond low tide.
Low tide line
PACIFIC OCEAN
Reservation Boundary
SITE NO. 6
SITE NO. 7 W. 7,416.53 S. 5,077.81
SITE NO. 8 W. 7,168.04 S. 4,929.55
W. 7,161.26 S. 4,938.23
SITE NO. 8a W. 7,174.82 S. 4,785.95
S.C.R. 296 BATTY. 241 W. 8,145.86 S. 3,864.38
AA-3" W. 7,114.08 S. 5,759.09
AA-3" W. 7,323.61 S. 5,677.12
AA-3" W. 7,147.35 S. 5,536.64
SCALE IN FEET

Fort MacArthur Lower Reservation 1937 (NARA)

Fort MacArthur Upper Reservation 1937 (NARA)

Battery Farley Angels Gate City Park (Mark Berhow)

Restored power room iin Battery Osgood-Farley Angels Gate City Park (Mark Berhow)

Using the experience of the widely spaced, adjacent/same-level magazines of the Panama 14-inch batteries, it utilized all of those features in its design. The two platforms were splayed out to give a wider field of fire, and between them was the traverse with shared magazine. Access was from a road that led to a tunnel on the right side, the rear of the traverse was covered for more protection from flank fire than Osgood-Farley. Leary was the western, No. 1 emplacement, Merriam the eastern, No. 2 mount. Battery names were conferred with General Orders No. 15 of April 25, 1916 for Brigadier General Peter Leary and Major General Henry C. Merriam, both recently deceased. Work was done from November 1915 to November 1917. It was transferred on October 10, 1919 for a cost of $251,363.63. Battery Leary was armed with one 14-inch gun Model M1910M1 on Model M1907M1 disappearing carriage (#14/#18). Battery Merriam was armed with one 14-inch gun Model M1910M1 on Model M1907M1 disappearing carriage (#16/#19). The batteries served with this armament until deleted by authority of December 1943. The actual dismounting of armament came with an order of January 14, 1944. Subsequent to this the new, permanent combined HECP/HDCP was built into the battery structure in early 1944. The emplacement still exists, though somewhat modified for the HECP, later Nike, and finally the Southern California Marine Exchange and Vessel Traffic Service. It is not generally open to the public.

- **BARLOW – SAXTON:** The emplacement for eight 12-inch mortars in four pits emplaced in a natural depression on the northern limit of the reservation. It fired to the south, southwest. The plans called for a complete parados in the rear of the pits to protect the battery from fire coming from the inner harbor and to screen it from public streets just beyond. Site plan was submitted on August 14, 1913, detailed engineering plans on June 18, 1914. It followed the type plans recently developed for similar batteries in Panama. It was a continuous battery with four pits each containing two mortars, split administratively into two named batteries. Magazines were under the parapet in front of the battery, and traverses between the pits held corridors for access and delivery of shell carts. The entire emplacement was accessed through a tunnel on the east side, in which also were located the latrines, power rooms, and storerooms. The retired position of the battery on the backside of the bluff allowed the front concrete to be reduced to a bare 3-feet in thickness. Work was done with an initial appropriation of $167,000 made on August 19, 1914 (and another $39,000 following that on March 4, 1915). Work was done from April 1, 1915 and completed for transfer of June 28, 1919 at a construction total cost of $290,300. It was named on General Orders No. 15 of April 25, 1916 for Brigadier General Rufus Saxton (two western pits) and Brigadier General John Barlow (two eastern pits). It was armed with eight Model 1912 12-inch, 15-caliber mortars on Model 1896MIII carriages (Saxton: #37/#29, #30/#30, #31/#31, #32/#32 and Barlow: #33/#33, #34/#34, #35/#35, and #36/#36). The mortars and carriages were shipped here on May 23, 1916. The battery had a long service life, and the armament was never changed. Removal and destruction came with the completion of more modern defenses in late 1943. The emplacement still exists and has been extensively used for filming television and movie projects. The emplacement is on property of the Los Angeles Unified School District, and not normally open for visits.

- **LODOR:** A battery for four 3-inch pedestal guns located on what was known as Dead Man's Island on the eastern side of the entrance to the San Pedro Harbor. Plans were submitted initially on June 17, 1916. The location on the low-lying spit of island on the east side of the current harbor was thought superior for coverage of the mine field. Considerably debate occurred around the proper form of the battery. At first it was decided to use mobile 3-inch guns to be stored at Fort MacArthur pending attack (decision of September 19, 1916). That was reversed on April 13, 1917 when permanent gun blocks for Model 1903 pedestal guns were authorized. The emplacement was always conceived as a minimal structure—just gun blocks (four, separated by 50-feet from each other) with no parapets or protection and a light magazine mainly intended for weather protection. Work was done from April 15 to July 31 1916. It was transferred on July 31, 1917 for a cost of just $1,948.91. Unfortunately, though, the Los Angeles harbor mine field was never authorized. Armament was received and mounted in early 1918. It consisted of four 3-inch Model 1903 guns and pedestal mounts (all Watervliet #105/#105, #106/#106, #107/#107, and #108/#108). The battery was named on General Orders No. 117 of August 30, 1917 for Brigadier General Richard Lodor, Civil War officer who later taught at West Point. The battery was disarmed in 1927, and the emplacement destroyed in 1928 when the island was dredged away for harbor improvements. The armament went into storage and was later used in the JAAN batteries in early 1942. No traces of the emplacement remain today.

• **Railway Battery:** Two 14-inch railway guns were a very important feature of the Los Angeles defenses in the 1920s and 1930s. At times they were referred to as Battery Erwin, but that name was never officially bestowed. As part of the project the distribute movable artillery in the 1920s, two Model 1920 railway carriages were allocated to Fort MacArthur. One circular firing platform was built on the lower reservation of the fort on a rail spur and the guns were normally kept in storage in sheds not far away. The first gun arrived in 1925, the second not until 1929. These were both Model 1920MII guns on Model 1920MII railway carriages (#7/#1 and #10/#4). Two similar guns were constructed and stationed in the defenses of the Panama Canal. The firing platform was built between October 1925 and February 1926. Transfer of the was made on May 10, 1926 for a cost of $9,344. Two new platforms were built in early 1937 on a fill area on the "Bottomside" reservation below the bluff of the Lower Reservation. The new firing platforms were transferred on April 21, 1942 for a cost of $27,844. There was damage from the concussion of firing the guns, so they were moved to other sites twice for training (one in Goleta, one at Camp Pendleton). The guns served well through World War II; they had the longest range in the defenses until the 16-inch guns at Battery Bunker were installed. The guns were removed in 1944 for projected use in Europe, and the platforms covered over or removed in the years following . Little remains today of the emplacement structures for these guns.

• **Battery #241:** A standard 1940 Program dual 6-inch barbette battery for Fort MacArthur. It was emplaced on the southern edge of the headlands, in front of and below the 14-inch disappearing batteries, and fired to the south. Construction required the removal of a 155mm Panama mount previously on the location. It was originally assigned national priority #17 and funded with the FY-1942 Budget. Construction started on May 4, 1943; concrete work being completed by the end of the year. It was of standard 200-series type, with 210-feet between guns. It was of the type where the rear power room and its entrance was folded to the left rather than straight behind the central traverse to fit it into the ridge. Work was suspended before completion, but it was provisionally transferred at a cost of $219,000. It was allocated two 6-inch guns Model M1(T2) and M4 barbette carriages (#28/#13 and #29/#14). In 1946 it was proposed to complete the battery, and apparently that was done, the armament being reported as being mounted in 1948, probably from Battery Harrison at Bolsa Chica. While probably deleted in 1948-1949, the armament remained in place under U.S. Navy ownership until scrapped in 1956. The emplacement still exists in Angels Gate Park and retains its power plant but the interior is closed to the public. The Korean Friendship Bell is located on top of the battery.

AMTB Batteries – HD Los Angeles (1942-1945)-In 1942 five batteries of AMTB guns protected the San Pedro Harbor. There were four batteries of 90mm guns and 37mm guns and one of only 37mm guns. These were fast-firing weapons, which were supplemented with .50-caliber machine guns and searchlights as needed. The 90mm batteries were four guns, two of which were fixed and two were mobile mounts. These were supplemented with two mobile 37mm guns. These batteries were located at the Gaffey Bulge where Gaffey Street winds around Battery Leary-Merriam; JAAN No. 1, Navy Field; JAAN No. 2, Bluff Park; and Terminal Island. The fifth battery was designated Dolphins. This was a line of 10 37mm guns on towers with searchlights located along the harbor breakwater— making a total of 34 guns.

Gaffey Bulge Gun Battery

- **AMTB #3A**: A 1943 Program AMTB battery for two 90mm fixed and two 90mm mobile guns for Fort MacArthur. It was emplaced at the bulge of Gaffey Street on the eastern side of the upper reservation. Consequently, it was informally known locally as battery "Gaffey Bulge." Although it was never officially named. Tactically it served as Tactical No. 3A. It was built from June to October 1943, for transfer on May 4, 1944 at a cost of $19,144. The battery was dismounted shortly after the end of the war. The two permanent gun blocks remain alongside the street. The battery site is open to the public.

Navy Field Gun Battery

- **AMTB-Navy Field**: A 1943 Program AMTB battery for two fixed and two mobile 90mm guns. It was built at Navy Field from June until September 1943, being transferred with the other three Los Angeles AMTB batteries on May 4, 1944 for a combined cost of $76,576.03. It was disarmed shortly after the end of the war. No traces of the emplacement are believed to have survived.

Cabrillo Beach Gun Battery

- **JAAN-1**: A battery for two 3-inch pedestal guns emplaced on the rise over Cabrillo Beach east of Point Fermin as an early AMTB battery. Two guns from old Battery Lodor at Fort MacArthur were used in this emplacement. Work was done on the emplacement from July 6, 1942 to January 11, 1943 and completed for transfer on April 12, 1943 at a cost of $8040.11. It consisted of simple gun blocks 100-feet apart, sheltered connecting tunnels and two magazines. It carried two 3-inch Model 1903 guns and pedestal mounts (#105/#105 and #106/#106). The guns received the newly developed box-type shields. The battery served during the war as Tactical No. 4A. Armament was probably removed in 1944-1946. One gun block (No. 2) still exists at a public park. The battery site is open to the public.

Terminal Island Gun Battery

- **AMTB-Terminal Island**: A 1943 Program AMTB battery for two mobile and two fixed 90mm guns erected at Terminal Island in Long Beach Harbor. Work was done from June until September 1943. Transfer was made on May 4, 1944 for an allocated cost of $19,144, It was armed as described with two fixed blocks holding 90mm guns on M3 carriages and temporary shelters and magazines. The armament was removed shortly after the end of the war. One gun block still exists, though covered currently by a building.

Bluff Park Gun Batteries

- **JAAN-2:** A battery for two 3-inch guns serving in the early AMTB role. It also serving as Tactical Battery No. 4B. It was built from August 1 1942 to January 11, 1943. Transfer was made on April 12, 1943 for a cost of $8122.16. It was located on the beach in Long Beach at a location known as Bluff Park. It was armed with two of the former guns from Battery Lodor at Fort MacArthur. It had two 3-inch Model 1903 guns and pedestal mounts (#107/#107 and #108/#108). The guns received the newly developed box-type shields. The battery served until removed in the 1944-1946 time period. One gun block (No. 2) still exists at Bluff Park. The battery site is open to the public.

- **AMTB-Bluff Park:** A 1943 Program AMTB battery consisting of two 90mm fixed and two 90mm mobile guns. It served as the permanent replacement of Jaan-2 and was sited just a short distance from that work. It consisted of two concrete gun blocks and two roll-away shelters for the mobile mounts, along with troop and ammunition shelters. Work was done from June to September 1943, for transfer of May 4, 1944 at a cost of $19,144. It was disarmed postwar, and no remains exist today.

Bolsa Chica Military Reservation (1942-1948) is located next to the Pacific Coast Highway (California Route 1) and adjacent to Warner Avenue in the town of Sunset Beach California. Bolsa Chica is about twenty miles due southeast of Fort MacArthur. Originally a sportsman club for hunting and fishing, the site was purchased in 1941. It received a set of 155 mm guns on Panama mounts. As part of the 1940 Modernization Program, Bolsa Chica was the site of 16-inch casemated battery and 6-inch battery. The construction of Battery #128 was deferred short of being armed, while Battery Harrison was essentially completed. Sometime after World War II, parts of the reservation were sold to private interests. In 1995, the concrete gun emplacements were demolished, and the area developed for private homes. Battery #128's PSR remains along with a reserve magazine. The two Panama mounts still remain in the Bolas Chica Ecological Preserve across the shallow bay from the highway.

Bolsa Chica Gun Batteries

- **Battery #128:** A standard 1940 Program dual gun, casemated 16-inch battery. It was emplaced at the Bolsa Chica Military Reservation on the southern flank of the Los Angeles sector in Huntington Beach. Within the reservation the battery was on a slight plateau in the center of the reservation, on the eastern side firing to the south, southwest. It was originally assigned national priority #24 of September 11, 1940 and then #25 in August 11, 1941. Although planned to have been started under the FY-1942 Budget, work was slow in getting started, authorization to start work was granted finally on November 13, 1942. Physical construction work was started on April 17, 1943 and the concrete work was essentially complete on January 21, 1944. It had been ordered deferred on November 26, 1943. The battery was never completed for service. It was provisionally transferred on January 5, 1945 for a cost to that date of $758,878.76. Fire control stations and a separate PSR to the rear of the battery were also built. The battery itself followed conventional design standards for the 100-series 16-inch emplacements. No armament was ever assigned or received. The emplacement was never named, it was simply known as Battery Construction No. 128 during construction. In later years the site was sold for commercial development, and the emplacement was destroyed for private homes in the mid-1990s (though the PSR currently survived with buried entrances).

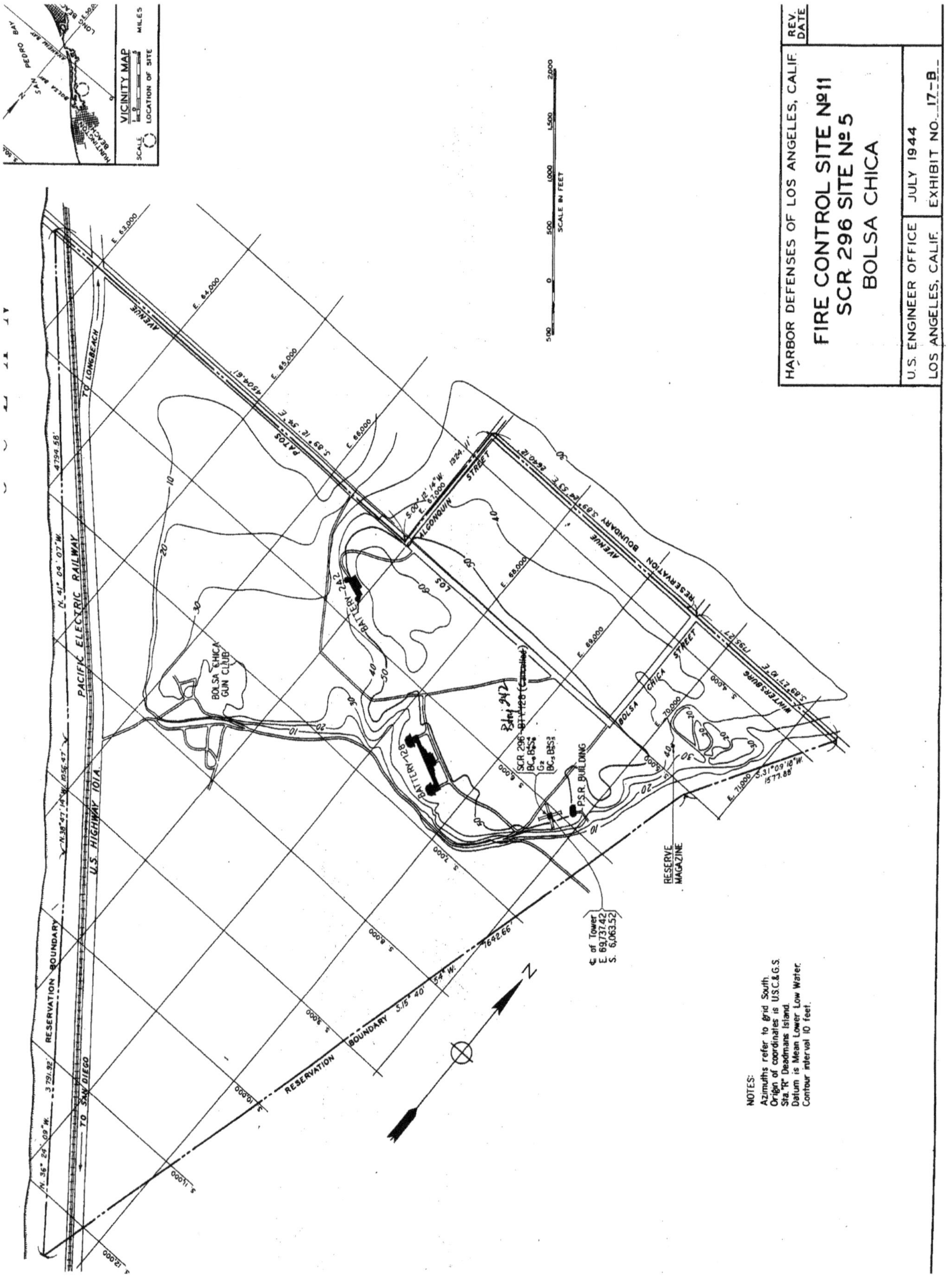

HARBOR DEFENSES OF LOS ANGELES, CALIF.
FIRE CONTROL SITE Nº11
SCR 296 SITE Nº 5
BOLSA CHICA
U.S. ENGINEER OFFICE
LOS ANGELES, CALIF.
JULY 1944
EXHIBIT NO. 17-B
REV. DATE
VICINITY MAP
LOCATION OF SITE
MILES
SCALE IN FEET
PACIFIC ELECTRIC RAILWAY
U.S. HIGHWAY 101A
TO LONG BEACH
TO SAN DIEGO
RESERVATION BOUNDARY
LOS PATOS AVENUE
ALGONQUIN STREET
BOLSA CHICA STREET
WINTERSBURG AVENUE
BOLSA CHICA GUN CLUB
BATTERY-242
BATTERY-128
P.S.R. BUILDING
RESERVE MAGAZINE
NOTES:
Azimuths refer to grid South.
Origin of coordinates is U.S.C.&G.S. Sta. "R" Deadmans Island.
Datum is Mean Lower Low Water.
Contour interval 10 feet.

- **HARRISON**: A 1940 Program dual 6-inch battery also built at the Bolsa Chica reservation. It was sited on the plateau behind the lagoon, to the west of the 16-inch battery roughly on the same axis, and also fired to the south, southwest. Under construction it was assigned the name Battery Construction No. 242. Although originally planned with as low priority project and funded under the FY-1943 Budget, the work was accelerated. Construction was started on March 31, 1943 and completed in late June of 1944. It was of standard 200-series design, but of the type with the power room and rear entry bent to the side, in this case to the right flank adjacent to the entry to the magazine corridor by gun No. 1. It was armed with two 6-inch Model M1(T2) guns on Model M4 barbette carriages (#28/#11 and #29/#12). The emplacement was named on General Orders No. 51 of June 10, 1946 for Major Harry J. Harrison, Coast Artillery Corps. The battery was retained in the 1946 Review, but was soon deleted and disarmed, probably in 1947-1948. In later years the fort property was sold to private developers, and the battery emplacement completely destroyed in the early 1990s for construction of private homes. No traces remain today.

Demolition of Battery #128 (above) and Battery Harrison (below) (Mark Berhow)

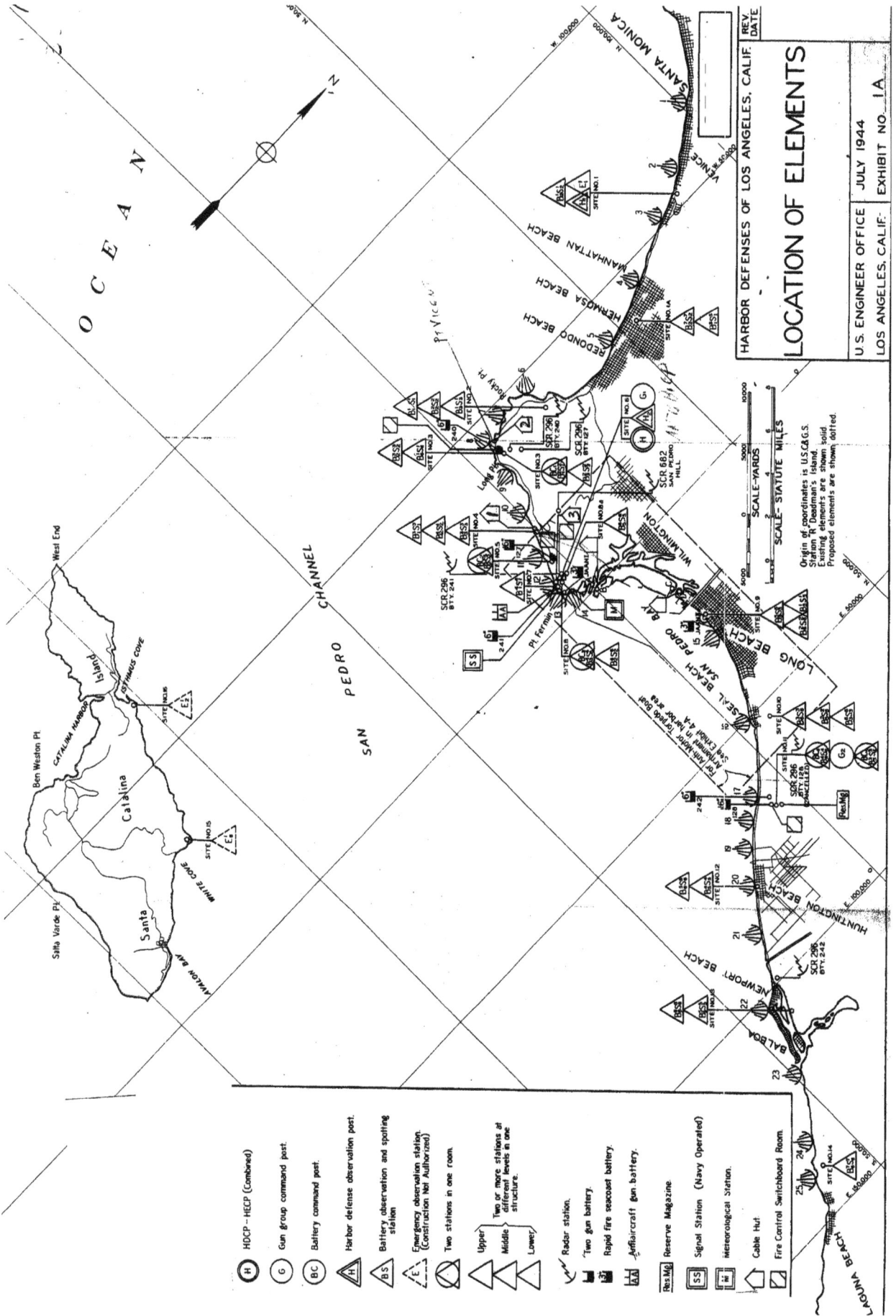

Harbor Defenses of Los Angeles, Calif.

LOCATION OF ELEMENTS

U.S. ENGINEER OFFICE — JULY 1944
LOS ANGELES, CALIF. — EXHIBIT NO. 1A

Los Angeles World War II-era Site Locations. Stations housed in a single structure are connected by dashes (-)

location	Loc#	Purpose
Pacific Palisades		
Playa Del Ray	1	HDOP1, E1/240, BS1/127
Hermosa Beach/ Redondo Beach	1A	BS5/240, BS7/127
Bluff Cove	2	BS1/240, BS2/127, BS1/241
Rocky Point		
Point Vicente MR	3	Batt. Tact. #1 BCN 240, BC-BS2/240, BS3/241, BS3/127, BS1/128, SCR296
San Pedro Hill	3C	SCR-682
West Seabench		searchlight
East Seabench	4	BS3/240, BS4/127, BS3/241
White Point MR	5	Batt. Tact. #2 BCN 127 Bunker, BC-BS5/127, PSR 127
Palos Verdes Hill		SCR296 Bunker
Fort MacArthur UR	6	HECP-HEDP, HDOP2, G1, SBR
Fort MacArthur UR	7	BS4/240
Fort MacArthur UR	8	Batt. Tact. #3 BCN 241, BS7/127, BC-BS4/241
Fort MacArthur UR	8A	BS6/242 (old Leary BC), SCR296-241
Cabrillo Beach		Batt. Tact. #4 JAAN #1
Hilton Hotel Cupula	9	BS5/241, BS1/242, BS2/128
Bluff Park		Batt. Tact. #5 JAAN #2
Landing Hill/ Seal Beach	10	BS6/241, BS2/242, BS3/128
Sunset Beach		searchlight
Bolsa Chica MR	11	Batt. Tact. #6 BCN 242, Batt. Tact. #7 BCN 128, BC-BS3/242, BC-BS4/128, G2, PSR 128
Huntington Beach	12	BS4/242, BS5/128
Costa Mesa	12A	SCR radar Panama Mount
Newport Beach	13	BS5/242, BS6/128
Abalone Point	14	BS7/128

Battery Barlow
Saxton Angels Gate
Park (Mark Berhow)

Berhow, Mark A. and David Gustafson. *The Guardian at Angels Gate, Fort MacArthur, Defender of Los Angeles,* Fort MacArthur Military Press. San Pedro, CA, 2001.

Small, Charles S. *California's Railway Guns.* Railhead Publications. Greenwich, CT, 1984.

THE HARBOR DEFENSES OF SAN FRANCISCO – CALIFORNIA

San Francisco Bay is one of the largest natural harbors on the west coast of the United States, an important commercial and military harbor from the 1700s. The lands around the entrance at the Golden Gate were reserved for military use by the United States government beginning 1849 and sections remained in military use though the 1990s. The military importance of San Francisco Bay resulted in a wide range of defensive construction programs around the Golden Gate. The U.S. Army reservations feature a wide range of American seacoast fortifications from the Third System brick Fort Point to the Nike defenses of the Cold War. Nearly all the old military reservations were transferred to public use, making up a great part of the Golden Gate National Recreation Area, which was established in 1972. The Presidio Trust operates much of the former presidio lands, and the State of California made Angel Island into a state park. A visit to the San Francisco seacoast defenses provides a comprehensive and accessible view of a wide range of fortification structures. Forts Mason, Winfield Scott, Baker, Barry, and Cronkhite, as well as the Presidio, have a superb collection of remaining garrison buildings.

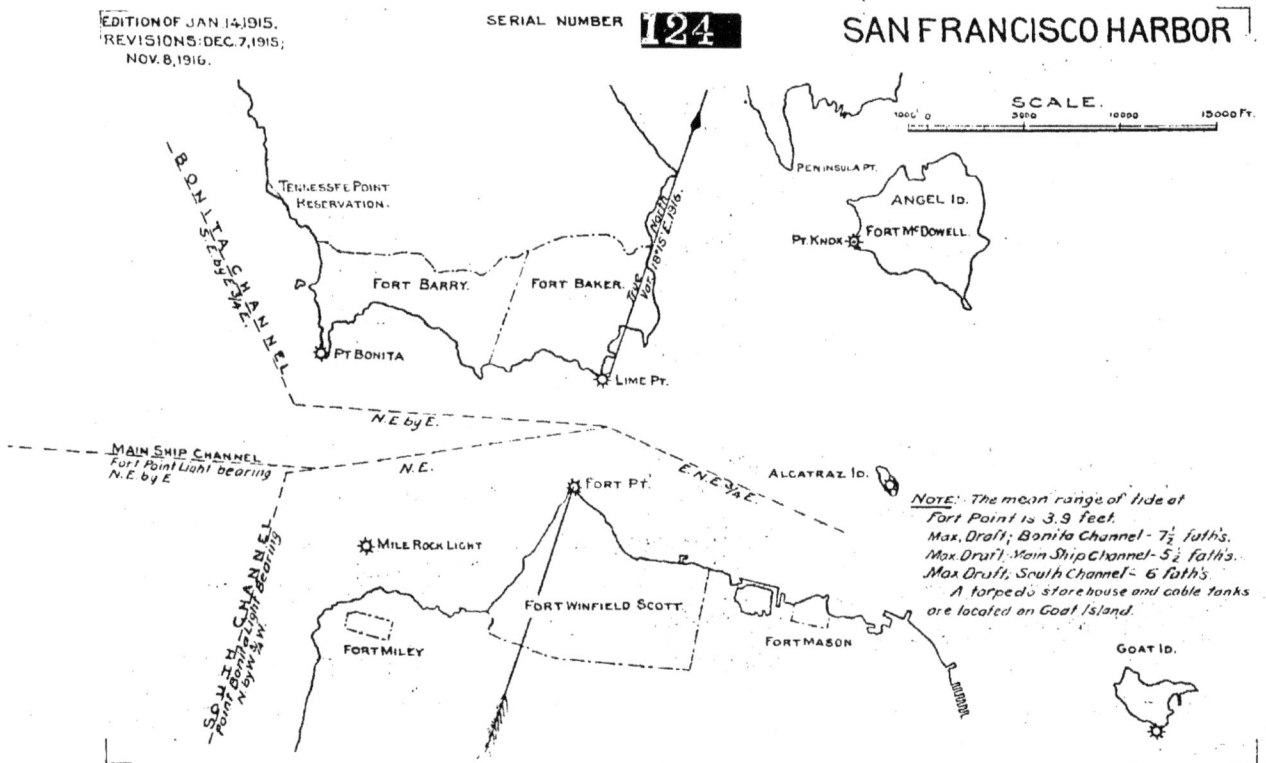

Fort Cronkhite (1937-1950) is located directly north of Fort Barry on the Pacific Ocean, about four miles west of Sausalito, California. The military reservation runs from Rodeo Lagoon to north of Tennessee Point and east to Wolf Ridge. Fort Cronkhite original 800 acres was acquired in 1937 to construct one of the first 16-inch casemated batteries. It was named in General Orders 9 of 1937 for Maj. Gen. Adelbert Cronkhite, U.S. Army. Battery Townsley, mounting two 16-inch guns on long-range barbette carriages in two reinforced concrete gun houses connected by a service gallery with magazines and power rooms. By 1942 the post contained a reserve magazine, an anti-aircraft battery, and a large barracks area. At end of World War II, it became a sub-post of the Presidio of San Francisco. In 1950 the fort hosted a 120 mm antiaircraft guns battery, followed by the Nike missile launch and radar control sites located at Fort Cronkhite/Fort Barry. Fort Cronkhite was transferred to National Park Service in 1972 to become part of the Golden Gate National Recreation Area. Fort Cronkhite hosts a variety of National Park activities and

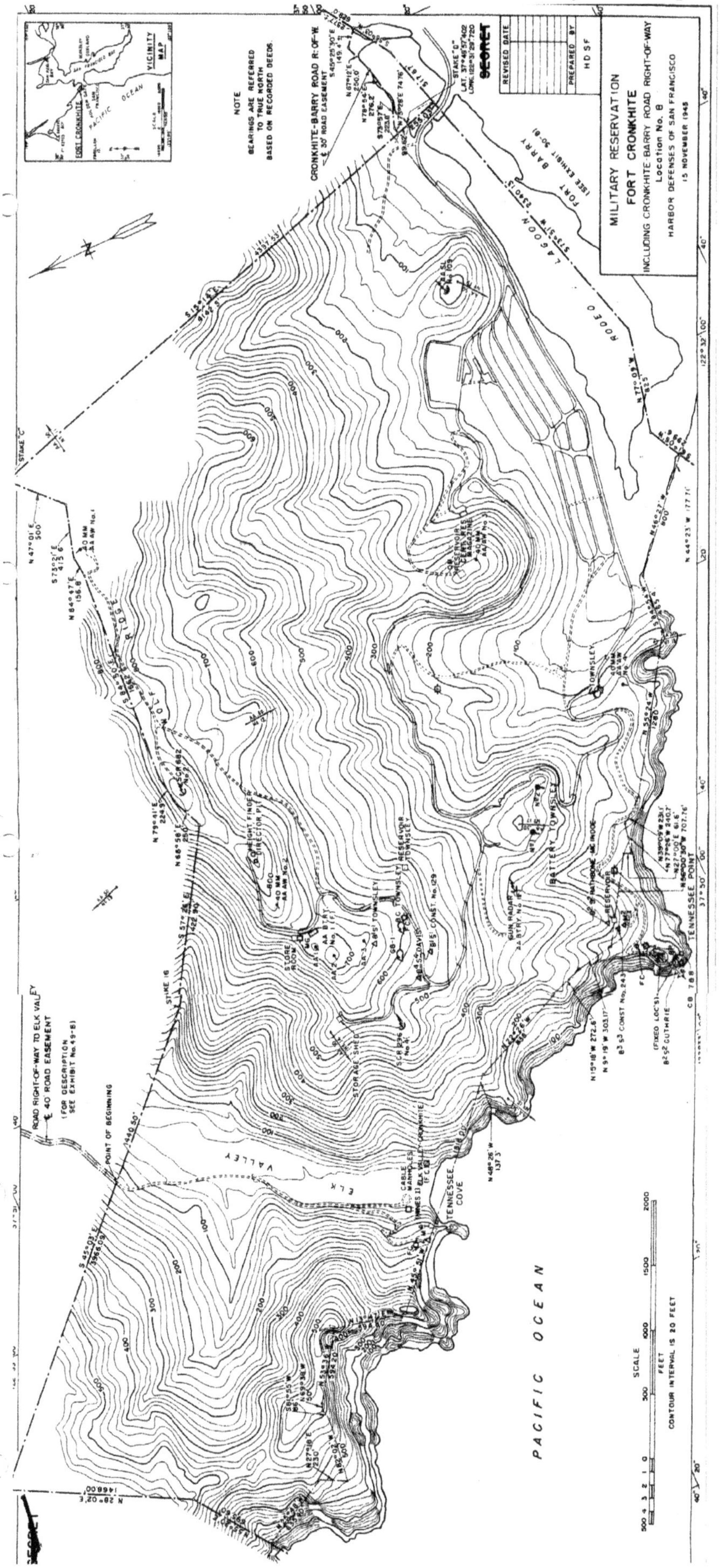
MILITARY RESERVATION
FORT CRONKHITE
INCLUDING CRONKHITE-BARRY ROAD RIGHT-OF-WAY
Location No. 8
HARBOR DEFENSES OF SAN FRANCISCO
15 NOVEMBER 1945
SECRET
REVISED DATE
PREPARED BY
H D S F
NOTE
BEARINGS ARE REFERRED TO TRUE NORTH BASED ON RECORDED DEEDS.
CRONKHITE-BARRY ROAD R-OF-W
VICINITY MAP
PACIFIC OCEAN
RODEO LAGOON
FORT BARRY
WOLF RIDGE
ELK VALLEY
TENNESSEE COVE
TENNESSEE POINT
BATTERY TOWNSLEY
POINT OF BEGINNING
ROAD RIGHT-OF-WAY TO ELK VALLEY
STORAGE SHED
SCALE
FEET
CONTOUR INTERVAL IS 20 FEET

Battery Townsley, Fort Cronkhite Marin Headlands Golden Gate National Recreation Area (Terry McGovern)

16-inch MkVII navy gun barrel at Battery Townsley
Marin Headlands Golden Gate National Recreation Area (Mark Berhow)

partners including the Marin Headlands Native Plant Nursery, the Golden Gate Raptor Observatory, the Marin Headlands Marine Mammal Center (in the old SF-87 launch area), and the Headlands Center for the Arts. Many of the World War II "temporary" wooden buildings still survive today. In 2010 the group of volunteers began cleaning up and restoring Battery Townsley. In 2012 a Navy Mk VII 16-inch gun barrel was placed on display behind the entrance to the southern gun emplacement. It is opened on scheduled days with guided tours.

Fort Cronkhite Gun Battery

- **TOWNSLEY**: A battery for two 16-inch barbette guns built on a new reservation to the north of Fort Barry on the northern side of the San Francisco defenses. This battery, along with sister Battery Davis at Fort Funston, were the prototypes for new 16-inch type batteries soon to be adopted in standardized form in the 1940 Program of new American coastal defense. While it has many features in common with these later emplacements, the arrangement of the gun rooms and connecting gallery make this emplacement quite unique. It was built into Rodeo Ridge, north of Rodeo Lagoon firing to the west. Due to the topography of the adjacent ridge points, the gun houses were relatively closely spaced at just 350-feet apart, and the connecting passageway bent to follow the ridge between the houses. It was built with cut and cover engineering technique. Construction was authorized in 1934, after several years of planning. Work was actually started in January 1938, about a year after the work at Fort Funston began. It was finished for transfer on July 24, 1940 at a cost of $595,000. It was armed with two 16-inch navy MkIIM1 on barbette carriages Model 1919M2 (#87/#15 and #86/#16). Also built were a battery commander's station on Wolf Ridge, an adjacent PSR room and an extensive, separate reserve magazine, along with numerous assigned base-end stations. It was named on General Orders No. 9 of December 17, 1937 for Major General Clarence P. Townsley. The battery served throughout the war, being finally disarmed in 1949. The emplacement still exists and is on public display at the Golden Gate National Recreation Area. On 1 Oct 2012, a surplus 16-inch naval Mark VII gun tube was delivered on site for use as a display gun behind the battery. The external elements of the Battery, PSR and Reserve Magazine are open for self-guided tours during park hours. Parts of the interior of the Battery are open to the public on one day of the month. The interior of the PSR and the Reserve Magazine are not open to the public.

Fort Barry (1904-1950) is located on the northern side of the Golden Gate Channel, the entrance to San Francisco Bay. The military reservation runs from Point Diablo to the west to Point Bonita and then north to Rodeo Lagoon. Original part of Fort Baker, Fort Berry's reservation of 1,344 acres created out of Fort Baker in 1904. It was named in General Orders 194 of 1904 for Bvt. Maj. Gen. William F. Barry, U.S. Army. A half-mile long tunnel connected Fort Barry and Fort Baker. The fort had a total of seven Endicott Program batteries. In 1928, Battery Wallace with two 12-inch guns on long-range barbette carriages, was constructed at Fort Barry. During World War II, Battery Wallace was casemated, and a new 16-inch casemated battery was built on the spine of the ridge dividing the Fort Barry and Fort Baker reservations. While Battery 129 was never completed, its location at 821 feet above sea level makes it the highest of the 100-series batteries in the 1940 Program. At end of World War II, the fort became a sub-post of the Presidio of San Francisco. During the Cold War, two Nike missile launch sites were located at Fort Barry and Fort Cronkhite. Fort Barry was transferred to National Park Service in 1972 to become part of the Golden Gate National Recreation Area. Fort Barry is the site of the Marin Headlands Visitor Center, located in the World War II-era chapel building. The garrison area is used as seasonal housing, a youth hostel and the Highlands Center for the Arts. The balloon hangar at Fort Barry is a surviving element of the U.S. Army's brief experimentations with using tethered balloons as part of the nation's system of coastal defenses. Con-

SAN FRANCISCO HARBOR

FORTS BARRY AND BAKER
Scale of Feet

1000 0 1 2 3 4 5 6 7 8000

SERIAL NUMBER

EDITION OF MAR. 3, 1919.
REVISIONS APRIL 17, 1920

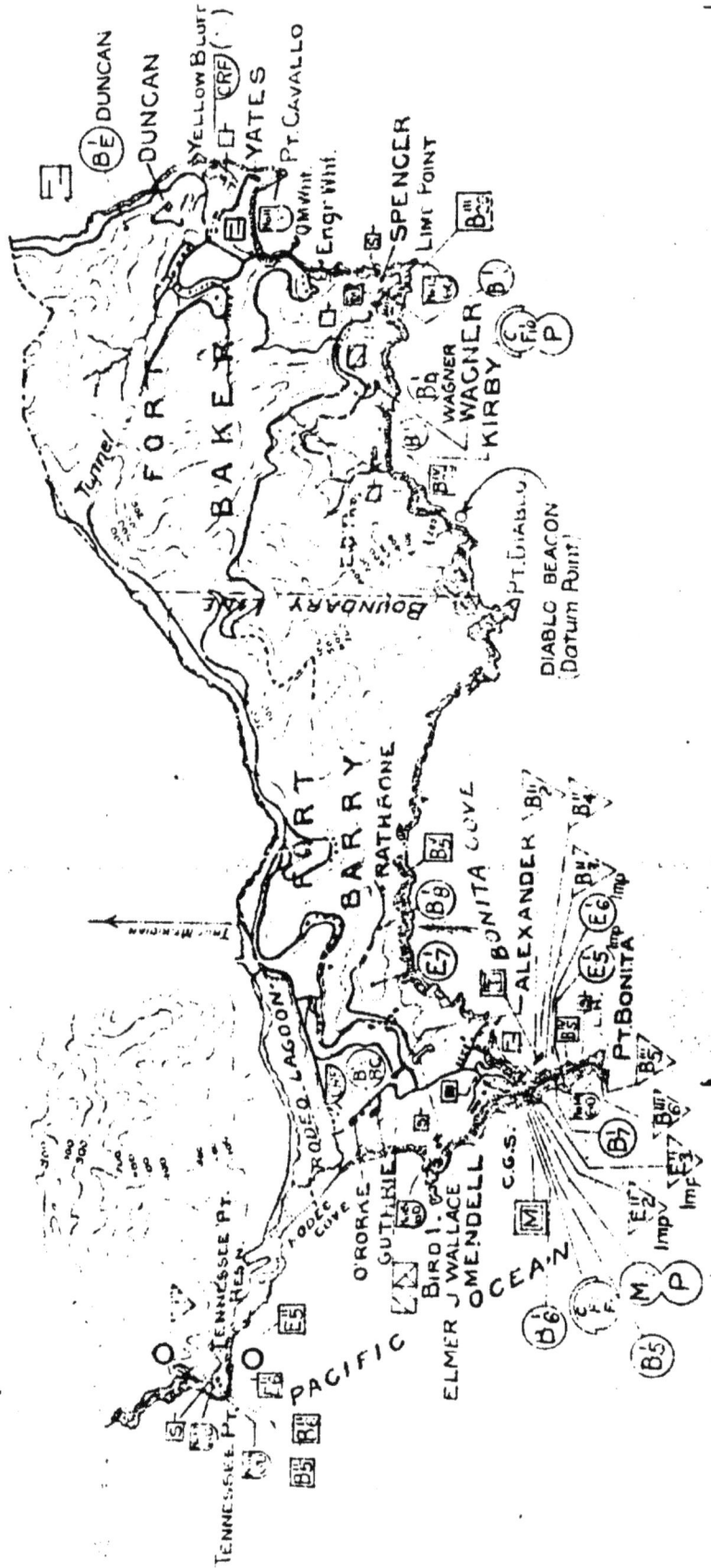

DUNCAN
B'ᴱ DUNCAN

YELLOW BLUFF
CRF
YATES
Pt. CAVALLO

Q Wht
Engr Whf

SPENCER
Lime Point

B'''

WAGNER
B D WAGNER
B' KIRBY

Pt. DIABLO

DIABLO BEACON
(Datum Point)

BOUNDARY LINE

TUNNEL

FORT BAKER

FORT BARRY

RATHBONE

BONITA COVE

ALEXANDER B'₂ B'₄

B₈
E₇
E₅ E₆ Imp
B₃
B₉
Pt. BONITA B₅

CGS
MENDELL
M

B₇

Imp E₁

B₆ P
M
B₅

RODEO LAGOON

RODEO COVE

ORORKE
GUTHRIE
BIRD I
ELMER J WALLACE

TENNESSEE PT. RES.

PACIFIC OCEAN

E₅
B₅

structed and abandoned the same year, the structure is the only surviving hangar of its type that actually housed an army balloon, and one of only two examples of its type known to survive in the country. One of the Nike battery integrated fire control sites was located on top of Battery 129, and one of the Nike Administration sites houses a youth activities center. Nike Launch area SF-88L was transferee to the park service pretty much intact. It has since been restored and is open for tours on specified days. The park is open to the public while most batteries are sealed.

Fort Barry Gun Batteries

- **MENDELL:** A battery for two 12-inch disappearing guns emplaced to the north of Point Bonita and the bluff firing to the west. Fort Barry was the final San Francisco fort armed during the Endicott Program and received mostly the later types of battery emplacements. This battery was submitted on September 12, 1900 for two emplacements for the Model M1897 disappearing carriage. There was some debate about how far south to locate the battery, but eventually this location somewhat to the north was selected. It was of type construction but was half-sunken to protect the rear from possible reverse fire from enemy ships already past the point. The emplacement had its own reserve power plant. Work was done from October 1901 to August 1902. While under construction it was found that the underlying soil of the bluff was too loose, and the emplacement concrete footings had to be substantially increased. The carriages arrive by barge at Rodeo Beach on June 6, 1902 and they were mounted early the following year. Transfer came on June 8, 1903 for a cost of $128,016. It was named in General Orders No. 120 of November 22, 1902 for Colonel George Mendell, and engineer of prominence in the San Francisco fortification history. It was armed with two 12-inch Model 1895M1 Bethlehem guns on Model 1897 disappearing carriages (#4/#30 and #5/#31). In about 1917 the gun carriages were given an additional 5-degrees of elevation in order to increase the maximum range. The battery served fully armed throughout the post and inter war period. It was not authorized for removal until July 7, 1943. The guns were shipped away on December 6, 1943 and the carriages were scrapped. The emplacement still exists at the Golden Gate National Recreation Area and is open to the public.

- **WALLACE:** A 1915 Program dual 12-inch barbette battery at Fort Barry on a ridge overlooking Battery Mendell, firing to the west. The site was selected by the local board in December of 1915, and it was begun in January 1917 as one of the first of this new type of emplacement in the United States. In February of 1917 a very simple plan was approved as a sort of engineering experiment. At that time there was a healthy debate between building heavily protected, but expensive fortifications vs. very open, dispersed, inexpensive emplacements relying on camouflage and natural features for protection. All the new 12-inch, long-range batteries were built with open, all-round fire platforms with a heavily protected magazine in the traverse between the two guns. But at Fort Barry it was decided to just build and arm the two-gun blocks and rely on a simplest wood magazines, plot, and equipment sheds. Work was begun on March 6, 1917 on this emplacement to test this concept. Cost was estimated (exclusive of the actual armament) at just $80,394. Work was done in a little over a week, completing on March 15, 1917. Inspection and debate soon decided that a full, protected magazine was a worthwhile investment, and in early 1918 approval was granted to build a standard traverse magazine between the two completed and armed gun platforms. Work was done in 1919-1920. Transfer of the battery was made on June 24, 1921 for a cost of $273,464.41. It was armed with two 12-inch Model 1895M1A4 Watervliet guns on long-range barbette carriages Model 1917 (#61/#2 and #68/#3). The emplacement was of standard, open-back type. The emplacement was named on General Orders No. 63 of May 12, 1919 for Colonel Elmer J. Wallace who was

SAN FRANCISCO HARBOR
FORT BARRY.

SERIAL NUMBER **108**

EDITION OF JAN.14.1915.
REVISIONS: DEC.7.1915;
NOV.8 1916

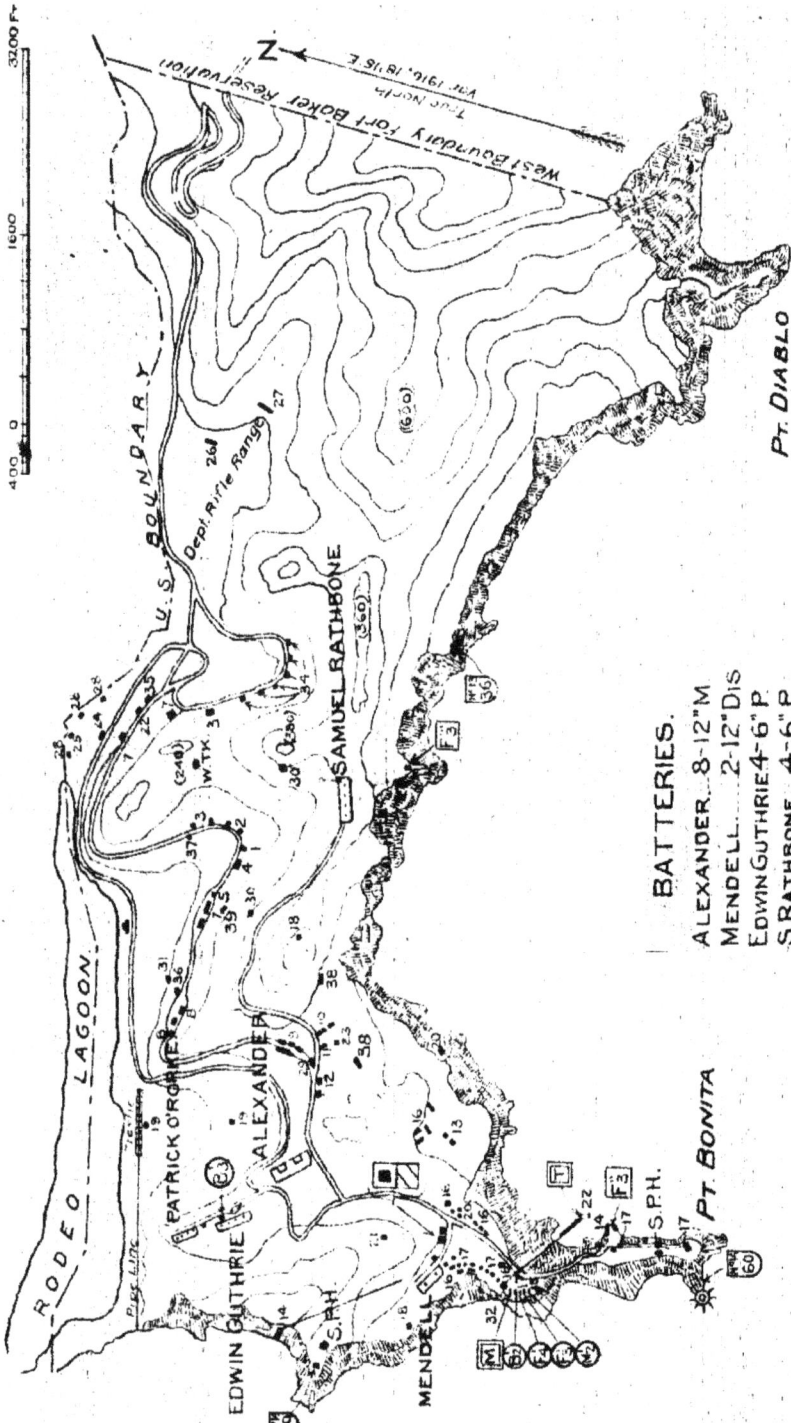

Var. 1916, 18:16 E.
True North

West Boundary Fort Baker Reservation

U. S. BOUNDARY

Dept. Rifle Range

RODEO LAGOON

PATRICK O'RORKE

EDWIN GUTHRIE

MENDELL

ALEXANDER

SAMUEL RATHBONE

Pt. BONITA

Pt. DIABLO

LEGEND.

1. ADMIN. BLDG.
2. COMDG. OFF. QRS.
3. OFFICERS QRS.
4. HOSPITAL.
5. HOSPITAL ST. QRS.
6. N.C.O.QRS.
7. BARRACKS.
8. GUARD HOUSE.
9. STOREHOUSES & SHOPS.
10. STABLE & WAGON SHED.
11. FIRE ENGINE HOUSE.
12. OIL HOUSE.
13. U.S.L.S. STATION.
14.   "   - BOATHOUSE.
15.   "   - LOOKOUT.
16.   "   - ENG.DEPT.BLDGS.
17.   "   - L.H. BLDGS.
18. ENG. DEPT. RESERVOIRS.
19.   "   "   - WELLS.
20. L.S.S. WELL.
21. L.H. WELL.
22. STOREHOUSES.
23. STABLE LAVATORY.
24. STABLES.
25. BLACKSMITH SHOP.
26. PISTOL RANGE BUTTS.
27. 600 YARD BUTTS.
28. CIVILIAN EMP. QRS.
29. COAL SHED.
30. RESERVOIR.
31. TEMP. CANTEEN.
32. HOIST.
33. PUMP HOUSE.
34. KITCHEN.
35. POST EXCHANGE.
36. GYMNASIUM.
37. FLAGSTAFF (75'HIGH)
38. DWELLING.
39. HANDBALL COURT.

BATTERIES.

ALEXANDER . . 8-12"M.
MENDELL . . . 2-12"D.S.
EDWIN GUTHRIE 4-6"P.
S.RATHBONE . . 4-6"P.
PATRICK O'RORKE 4-3"P.

SAN FRANCISCO HARBOR
FORT BARRY-D1.

SERIAL NUMBER

EDITION OF MAR.3,1919.
REVISION: APR.17,1920.

True North.
Var. 1916, 18°15'E.

Scale of Feet.
400 0 400 800 1200 1600 2000 2400

P A C I F I C   O C E A N

Rodeo Lagoon

Pt. Bonita

Bird Island

ALEXANDER
MENDELL
EDWIN GUTHRIE
PATRICK O'RORKE
ELMER WALLACE

LEGEND

5 N.C.O. QRS.
7 BARRACKS.
8 GUARD HOUSE.
9 POST EXCHANGE.
10 GYMNASIUM.
11 GARAGE.
12 STOREHOUSE.
13 MESS HOUSE.
14 COMMISSARY.
17 "
21 Q.M. STORE HOUSE.
22 Q.M. STABLE.
23 Q.M. WAGON SHED.
24 Q.M. BUNK & FIRE HO.
25 Q.M. COAL SHED.
26 Q.M. OIL HOUSE.
27 Q.M. CARPENTER AND
   BLACKSMITH SHOP.
28 Q.M. BAKERY.
29 Q.M. SCALES.
31 ORDNANCE & ARTY.
   ENGR'S. ST. HO.
41 ENGR. DEPT. BLDGS.
42 ENGR. DEPT. ROCK BIN.
43 ENGR. DEPT. SAND BIN.
44 ENGR. DEPT. TRAMWAY.
45 ENGR. DEPT. DWELLING.
70 Y.M.C.A.
71 SCHOOL HOUSE.
80 NAVY DEPT. RADIO STA.
81 " " COMPASS HO.
82 " " RADIO QRS.
83 U.S. COAST GUARD STA.
84 " " BOAT HO.
85 LT. HO. KEEPERS DWELL.
86 LT. HO. " ASST. "
87 " " DEPT. WIND MILL.

BATTERIES.

ALEXANDER_4-12" M.
MENDELL_2-12" Dis.
EDWIN GUTHRIE 4-6" Bar.
PATRICK O'RORKE 4-3".
ELMER J. WALLACE_2-12".

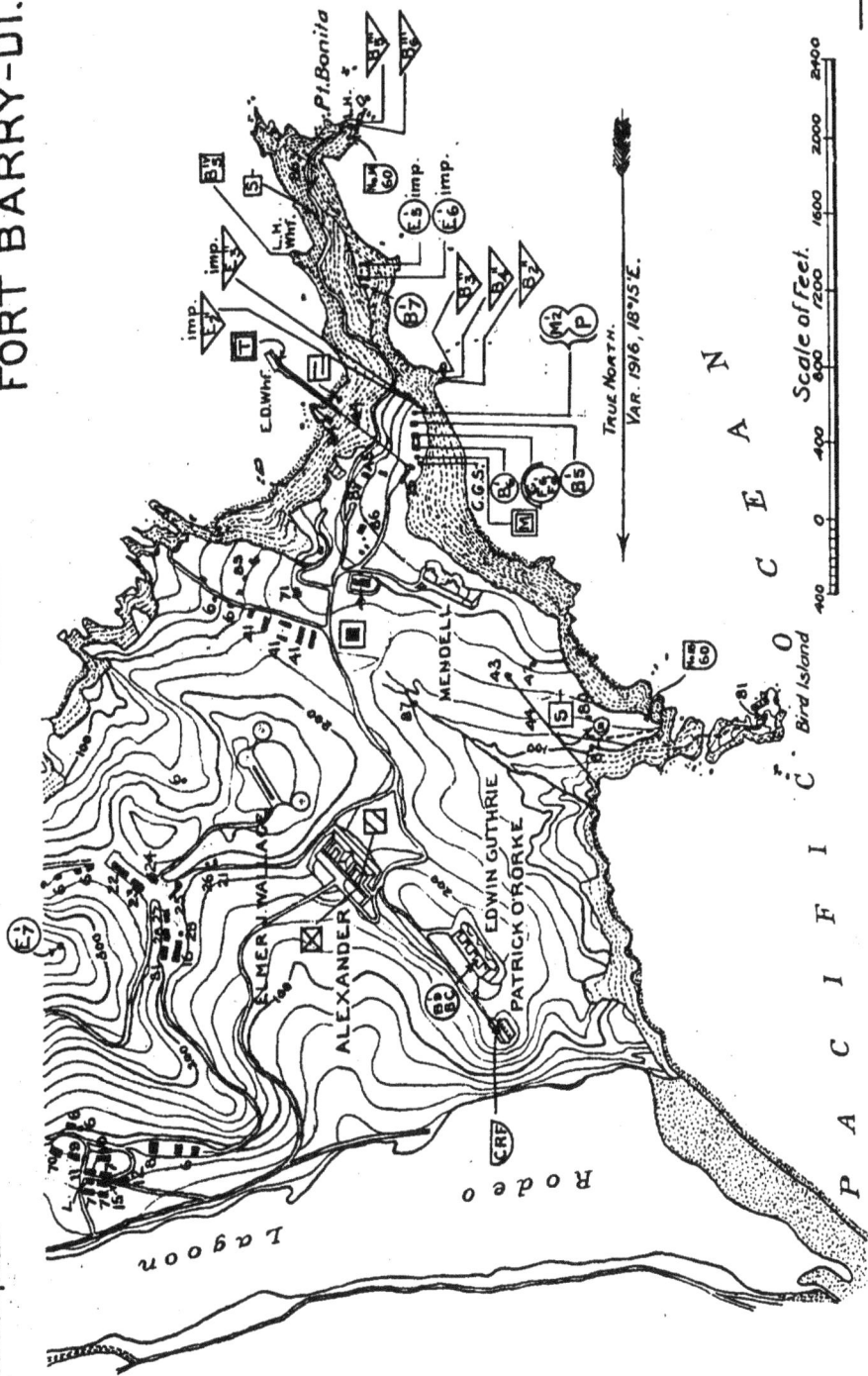

SAN FRANCISCO HARBOR
FORT BARRY–D2.

SERIAL NUMBER

EDITION OF MAR. 3, 1919.
REVISION: APR. 17, 1920.

TRUE NORTH.
VAR. 1916, 18°½'E.

Scale of Feet.
100 0 400 800 1200 1600 2000 2400

SAMUEL RATHBONE

LEGEND.

1 ADMIN. BLDG.
2 COMDG. OFF. QRS.
3 OFFICERS QRS.
4 HOSPITAL.
5 HOSPITAL STO. QRS.
6 N.C.O. QUARTERS.
7 BARRACKS.
8 GUARD HOUSE.
9 POST EXCHANGE.
10 GYMNASIUM.
11 GARAGE.
12 STORE HOUSE.
13 COAL SHED.
14 FIRE HOUSE.
15 MESS HOUSE.
16
17 RESERVOIR.
18 CAMP KITCHEN.
22 Q.M. STABLE.
23 " " WAGON SHED
70 Y.M.C.A.
72 HANDBALL COURT.

BATTERIES.
S. RATHBONE 4–6" BAR.

SAN FRANCISCO HARBOR

FORT BARRY-D3.

Scale of feet.

-400   0   400   800   1200   1600

EDITION OF MAR. 3 1919.
REVISIONS: APRIL 17, 1920.

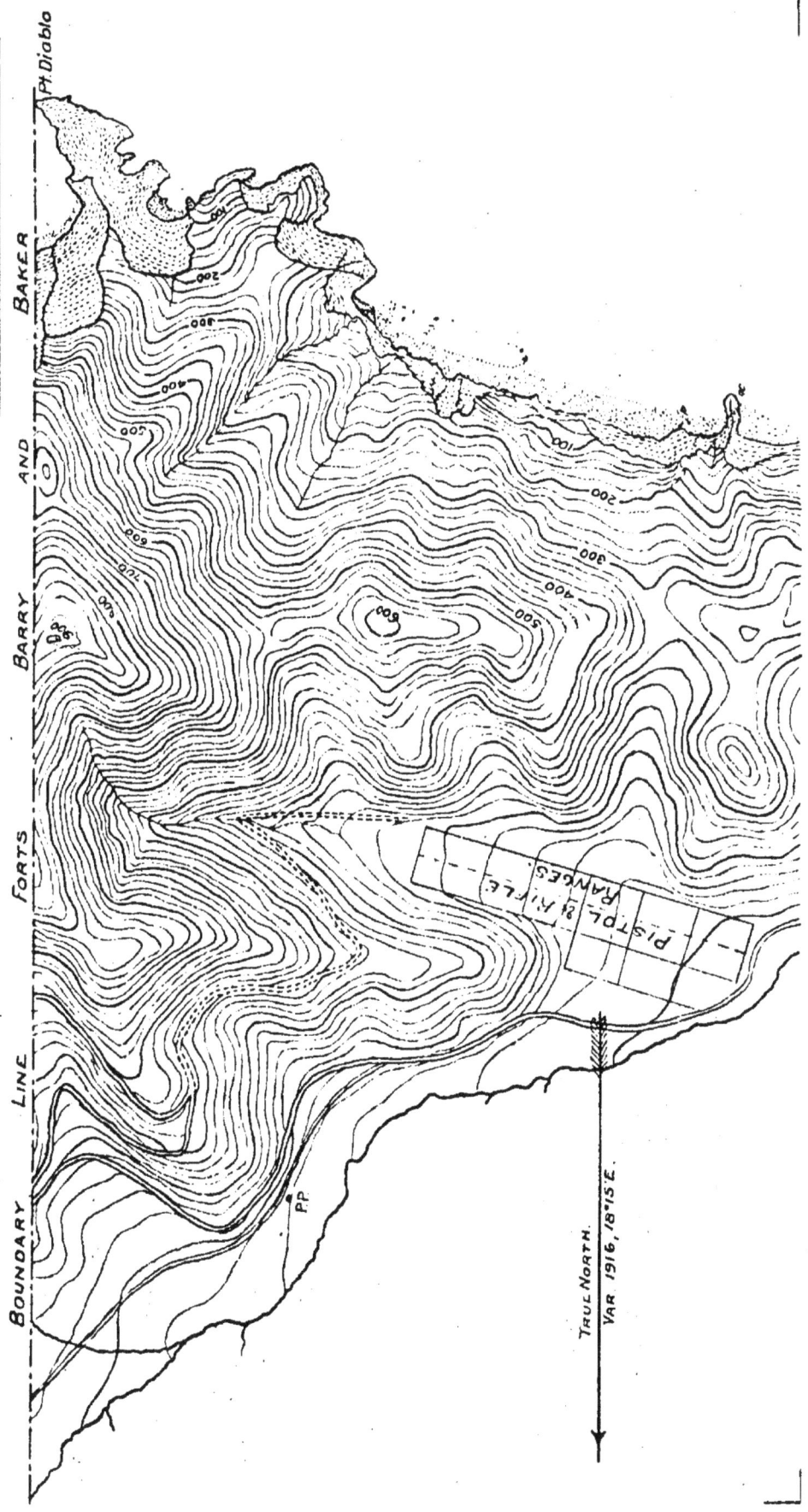

SERIAL NUMBER

BOUNDARY   LINE   FORTS   BARRY   AND   BAKER

Pt. Diablo

PISTOL & RIFLE RANGES

P.P.

TRUE NORTH.
VAR 1916, 18°15' E.

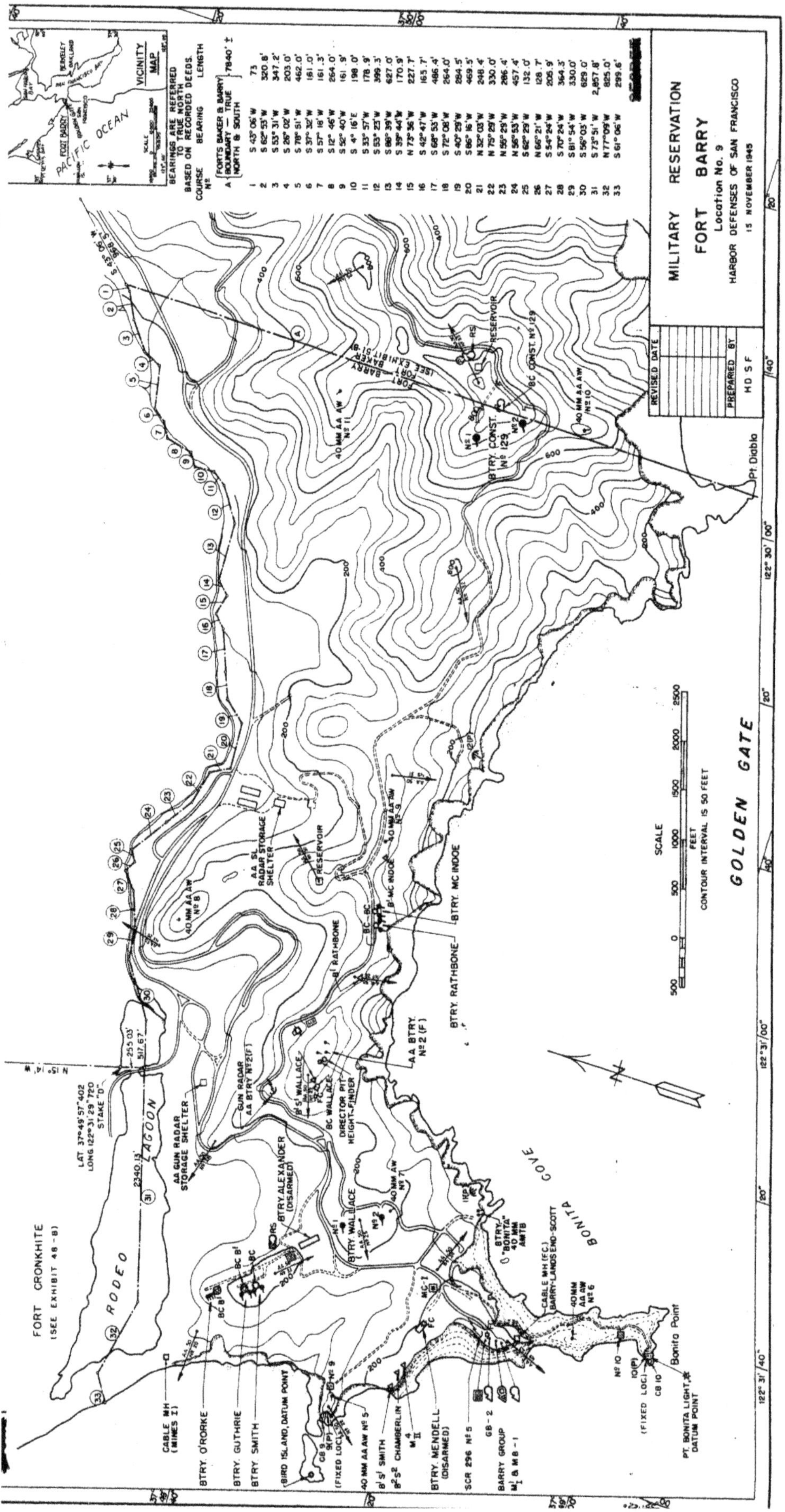
MILITARY RESERVATION
FORT BARRY
Location No. 9
HARBOR DEFENSES OF SAN FRANCISCO
15 NOVEMBER 1945
GOLDEN GATE
RODEO LAGOON
BONITA COVE
PACIFIC OCEAN
VICINITY MAP
SCALE
FEET
CONTOUR INTERVAL IS 50 FEET
FORT CRONKHITE
(SEE EXHIBIT 48-B)
Pt Diablo
Bonita Point
114 EXHIBIT 50-B

Fort Barry 1939 (NARA)

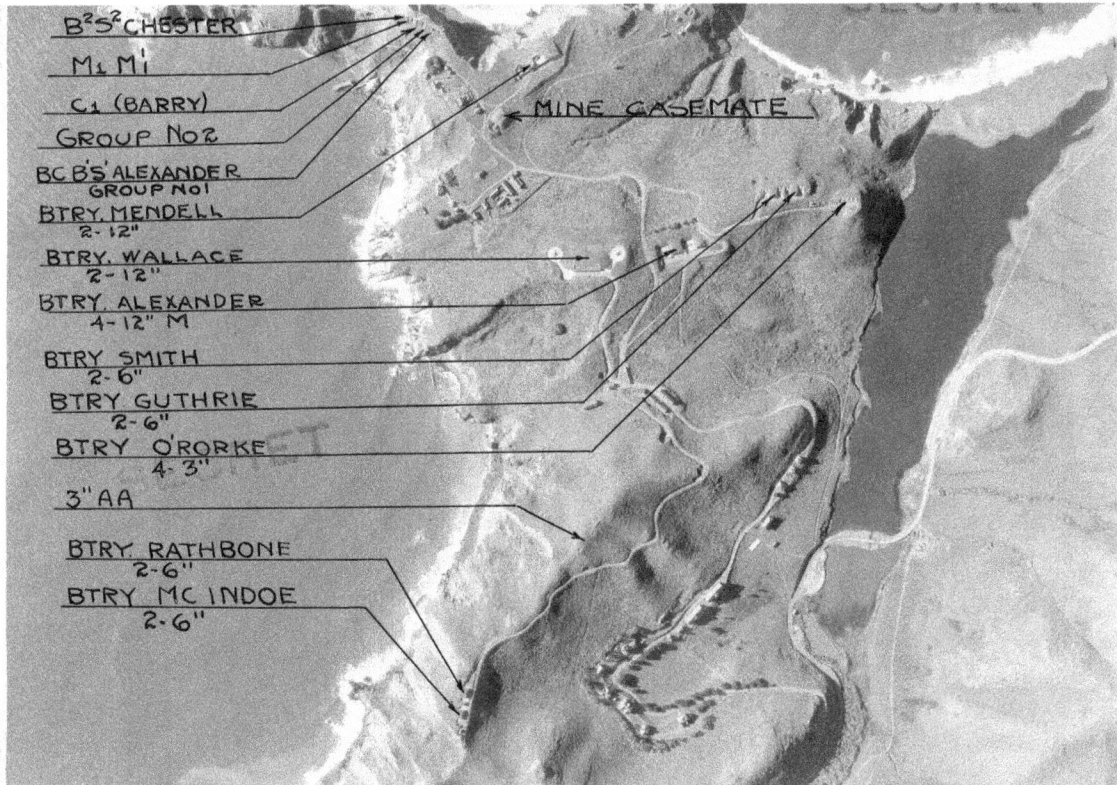

B²S² CHESTER

M₁ M¹

C₁ (BARRY)

GROUP No 2

BC B'S ALEXANDER
GROUP No 1

BTRY. MENDELL
2 - 12"

BTRY. WALLACE
2 - 12"

BTRY. ALEXANDER
4 - 12" M

BTRY SMITH
2 - 6"

BTRY GUTHRIE
2 - 6"

BTRY O'RORKE
4 - 3"

3" AA

BTRY. RATHBONE
2 - 6"

BTRY MC INDOE
2 - 6"

MINE CASEMATE

Fort Barry 1938 (NARA)

Forts Barry and Cronkhite
Marin Headlands Golden Gate National Recreation Area (Terry McGovern)

Battery Mendell and Bonita Head
Marin Headlands Golden Gate National Recreation Area (Terry McGovern)

Battery Alexander, Battery Rathbone-McIndoe and Battery O'Rorke
Marin Headlands Golden Gate National Recreation Area (Terry McGovern)

Battery Wallace Marin Headlands Golden Gate National Recreation Area (Terry McGovern)

Battery #129 Marin Headlands Golden Gate National Recreation Area (Terry McGovern)

mortally wounded in France during World War I. The gun tubes received were previously in storage received as spare guns. During proofing of the armament in 1928 tube #68 was severely damaged. It was removed and replaced in 1929 with Watervliet tube #75. In May of 1942 work was begun on battery modernization featuring new overhead casemates for the gun platforms. This work was accomplished for transfer on March 31, 1944 for a cost of $761,000. The battery was phased out at the end of the San Francisco defenses in 1948, and the armament was scrapped. The emplacement still exists at the golden Gate National Recreation Area. The battery is open to the public, but the service gallery and service rooms are closed.

- **ALEXANDER:** The mortar battery emplaced on the Fort Barry reservation, in a small, protected valley inland of the western shoreline. It was not too far southeast of Battery Mendell. Submission was made on September 12, 1900 for an eight mortar, two-pit design with a construction cost estimate of $89,362. It was of typical late Endicott mortar type design, with wide pits and a "T-shaped" magazine located solely in the traverse between the two pits. Excavation was done in September 1901, and actual concrete work was done from October 1901 to July of 1902, sharing the construction plant with Battery Mendell. It was transferred on June 8, 1905 for $100,382. Naming was designated in General Orders No. 120 of November 22, 1902 for Colonel Barton S. Alexander of Civil War engineer service. The armament was transported by barge to Rodeo Beach but lost in a storm on April 6, 1902. While all the major parts were subsequently recovered, arming was delayed for a year. It was finally armed with eight 12-inch Model 1890M1 Watervliet mortars on Model 1896 carriages (#145/#279, #147/#277, #148/#281, #150/#283, #151/#282, #155/#284, #159/#278 and #160/#280). In the summer of 1918 four mortars were removed from the forward two positions in each pit (tubes #145, #148, #150, and #159). The final four guns and carriages continued to serve until authorized for removal on January 23, 1943. The emplacement was subsequently used for TNT storage until the reservation was shut down. The emplacement still exists in good condition at the Golden Gate National Recreation Area. The battery has been used as a youth group campsite over the years,

- **RATHBONE – McINDOE:** A battery for four 6-inch rapid-fire guns emplaced along the bluff line mid-way between Lime Point and Bonita Point, firing to the south. By the time it was submitted, the type plans for 6-inch Model 1900 batteries had been substantially modified so that they were more like the type disappearing batteries—with individual gun platforms and wide traverses holding the magazines. The final plan was submitted on November 2, 1902, which was slightly revised on January 31, 1903. Concrete construction had to await the moving of the concrete plant from work at mortar battery Alexander. Transfer was made on June 8, 1905 for a cost of $92,511. It was of typical 1903 type design, although the No. 1 position (the most westerly) was configured so as to be able to fire all-round. It was named on General Orders No. 194 of December 27, 1904 for Lieutenant Samuel B. Rathbone who was mortally wounded during the War of 1812. It received four 6-inch Model 1900 guns on M1900 pedestal mounts (#19/#26, #29/#42, #33/#43, and #34/#44). As the battery was built and armed after the gun recall for breech replacement, these serial numbers were for the original armament. In December 1917 the No. 3 and No. 4 emplacements were dismounted, the tubes being taken for field artillery use, and the carriages left in place. In May 1919 these exact same tubes were returned to the battery and remounted. On General Orders No. 13 of March 22, 1922, the battery was tactically split, the first two emplacements remaining Battery Rathbone, but No. 3 and No. 4 became Battery McIndoe, named for Brigadier General James F. McIndoe, killed in France in 1918. The four guns continued to serve until deleted in 1948. The emplacement still exists in good condition at the Golden Gate National Recreation Area. The battery is open to the public, but the magazines are closed.

- **GUTHRIE – SMITH**: A second 6-inch, four-gun battery for Fort Barry, this one emplaced on the western side of the reservation, north of Point Bonita, firing to the west. It was situated on a ridge just north of Rodeo Lagoon, positioned to cover any attempted landing at vulnerable Rodeo Beach. It was planned and submitted simultaneously with Rathbone. The two batteries shared the same late, type 1903 plan with separate platforms. It had the primary magazines shared by two platforms between No. 1 and No. 2 and then between No. 3 and No. 4. The central traverse had small service rooms, but no ammunition storage. Ammunition service was by hand, with no hoists. The four platforms all faced south with no ability to fire to either flank. Work was done in 1904-1905. It was transferred on June 8, 1905 for a cost of $69,193. The battery was named in General Orders No. 194 of December 27, 1904 for Captain Edwin Guthrie who was mortally wounded during the Mexican War. It was armed with four 6-inch Model 1900 guns and pedestal carriages, the latter received in late 1905. The tubes were shipped here in January and March of 1906 after their recall and modification of breechblocks at Watervliet Arsenal. It was armed with #2/#13, #3/#14, #5/#15, and #12/#16. In December 1917 the guns from emplacements No. 3 and No. 4 were removed for use on field carriages destined for France. However, they were returned and re-emplaced in their previous assignments in May of 1919. In 1922 the battery was tactically split, emplacement No. 1 and No. 2 remaining as Battery Guthrie, but No. 3 and No. 4 became Battery Smith. This was named on General Orders No. 13 of March 22, 1922 for Colonel Hamilton A. Smith, who died in France in 1918. These four guns continued to serve until authorized for removal after World War II in 1948. The emplacement still exists at the Golden Gate National Recreation Area. The battery is open to the public, but the magazines are closed.

- **O'RORKE**: A battery for four 3-inch pedestal guns emplaced on the right flank of Battery Guthrie overlooking Rodeo Beach. An early project for a two-gun, masking parapet emplacement actually at Point Bonita was not approved. Submission was made on June 27, 1902 from funds made available by the Fortification Act of June 6, 1902. Excavation started immediately in the summer of 1902, and by mid-1903 the construction was complete. It was of conventional design, with the platforms in line, and built for the Model 1898 masking parapet mount. However, during construction, the decision was made to complete the battery for new pedestal mounts. New platform bolts were installed and the barrel niches for the balanced pillar guns filled in. It was transferred on June 1, 1905 for a cost of $24,462.76. It was named on General Orders No. 194 of December 27, 1904 for Colonel Patrick Henry O'Rorke killed in action at Gettysburg on July 2,1863. It was armed with four 3-inch Model 1903 guns and pedestal mounts (#90/#68, #91/#69, #92/#70, and #94/#71). This armament remained in place until finally removed on March 7, 1946. The emplacement still exists at the Golden Gate National Recreation Area. The battery is open to the public.

- **Battery #129**: A 1940 Program battery for two 16-inch casemated barbette guns at Fort Barry. Due to the recent construction of the new 16-inch batteries at Fort Funston and Cronkhite, additional 1940 Program batteries of this size for San Francisco were considered of relatively low priority. One additional battery to the north and one to the south were authorized with the program. Battery Construction No. 129 (it was never named) was approved on September 26, 1940 for emplacement on Diablo Ridge, which though it is on the far eastern extreme of Fort Barry, is also the highest elevation available (at 821-feet, this was the greatest height used by the U.S. for a seacoast battery location). Siting was authorized on June 7, 1941 and work actually began in May of 1942. The original national priority of September 11, 1940 was #23, being slightly increased to #21 in the August 11, 1941 list. First construction funding of $625,000 was part of the FY-1942 Budget. The design plan was modified from the standard 100-series type to fit the tight topography of the site.

Generally firing to the west, the gun corridors were literally tunneled out of the ridge peak. The central rear entry was reduced in size, and a number of other modifications were made to the room arrangements. Work was stopped on November 23, 1943. The status was changed to "suspended" by authority of December 8, 1943. Though incomplete, the battery structure was conditionally transferred on January 12, 1944 at a cost of $2,076,000. No armament was ever mounted, though it was intended for two 16-inch guns MkIIM1 on M1919M5 barbette carriages. Interestingly the hill where the battery was constructed is now named "Hill 129" reflecting the construction number of the battery. The emplacement still exists at the Gold Gate National Recreation Area. The battery is open to the public, but the service gallery and rooms are closed.

Fort Baker (1852-1950) is located on the northern side of the Golden Gate Channel, which is the entrance to San Francisco Bay, and about two miles south of Sausalito, California. In the 1850s plans were made to build a masonry fort on the northern side of the Golden Gate at Lime Point. Some preparation work was undertaken, but a fort was not built. During the 1870s Period an earthen battery was built at Yellow Bluff, called Battery Cavallo. It was named in General Orders 32 of 1897 for Col. Edward D. Baker, USV, who had served as a Senator from California and was killed in action during the Civil War in 1861. Construction of one of the first Endicott Program batteries in the United States begun on a high bluff above Lime Point in 1893. Battery Spencer, which mounted three 12-inch guns on barbette carriages was the first completed. By 1904, the fort held a collection of seven Endicott Program batteries with five more under construction. The fort also had a controlled submarine mine complex. In the same year, Fort Baker's reservation was divided into two forts. The section from Point Diablo to the east remained Fort Baker, while west of Point Diablo became Fort Barry. The coming of World War II resulted in a major effort to modernize the defense of San Francisco, but these new coast artillery batteries were built to the north and south of the Fort Baker's reservation. The only addition to Fort Baker during World War II was a AMTB battery. At end of World War II, the fort became a sub-post of Fort Scott and then a sub-post of the Presidio of San Francisco. A portion of Fort Baker was transferred to National Park Service in 1972 to become part of the Golden Gate National Recreation Area, while the rest of the fort continued to be under U.S. Army control. In 1995, the military transferred its land to Golden Gate National Recreation Area. The NPS area around Horseshoe Cove is home to the Bay Area Discovery Museum, U.S. Coast Guard station, and the Travis AFB Sailing Marina. In January 2005, an agreement was reached by the city of Sausalito and the National Park Service with developers for a retreat and conference center. Thirteen historic lodging buildings and seven historic commons buildings were renovated; and thirteen new lodging buildings were built, the facility opened to the public in 2008 as the Cavallo Point, The Lodge at the Golden Gate. It features lodging, dining, conference space, and a spa. The restored area around the parade ground still retains the look and feel of the military years of Fort Baker, and its batteries are in excellent condition. The fort is open to the public, but most structures themselves are closed to the public.

Fort Baker Gun Batteries

- **SPENCER:** A battery for three 12-inch guns on barbette carriages authorized on July 23, 1892 as the first new Endicott emplacement for the Lime Point reservation of Fort Baker. Local engineers submitted their plans on October 5, 1892—originally for just the first two emplacements known at the time as emplacements No. 3 and No. 4. The site was the point of the ridge overlooking the Golden Gate, which had been the site of an earthen 1870s battery for Rodman guns known as Ridge Battery. In order to fit it into the constrained space, the three emplacements were fanned-out from the center. Also, they used steel beams in the roofs of the magazines to help reduce the required thickness of cover. They were equipped with ammunition hoists. Understandably the construc-

SAN FRANCISCO HARBOR
FORT BAKER.

SERIAL NUMBER 103

EDITION OF JAN.14,1915.
REVISIONS: DEC 7,1915;
NOV 5, 1916.

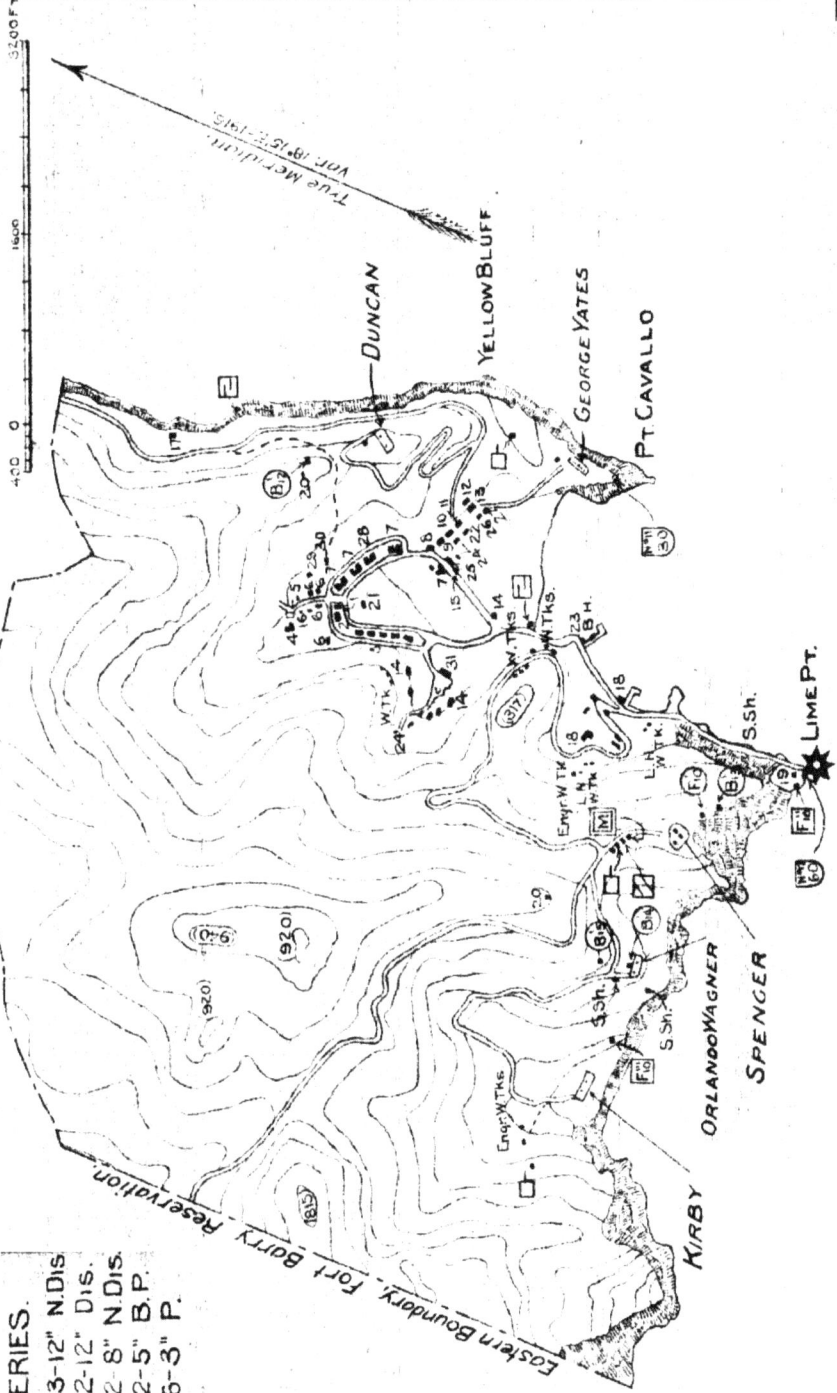

TRUE MERIDIAN
VAR. 18°15'E.-1915.

Duncan
Yellow Bluff
George Yates
Pt. Cavallo
Lime Pt.
Spencer
Orlando Wagner
Kirby

Eastern Boundary, Fort Barry Reservation.

BATTERIES.

Spencer	3-12" N.Dis.
Kirby	2-12" Dis.
Duncan	2-8" N.Dis.
O.Wagner	2-5" B.P.
G.Yates	6-3" P.

LEGEND.

1. ADMINST. BLDG.
2. COMDG. OFFS. QRS.
3. OFFICERS QRS.
4. HOSPITAL.
5. HOSPITAL QRS.
6. N.C.OFFICERS QRS.
7. BARRACKS.
8. GUARD HOUSE.
9. BAKERY.
10. COAL SHED.
11. Q. M. SUB STOREHOUSE.
12. STABLE.
13. WAGON SHED.
14. DWELLINGS.
15. ORD.STORE HOUSE.
16. TENNIS COURT.
17. TORPEDO CASEMATES ABANDONED.
18. ENGR. DEPT. BLDGS.
19. U.S.FOG SIGNAL.
20. RESERVOIRS.
21. FLAG POLE.
22. STORE HOUSE.
23. WAITING ROOM.
24. OUT HOUSES.
25. SQUAD ROOM.
26. CARPENTER & PT.SHOP.
27. BLACKSMITH SHOP.
28. POST EX. & GYM.
29. FIRE HOUSE.
30. SUB. STATION.
31. ENG. & SIG. CORP. ST.HO.

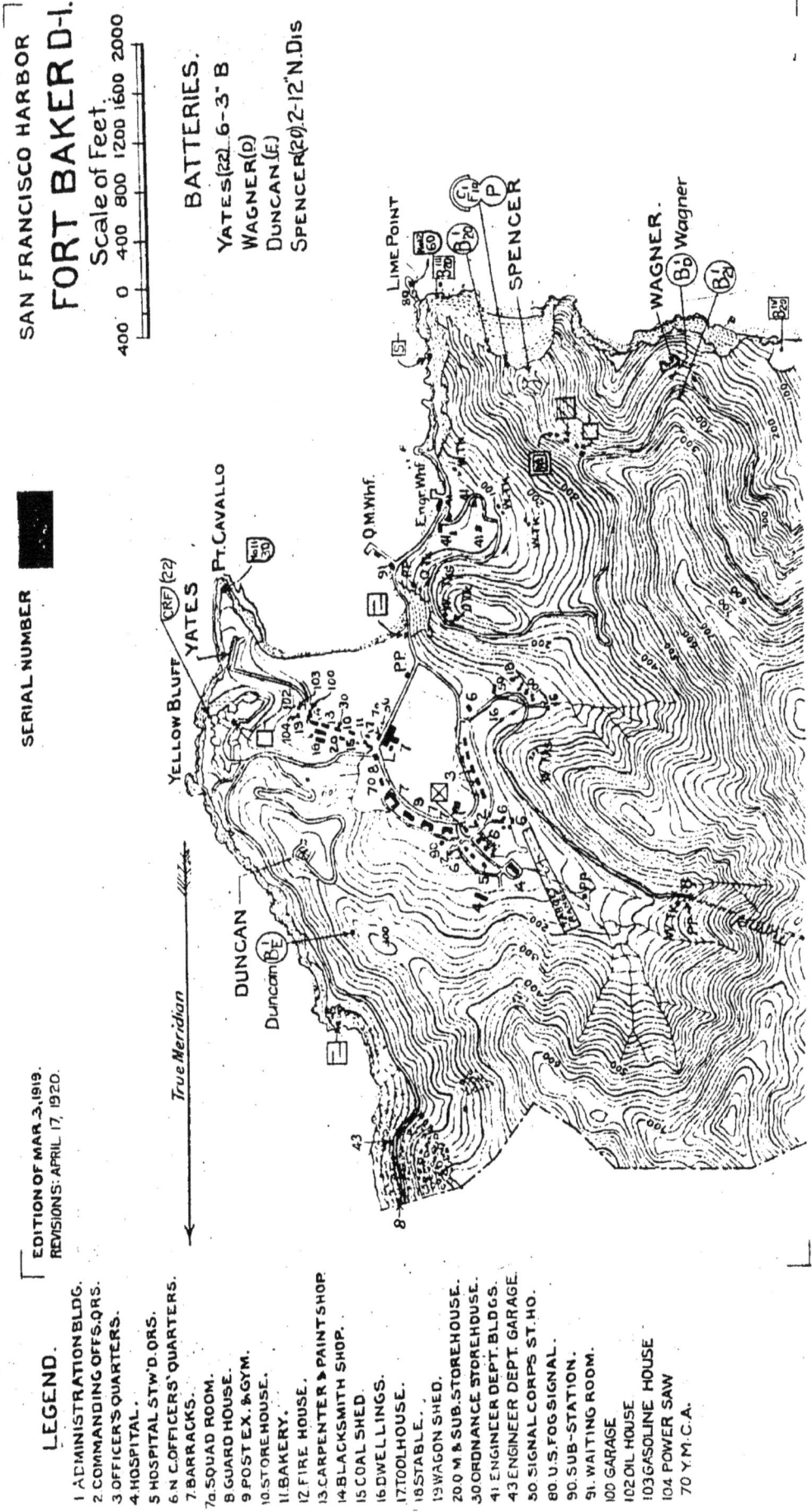

SAN FRANCISCO HARBOR
FORT BAKER D-I.
Scale of Feet
400 0 400 800 1200 1600 2000

BATTERIES.
YATES(22) 6-3" B
WAGNER(D)
DUNCAN(E)
SPENCER(20) 2-12" N.Dis

SERIAL NUMBER

EDITION OF MAR. 3, 1919.
REVISIONS: APRIL 17, 1920.

True Meridian

LEGEND.
1 ADMINISTRATION BLDG.
2 COMMANDING OFFS.QRS.
3 OFFICER'S QUARTERS.
4 HOSPITAL.
5 HOSPITAL STW'D.QRS.
6 N.C.OFFICERS' QUARTERS.
7 BARRACKS.
7a SQUAD ROOM.
8 GUARD HOUSE.
9 POST EX. & GYM.
10 STOREHOUSE.
11 BAKERY.
12 FIRE HOUSE.
13 CARPENTER & PAINT SHOP.
14 BLACKSMITH SHOP.
15 COAL SHED.
16 DWELLINGS.
17 TOOLHOUSE.
18 STABLE.
19 WAGON SHED.
20 Q.M.& SUB.STOREHOUSE.
30 ORDNANCE STOREHOUSE.
41 ENGINEER DEPT. BLDGS.
43 ENGINEER DEPT. GARAGE.
50 SIGNAL CORPS ST.HO.
80 U.S.FOG SIGNAL.
90 SUB-STATION.
91 WAITING ROOM.
100 GARAGE
102 OIL HOUSE
103 GASOLINE HOUSE
104 POWER SAW
70 Y.M.C.A.

SAN FRANCISCO HARBOR

FORT BAKER D-2.

Scale of Feet.

400 0 400 800 1200 1600 2000

SERIAL NUMBER

True Meridian

EDITION OF MAR. 3,1919.
REVISIONS: APRIL 17, 1920.

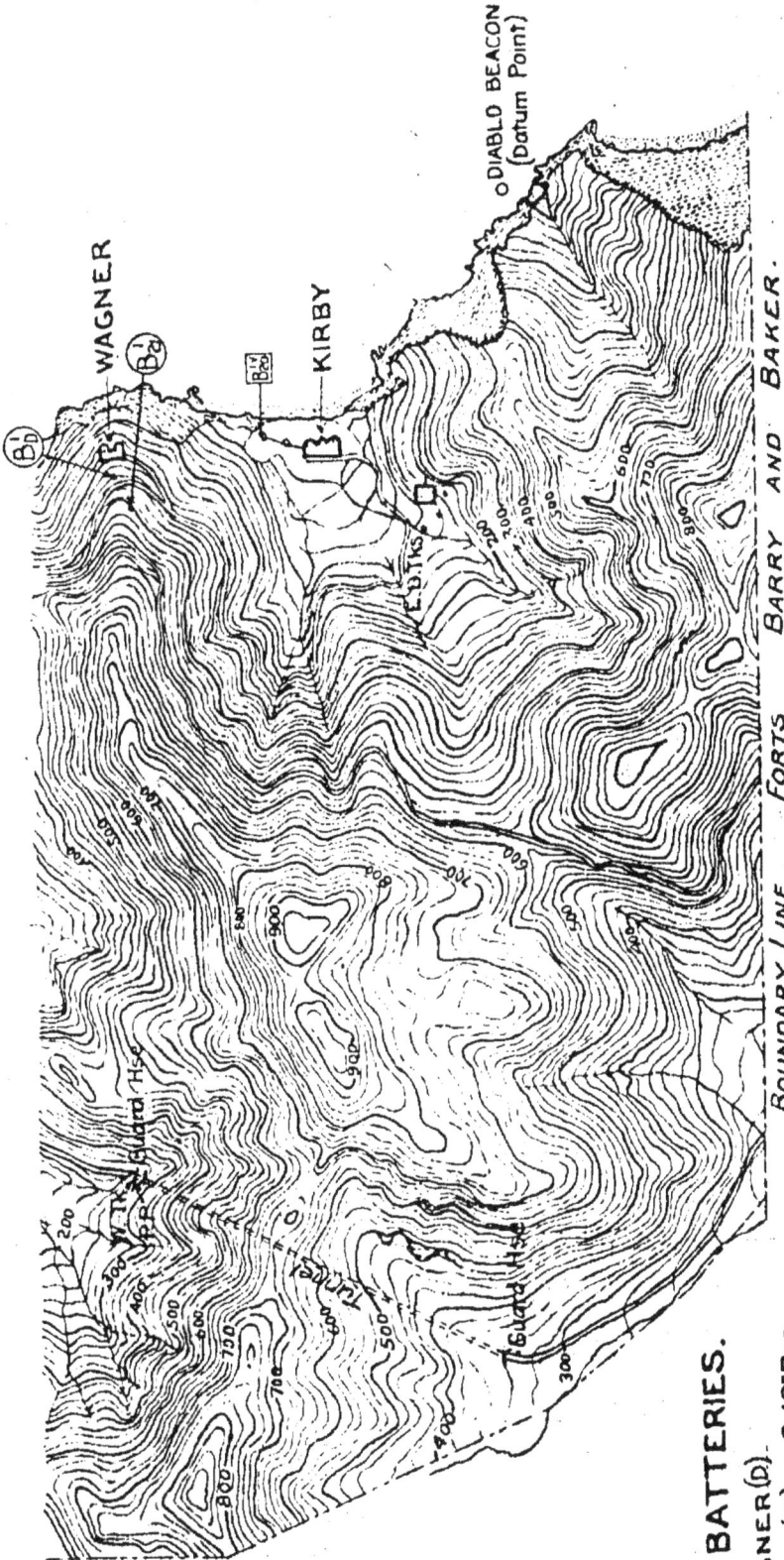

WAGNER

KIRBY

○ DIABLO BEACON
(Datum Point)

Guard Hse

Guard Hse

Boundary Line

Forts   Barry and Baker.

BATTERIES.

WAGNER (1).
KIRBY (2)....2-12"DIS.

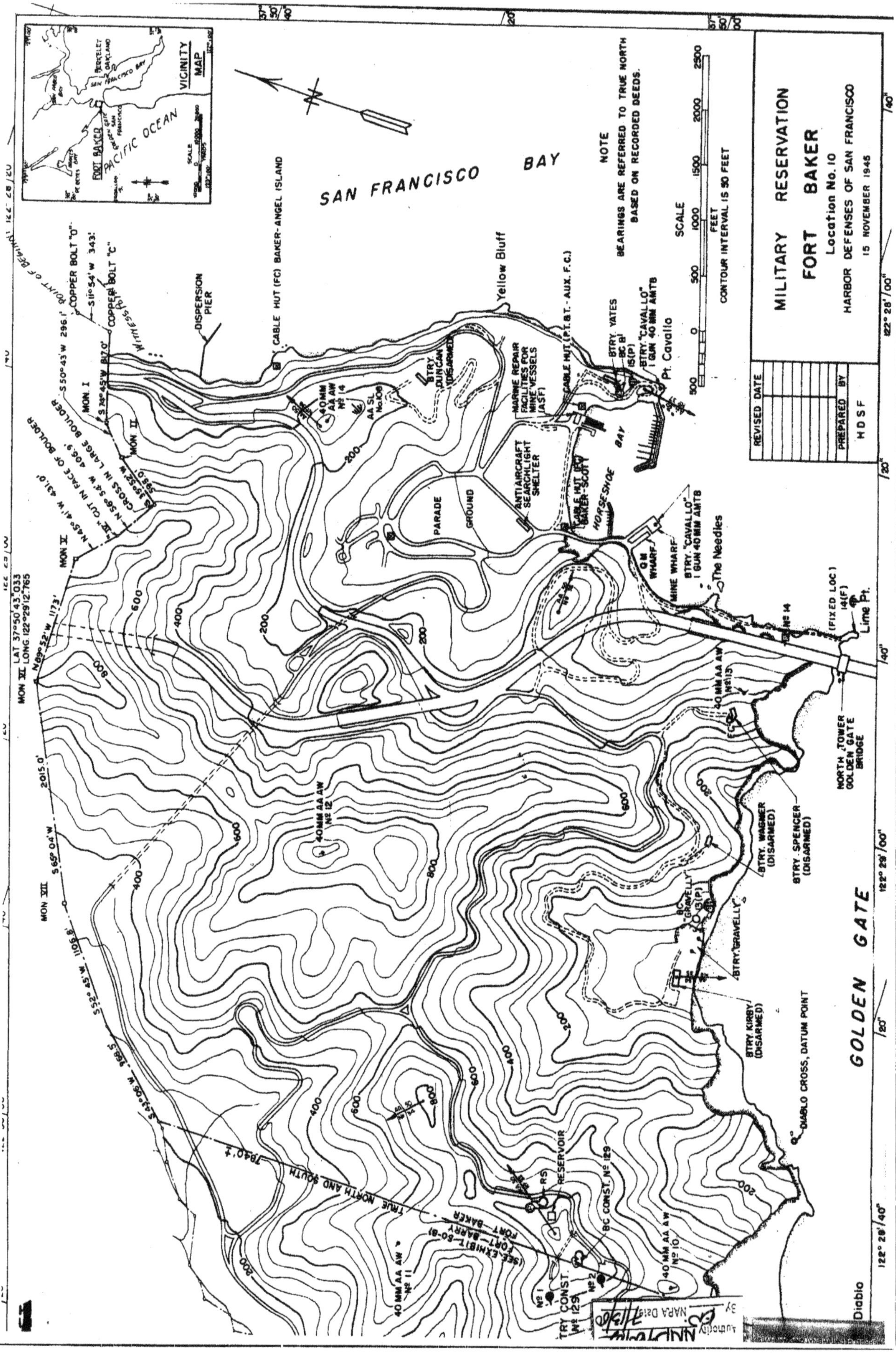
MILITARY RESERVATION
FORT BAKER
Location No. 10
HARBOR DEFENSES OF SAN FRANCISCO
15 NOVEMBER 1945
REVISED DATE
PREPARED BY
HDSF
NOTE
BEARINGS ARE REFERRED TO TRUE NORTH BASED ON RECORDED DEEDS.
SCALE
FEET
CONTOUR INTERVAL IS 50 FEET
VICINITY MAP
PACIFIC OCEAN
SAN FRANCISCO BAY
GOLDEN GATE
Yellow Bluff
Pt. Cavallo
HORSESHOE BAY
PARADE GROUND
ANTIAIRCRAFT SEARCHLIGHT SHELTER
MARINE REPAIR FACILITIES FOR MINE VESSELS (ASF)
BTRY. DUNCAN (DISARMED)
BTRY. YATES
DISPERSION PIER
CABLE HUT (FC) BAKER-ANGEL ISLAND
Q M WHARF
MINE WHARF
The Needles
Lime Pt.
NORTH TOWER GOLDEN GATE BRIDGE
BTRY. WAGNER (DISARMED)
BTRY. SPENCER (DISARMED)
BTRY. "GRAVELLY"
BTRY. KIRBY (DISARMED)
DIABLO CROSS, DATUM POINT
Diablo
RESERVOIR
TRUE NORTH AND SOUTH
FORT BARRY
FORT BAKER

Fort Baker 1938 (NARA)

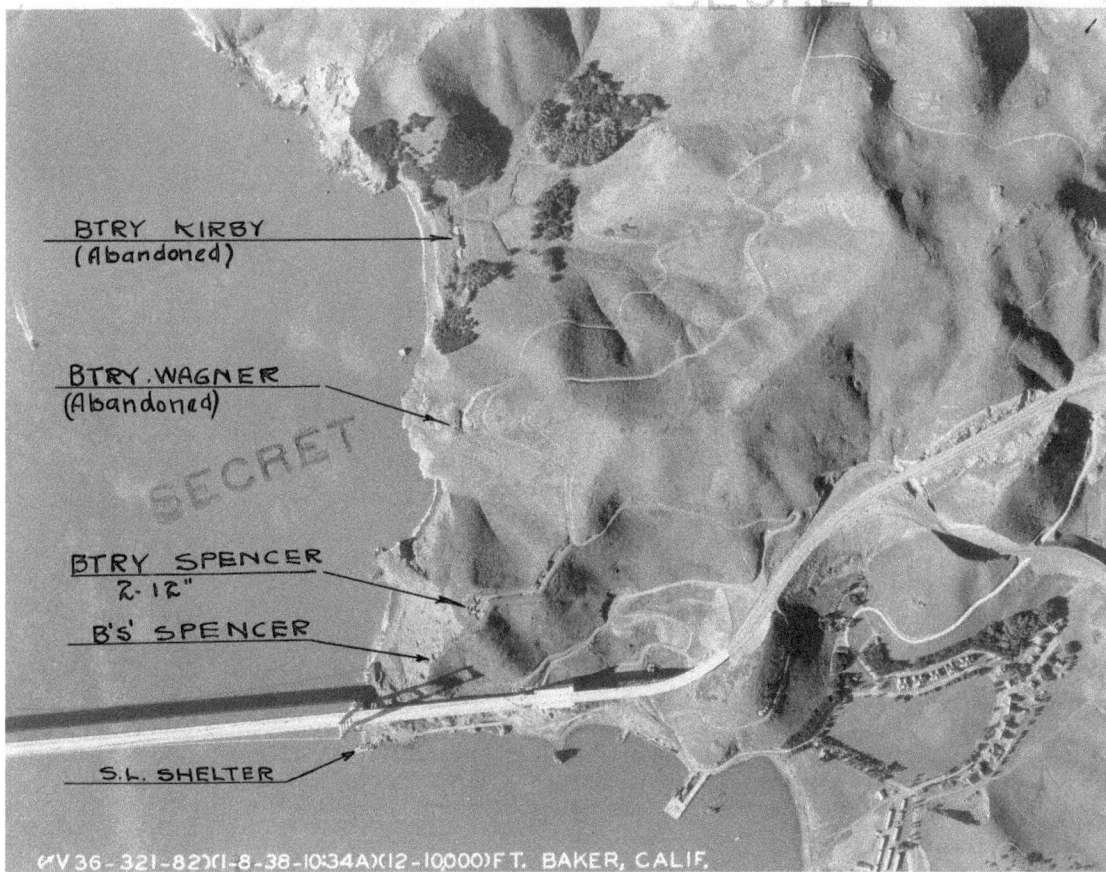

Fort Baker 1938 (NARA)

Fort Baker Marin Headlands Golden Gate National Recreation Area (Terry McGovern)

Battery Spencer Fort Baker Marin Headlands Golden Gate National Recreation Area (Terry McGovern)

tion time was prolonged, starting in 1893 and still going on as late as 1895 and 1896. Guns were mounted between 1895 and 1897. The battery was transferred on September 24, 1897 for a cost of $106,000. It was named in General Orders No. 16 of February 24, 1902 for Major General Joseph Spencer of the Continental Army. It was armed with three 12-inch Model 1888 Watervliet guns on barbette carriages Model 1892 (#10/#4, #16/#1 and #17/#5). As the emplacement was somewhat experimental in design, by 1904 the battery was deemed in need of major renovation. At that time $38,000 was spent to repair and enlarge the loading platforms, modify the location and type of ammunition hoists, improve communications access and add battery commander's station and latrines. In 1918 the gun from the No. 3 emplacement (Watervliet #17) was dismounted in common with the abandonment of most of the inner harbor covering batteries. This tube was subsequently used at Battery Chester at Fort Miley. The final two guns served until removed under authority of January 23, 1943. The emplacement still exists and is on public display at the Golden Gate National Recreation Area. The battery site is open, but the interior is closed to the public.

- **ORLANDO WAGNER:** A battery for two 5-inch, rapid-fire guns authorized for the Fort Baker reservation. In October 1898 plans were submitted for a battery of three 5-inch guns to go on Yellow Bluff on the western side of the reservation, as part of the defenses of the inner harbor, but the plan was rejected in November as Washington wanted the site saved for an 8-inch battery. Plans were re-submitted for another 5-inch battery on April 27, 1899 on a ridge lying midway between Lime Ridge and Gravelly Beach. However, the site was too narrow for three guns, plans were amended to just two guns. The plan closely followed the type recommendations and also was very similar to that submitted for Battery Sherwood at Fort Winfield Scott. The field of fire was to the south. Construction work was done in 1899, transfer coming on August 12, 1901 for a cost of $25,000. It was named in General Orders No. 194 of December 27, 1904 for 1st Lieutenant Orlando G. Wagner mortally wounded in 1862 during the Civil War. It was armed with two 5-inch Model 1897 Bethlehem guns on balanced pillar mounts Model 1896 (#19/#10 and #21/#11). An accident occurred here on September 27, 1910. During target practice– the spindle was blown off the No. 2 gun, wounding one of the gun's crew seriously. It was found that the piece had been manufactured with the defect and was soon repaired. The armament was finally removed in November 1917 for use on army field carriages, the pillars being scrapped in 1920. The emplacement was later used for TNT storage. It still exists on the park land of the Golden Gate National Recreation Area and is open to the public.

- **KIRBY:** The second 12-inch battery for Fort Baker, this one designed to be emplaced near the shore at Gravelly Beach. This was a small beach at the bottom of a large ravine descending from the headlands, facing directly south with an excellent field of view over the direct approach to the Golden Gate. In the 1870s an extensive linear battery for 15-inch Rodman guns had been built there. The original Endicott San Francisco plan called for a battery of four 12-inch gun lifts here, but by 1898 these plans had changed to a battery armed with disappearing guns. The directly exposed guns, without the advantage of bluff height to protect them, required disappearing mounts in order to protect the guns and crews. With the August 1898 appropriations, $60,000 was allocated to start this work. The plan for four guns entirely filled the beach from the hill on one flank to the next. Only a 60-degree lateral field of fire was required by the guns, the confines of the position being that restrictive. On August 4, 1898 plans were re-submitted for just three guns. Authorization was given to build initially a two-gun battery, with the ability to add a third emplacement, which was never granted. The position dictated a unique plan. As enemy fire could only come from directly in front, no side or rear protection was needed. Also, as enemy ships would only be ¾ of mile away, shot would not be plunging, and could be deflected by carefully shaped roof cover and front earthen protection. As these were low structures, they used adjacent ammunition service without lifts. Work was done in the fall of 1898 to early 1899—finished except for base rings, doors, and electrical plant and wiring. The guns arrived by barge in the fall of 1899 and were soon mounted. The battery was transferred on August 5, 1900 for a cost of only $70,334.18. It was named in General Orders No. 16 of February 14, 1902 for Lieutenant Edmund Kirby, mortally wounded in 1863 during the Civil War. The armament consisted of two 12-inch Model 1895 guns on disappearing carriages Model 1897 (Watervliet guns #12/#14 and #16/#15). The battery received several modifications in the coming years. In 1910 a new BC station with a plotting room underneath was added. One modification not made was to increase the elevation and thus range of the carriages. With such a short range required across the Golden Gate and any loss of depression being unacceptable, this change was not made like it was at most other disappearing 12-inch gun batteries. The guns were too important to remove in World War I and were finally taken out in 1933 and 1941. Official abandonment of the battery was authorized on January 23, 1943. The emplacement still exists on the property of the Golden Gate National Recreation Area. While the batterythe interior is closed to the public, the site is near a large group camp area, but

- **DUNCAN:** A battery for two 8-inch barbette mounts emplaced as part of the inner harbor defenses placed above Horseshoe Bay on Yellow Bluff. It was funded from the National Defense Act of March 1898 submission being made on June 4, 1898. It closely resembles the half battery built on Angel Island. It was of low profile, with platform adjacent to a protected magazine on the same level and no need for hoists. With the war emergency, work actually began before the formal approval of the submission plan, and was well underway in April of 1898. By the end of September, it was essentially completed. Carriages didn't arrive until July 1899. The emplacement was armed and transferred on May 5, 1900 for $57,535.40. The battery was armed with two 8-inch Model 1888 guns on Model 1892 barbette carriages (West Point Foundry #8/carriage #2 and Bethlehem M1888MII #24/carriage #4). It was named on General Orders No. 16 of February 14, 1902 for Colonel James Duncan of Mexican War service. In 1910 a new battery commander's station and plot was added atop the central traverse. This armament was removed in 1917 with the abandonment of the inner bay defenses. The guns were taken out in November 1917 and the carriages scrapped in May of 1918. The emplacement was subsequently used for ammunition storage. It still exists at the golden Gate National Recreation Area. The battery is fenced off and not open to the public.

- **YATES:** A rapid-fire battery of six 3-inch guns emplaced as part of the San Francisco inner defenses at Fort Baker. It was located on a small peninsula at the southeast corner of the reservation, at Point Cavallo. It fired into the inner harbor, to the southeast. The plan for construction was submitted on March 6, 1903. As a series of six emplacements, the final four (No. 3-6) were canted further to the east than the first two. It was of typical 1902 type design, with 42-foot gun centers between platforms. Work was done in 1903-1904. Transfer was made on June 8, 1905 for a construction cost of $41,406.82. It was named on General Orders No. 194 of December 27, 1904 for Captain George Yates of the 7th Cavalry who died at the Little Big Horn in 1876. It was armed with six 3-inch Model 1902 Bethlehem guns and pedestal carriages (#18/#18, #19/#19, #20/#20, #21/#21, #22/#22, and #23/#23). A new plot and BC station were added in 1910. The battery served until gradually disarmed in the 1940s. Two guns and carriages were moved in 1940, going to Battery Kirby Beach at Fort Baker until replaced by 90mm Battery Gravelly and then finally going to Fort Point in 1944. Two other guns were removed under authority of February 21, 1944 and moved directly to old Fort Point (guns and carriages #20 and #21). The final two guns and carriages (#18 and #19) stayed in place until removed under authority of March 7, 1946. The emplacement still exists on property of the Golden Gate National Recreation Area. The battery is open to the public.

- **Battery Kirby Beach:** A 1942 expedient AMTB emplacement for two relocated 3-inch pedestal guns. The two Model 1902 guns and pedestal mounts (Bethlehem #22 and #23) were moved here from Battery Yates and emplaced on simple concrete gun blocks and the slope to the east of old Battery Kirby in 1942. Work was done from August 4 to August 12 of 1942, but official transfer was not made until June 1, 1943 for a cost of $955.78. The emplacement consisted of only simple blocks and an adjacent wooden magazine. While never officially named, it was informally known as "Kirby Junior." The battery served only a short while, for with the completion of the 90mm AMTB battery Gravelly, it became redundant. The guns were moved in turn to a new site atop old Fort Point in mid-1943 (where they were joined by two other guns formerly at Battery Yates). At least one of the gun blocks still exists at the Golden Gate National Recreation Area. The battery is open to the public.

- **Battery Gravelly:** A 1943 Program 90mm AMTB battery. It consisted of two 90mm mobile and two 90mm fixed guns on simple concrete blocks. It was located to the east of Battery Kirby, not far from Battery Kirby Beach (which it replaced in function in 1944. Work was started in June 1943 and transfer made on January 18, 1944 for a cost of $4,908. The gun blocks were separated by 120-feet. The battery was disarmed postwar, but the blocks still exist at the Golden Gate National Recreation Area. The battery is open to the public.

Fort McDowell (1852-1928) is located on Angel Island in San Francisco Bay, near the town of Tiburon, California. This 640-acre island was first used by the military during the Civil War when Camp Reynolds was established. Earthen batteries were constructed to cover the Golden Gate channel, along with the guns at Fort Point, Lime Point, Alcatraz, and Black Point. During the 1870s Period, a battery mounting Rodman cannons on barbette carriages were installed at Point Knox. In 1899, a construction program begun at Quarry Point to build a large U.S. Army post, known as East Garrison, while the former Camp Reynolds was upgraded into the West Garrison. These two cantonment areas, along with a small North Garrison area, was named in General Orders 43 of 1900 for Maj. Gen. Irwin McDowell, U.S. Army. Constructed near West Garrison at Point Knox were three Endicott Program batteries. These batteries, along with similar 8-inch and 5-inch batteries at Fort Mason, Fort Baker, and Fort Scott, served as an inner-bay defense. This defensive scheme was abandoned by 1915, and Fort McDowell's Endicott batteries were disarmed. The fort continued in U.S. Army service, primarily as an induction center and also as an immigration and quarantine center. After World War II, Fort McDowell was declared surplus. Most of fort was transferred to the State of California in 1965 as a state park. A Nike missile launching site was located on the island until 1975. The island remains a state park with ferry service from Tiburon. Many buildings remain and the Endicott Program batteries survive today.

Fort McDowell Gun Batteries

- **DREW:** An emplacement for a single 8-inch gun on barbette mount approved for the southern side of Angel Island as part of the San Francisco inner harbor defenses. Work was authorized on March 13, 1898, and was underway by April 25, 1898. The work was emplaced on a rise known as Mortar Hill. The plan anticipated a second emplacement for the right flank, with two platforms for the Model 1892 barbettes separated with a single protected traverse containing the magazines on the same level. The plan strongly resembled that completed for Battery Duncan at Fort Baker. However, the No. 1 emplacement was never authorized, and the battery was built and armed with

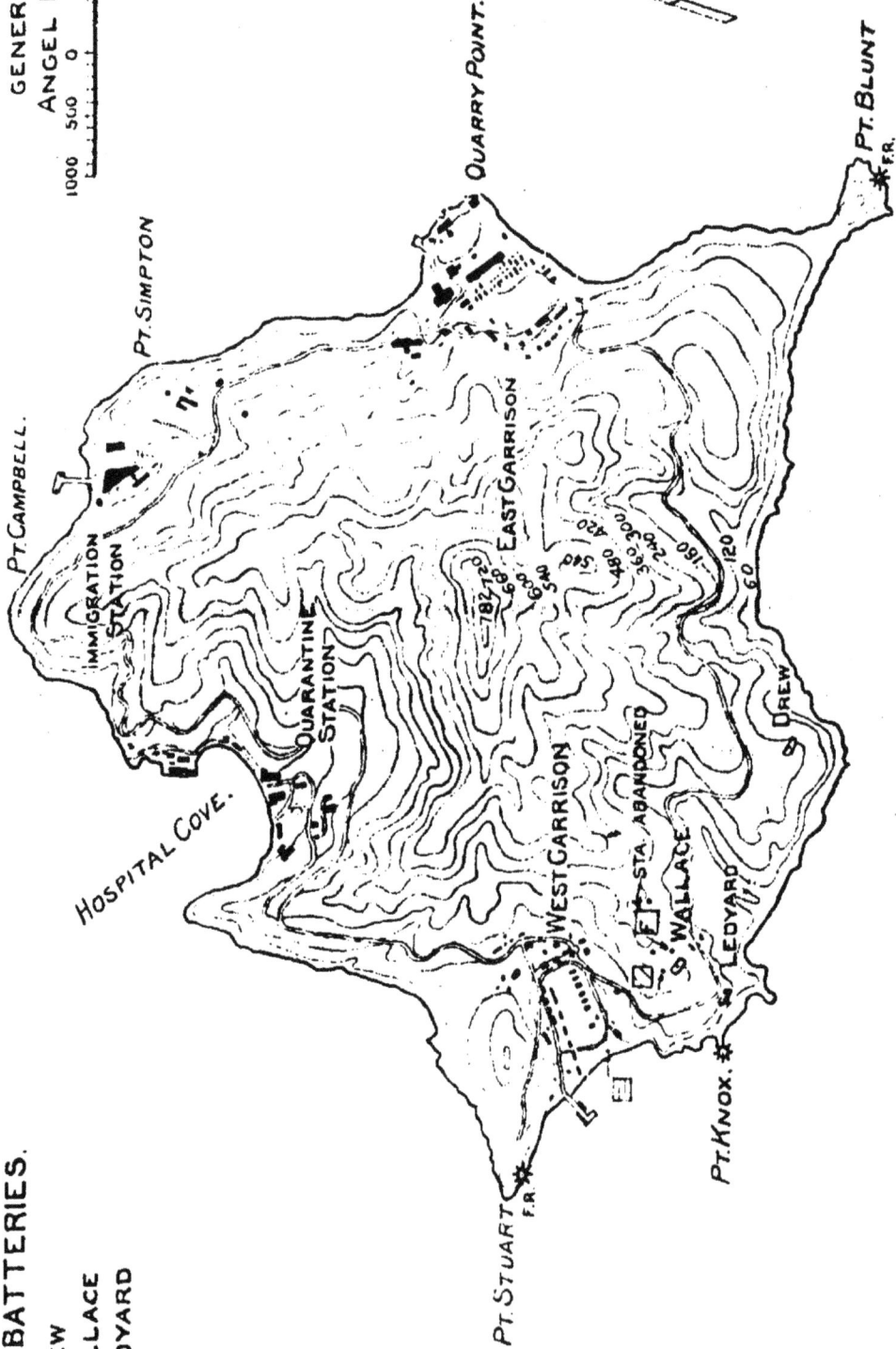

SAN FRANCISCO HARBOR

FORT McDOWELL

GENERAL MAP
ANGEL ISLAND

1000   500   0   500   1000   2000 Ft.

SERIAL NUMBER

EDITION OF DEC. 7, 1915,
REVISIONS NOV. 9, 1916 · APRIL 17, 1920.

BATTERIES.

DREW
WALLACE
LEDYARD

Pt.Campbell.

Pt.Simpton

Immigration Station

Quarantine Station

Hospital Cove.

Quarry Point.

East Garrison

West Garrison

Sta. Abandoned

Wallace

Ledyard

Drew

Pt.Stuart F.R.

Pt.Knox.

Pt.Blunt F.R.

SAN FRANCISCO HARBOR
FORT McDOWELL D-1.
ANGEL ISLAND.
EASTERN PORTION
Scale of Feet.

SERIAL NUMBER

LEGEND
EDITION OF DEC 7 1911
REVISIONS NOV 6 1916
APR 17, 1920.

1. ADMINISTRATION BLDG
2. COMMANDING OFF.QRS.
3. OFFICERS QUARTERS.
4. HOSPITAL.
5.
6. N.C.OFFICERS QRS.
7. BARRACKS.
8. GUARD HOUSE.
9. POST EXCHANGE.
10. MESS AND DRILL HALL.
14. BAND STAND.
16. PHOTOGRAPH GALLERY.
100. OIL HOUSE.
20. Q.M.OFFICE.
21. STOCKADE AND ST.HO.
22. ROCK CRUSHER.
23. COAL BINS.
24. FREIGHT SHED.
25.
26. COAL SHED.
27. SCALES.
71. TENNIS COURTS.
72. HAND BALL COURT.
11. NURSES' QUARTERS.
12. STORE HOUSE.

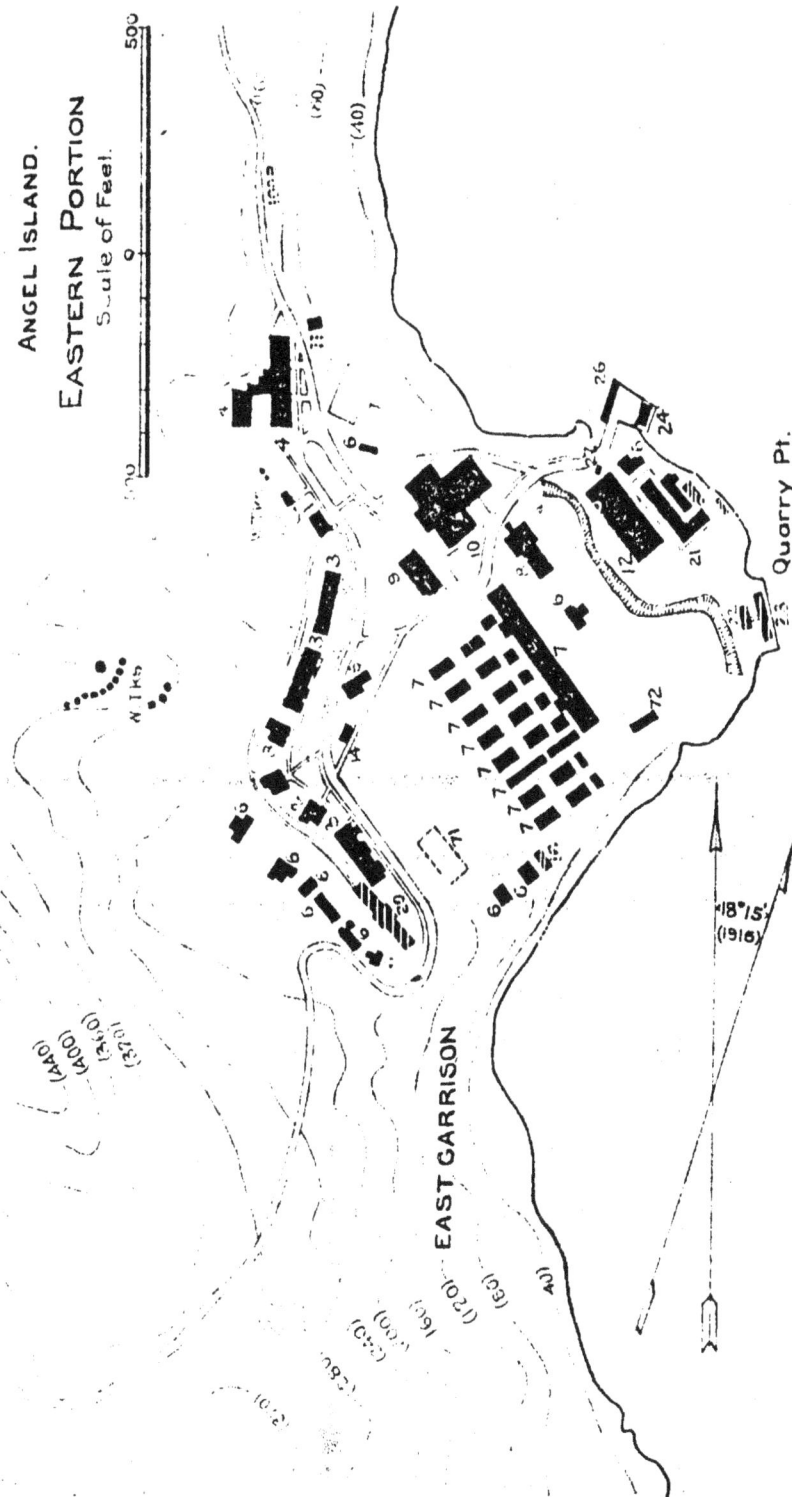

EAST GARRISON

Quarry Pt.

18°15' (1916)

SERIAL NUMBER 124

SAN FRANCISCO HARBOR
FORT McDOWELL D-2.
ANGEL ISLAND
WESTERN PORTION
Scale in Feet.

500   0   500

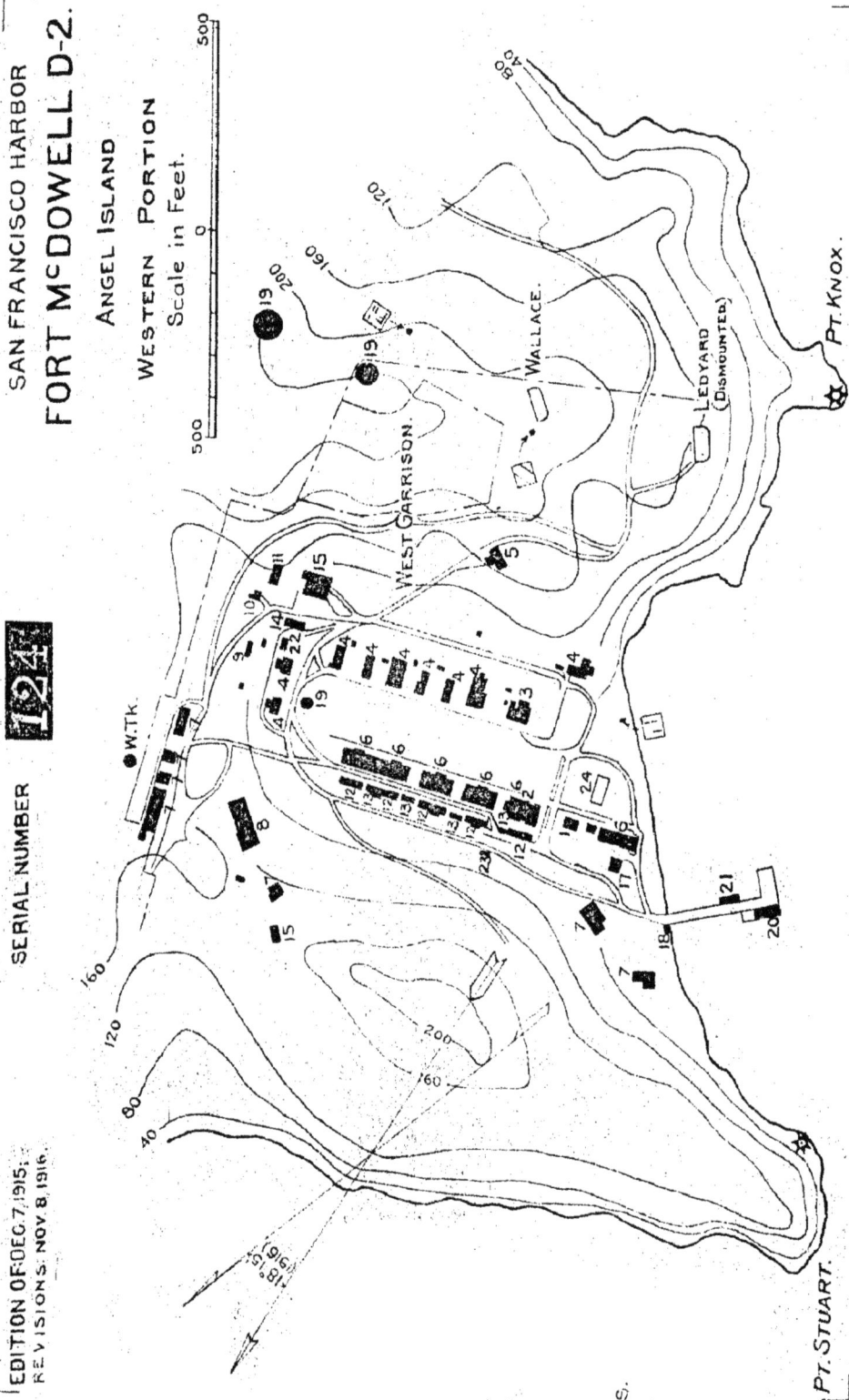

EDITION OF DEC. 7, 1915.
REVISIONS: NOV. 8, 1916.

LEGEND.
1. ADMINISTRATION BLD.
2. POST EXCHANGE.
3. COMDG. OFFS. QRS.
4. OFFICERS QRS
5. CHAPEL.
6. BARRACKS.
7. N.C. OFFICERS QRS.
8. HOSPITAL.
9. BLACKSMITH SHOP.
10. PAINT SHOP.
11. PLUMBING SHOP.
12. KITCHENS.
13. LATRINES.
14. WAGON SHED.
15. STABLES.
16.
17. ORD.
18. OIL HOUSE.
19. CISTERNS.
20. COAL SHED.
21. BOAT HOUSE.
22. BAKERY.
23. PHOTOGRAPHIC GALLERY.
24. TENNIS COURT.

BATTERIES.
WALLACE.......1-8"DIS.
LEDYARD.......2-5"P.
(DISMOUNTED)

PT. KNOX.

WALLACE.

LEDYARD
(DISMOUNTED)

WEST GARRISON.

PT. STUART.

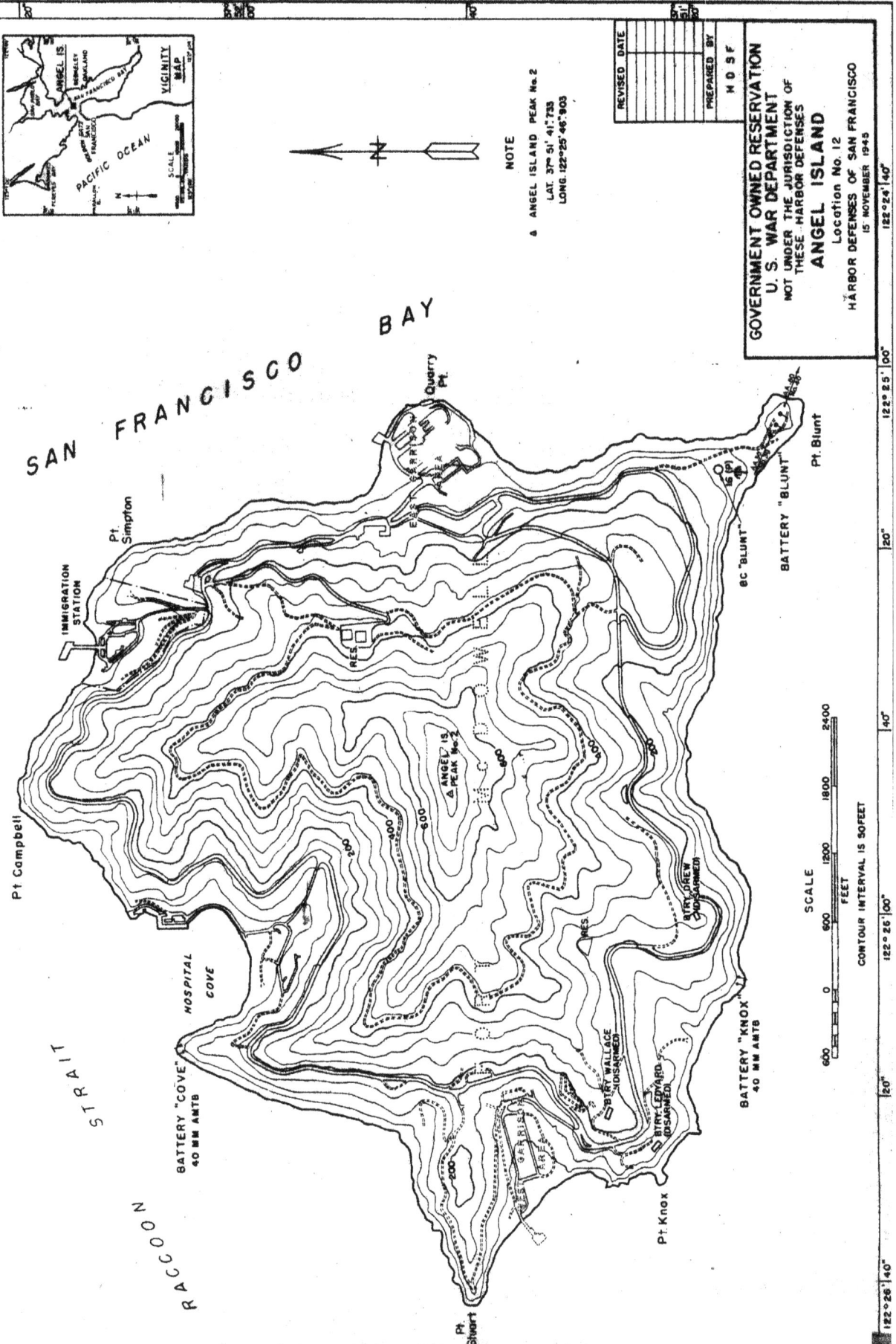
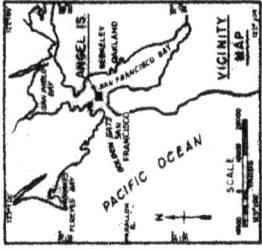

just a single gun platform. Work was done in 1898, the war emergency and relatively isolated location of Angel Island considerably adding to the cost overruns. The emplacement was done by June 1899, and the base ring mounted in November 1899 and the carriage and gun the following February. It was transferred on May 1, 1900 for a cost of $34,836.82. It was armed with one 8-inch Model 1888 Watervliet gun on Model 1892 barbette carriage (#36/#6). It was named on General Orders No. 16 of February 24, 1902 for Lieutenant Alfred W. Drew killed in action during the Philippine Insurrection. The gun was subsequently removed in 1918, though the battery had been declared as obsolete as early as 1915. Subsequently the emplacement was used for explosives storage. It still exists at Angel Island State Park. The battery is open to the public.

- **WALLACE:** An emplacement for a single 8-inch disappearing mount located on higher ground on the southwestern side of Angel Island, not far from the western garrison camp. Its plan was submitted on March 23, 1899 for this position above Knox Point. It had a conventional design, closely resembling Battery Burnham at Fort Mason. There was only a single gun, with adjacent, lower magazine on the right flank and hoist for lifting ammunition. Work was done from October to December 1899. A large 150,000-gallon reservoir already occupying the site had to be moved. It was reported completed on June 30, 1900 (without armament but with an exterior tool house and latrine on the left flank). Transfer was made on August 1, 1900 for a cost of $40,420.42. It was named in General Orders No. 16 of February 24, 1902 for Lt. Robert B. Wallace of the 2nd Cavalry, mortally wounded during the Philippine Insurrection. It was armed with one 8-inch Model 1888 Bethlehem gun on a Model 1896 LF disappearing carriage (#1/#28). This gun had been mounted in late 1901. The loading platform was widened in 1910. It was declared obsolete with the reductions of the inner harbor defenses, and the gun removed in 1917. The emplacement still exists at Angel Island State Park. The battery is open to the public.

- **LEDYARD:** An emplacement for two 5-inch, rapid-fire guns emplaced on the southwestern side of Angel Island, below and to the southwest of Battery Wallace. San Francisco engineers submitted plans for four 5-inch guns on October 30, 1899—two here and two for Fort Winfield Scott. It was emplaced 450-feet forward of Battery Wallace on the edge of the coastal cliff. It was at the site of an old 8-inch converted rifle, earthen battery and designed to cover the inner minefield. The design followed the type plans, but to manage costs it had no latrine nor power facilities. This site had been selected on December 20, 1899, excavation at the site beginning the following January. Work was completed, except for mounting armament and whitewashing rooms by June 30, 1900. The older, earthen Battery Knox was entirely removed during construction. It was not armed until 1907, at that time receiving two 5-inch Model 1900 guns on Model 1903 pedestal mounts (#1/#14 and #2/#15). It was named on General Orders No. 16 of February 14, 1902 for Lieutenant August C. Ledyard killed in action in 1899 in the Philippines. The armament served until removed in early 1917 and relocated to the new battery being built at Fort Miley. The emplacement was not subsequently used but still exists though somewhat damaged by earthquake and subsidence problems at the Angel Island State Park. The battery is open to the public.

- *Battery Blunt* (planned): A projected 1943 Program AMTB battery planned for Blunt Point on the southeastern point of Angel Island. It was authorized on June 1, 1943 and was to consist of two 90mm fixed and two 90mm mobile guns. However, the actual construction was deferred until early 1944. Then the project was entirely cancelled. No physical construction on the gun blocks was ever accomplished.

Fortress Alcatraz (1853-1907) is located on Alcatraz Island in San Francisco Bay. Alcatraz had been an active fortification site since the late 1840s. Extensive Civil War batteries were rebuilt in the 1870 Program, to eventually consist of sites for numerous 15-inch Rodmans. In the late 1880s the island was used as a mine storage depot and received an Endicott mine casemate in 1890. By early 1892 two 8-inch converted rifles were mounted as a practice battery for the San Francisco defenses, at the southeast end of Battery No. 13 (along with nine 15-inch Rodman cannons). Early Endicott plans called for five 12-inch lift guns in two batteries for the island, later changed to an equal number of disappearing guns. By the early 1900s plans for big guns had been dropped, but it was still envisioned to have a 6-inch disappearing battery here. While none of these were ever built, during the crisis of the Spanish American War two platforms were rebuilt to carry two 8-inch converted rifles and actually transferred on December 10, 1900 for a cost of $404.24. All remains were heavily modified for the later prison construction, but some significant parts of the earlier 1870s batteries remain. Alcatraz is a very popular tourist attraction today with hourly ferries from the waterfront of San Francisco. Only part of the island is open to the public.

Fort Mason (1845-1972) is located at Black Point near downtown San Francisco. Home of the commanding officers of Department of the Pacific, it was fortified during the Civil War (12-gun battery) and with a single emplacement during the Endicott Program. It was named in General Orders 133 of 1882 for Bvt. Brig. Gen. Richard B. Mason, U.S. Army. The famed officers club operated until the base was closed. The wharves of Fort Mason were used as a port of embarkation for both men and supplies during the wars of the 20th century. Fort Mason is the headquarters of Golden Gate National Recreation Area. As a whole, the former post is now a mix of parks and gardens and late nineteenth and early twentieth century buildings that are still in use. The old port of embarkation warehouses is home to the Fort Mason Center for Arts and Culture which features a variety of art shops, restaurants and a wide range of cultural activities and events. The park is open to the public.

Fort Mason Gun Battery

- **BURNHAM:** This was the only Endicott emplacement located at Fort Mason. Originally the Board had called for a dual 8-inch disappearing battery to be built here, but only a single emplacement was ever constructed. It was part of the inner defenses for San Francisco, located on the top of the Point San Jose bluff, above and behind the line of old Civil War earthen batteries. Plans were submitted on January 17, 1899. It was initially funded with $24,000 from the Fortification Act of May 7, 1898. Plans followed type mimeographs, with just a single platform and a lower-level magazine on the left flank. Ammunition service was by hand-operated lift. Work was done during the year of 1899, for transfer on August 21, 1900 at a cost of $32,137.62. It was armed with one 8-inch Model 1888 Watervliet gun tube on Model 1896 LF disappearing carriage (#4/#32). It was named on General Orders No. 16 of February 14, 1902 for Lieutenant Howard M. Burnham of killed in action in 1863 during the Civil War. For several reasons the emplacement was never considered favorably. By May 1908 it was already omitted from the district's active armament. It was considered to have an unimportant field of fire, to be of obsolete design, and at a post without much other coast artillery presence. In November 1908 it was reduced to just a caretaker status and the ammunition transferred for use elsewhere. In 1909 the gun was removed, and the carriage sent for re-emplacement at Fort Columbia. In subsequent years it was used for several purposes—in 1927 it was a plumbing shop, and in the 1940s it became the post air raid shelter. The emplacement still exists, but a structure is built on part of the emplacement. The battery exterior is open to the public.

FORT MASON

SAN FRANCISCO BAY.

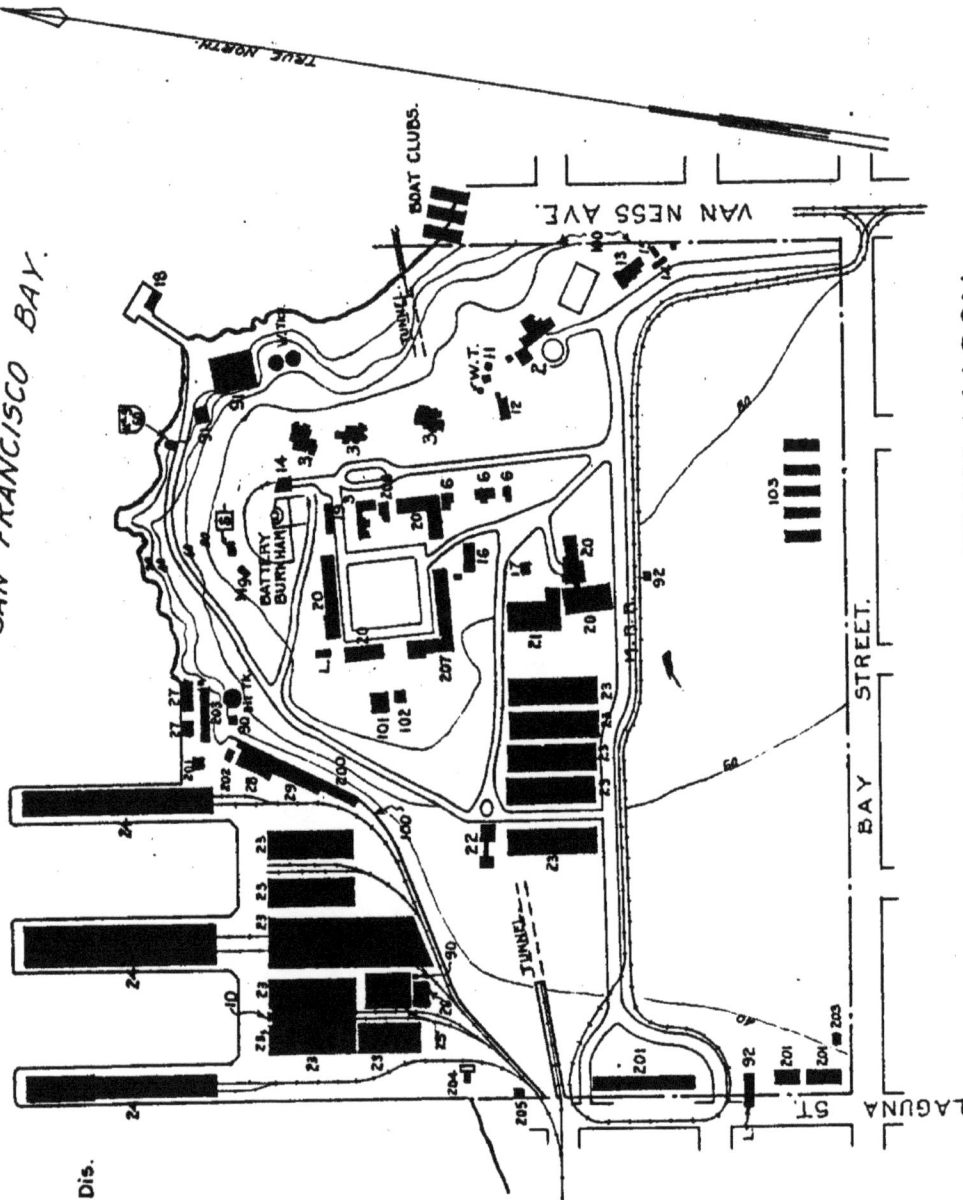

TRUE NORTH

VAN NESS AVE.

BOAT CLUBS.

BAY STREET.

LAGUNA ST.

BATTERY BURNHAM

TUNNEL

1000 FEET.

BATTERIES   1~8" Dis.

BURNHAM (Dismounted)

1
2 COMMANDING GEN. QRS.
3 OFFICER'S QRS.
4
5
6 N.C. OFFICER'S QRS.
7
8
9
10 FIRE DEPT.
11 BAND STAND.
12 HOT HOUSE.
13 ' ...NDING GEN. STABLE.
14 ...GE.
15 PIGEON HOUSE.
16 COAL SHED.
17 CARETAKER'S DWELLING.
18 BOAT WAITING ROOM.
19 CARPENTER & PAINT SHOP
100 RETAINING WALL.
101 STABLE.
102 GRAIN SHED.
103 STOREHOUSES.
20 Q.M. OFFICE
21 MESS BLDG. AND DINING ROOM.
22 CONSTRUCTING Q.M.
23 STOREHOUSE.
24 DOCK SHED.
25 CARPENTER SHOP.
26 SHEET METAL SHOP.
27 PAINT SHOP.
28 MACHINE SHOP.
29 ELECTRICIAN AND BOX MAKING SHOP.
200 CEMENT SHED.
20' 'AGE.
2(.. AND OIL STATION.
203 OIL SHED.
204 SCALES.
205 WHARFINGER AND TIME-KEEPER.
206 Q.M. SARGENT.
30 SUB. STATION.
91 CITY PUMPING STATION.
92 WAITING STATION.
207 SHOE REPAIR BLDG.

Fort Winfield Scott (1852-1950) is located on the southern side of the Golden Gate Channel, which is the entrance to San Francisco Bay. The Spanish built the first defense work, called El Castillo de San Joaguin or Fort Blanco, at Fort Point. In 1852, the U.S. Army started construction of a Third System fort on the same location, which was known as Fort Point until General Order 133 of 1882 named the masonry fort and its surrounding military reservation Fort Winfield Scott. Fort Scott was developed as a separate post on the western portion of the Presidio of San Francisco, formally established as a separate post after 1900. The post provided the primary harbor defense garrison to the City of San Francisco and the support facilities throughout the Bay area. Two 1870s earthwork batteries mounting 15-inch Rodman cannon were installed on bluffs behind the Third System fort. Construction of the first Endicott Program batteries was undertaken in 1892 on the high bluffs south of the old fort. These early batteries consisted of Batteries Godfrey and Saffold, a total of five 12-inch guns on barbette carriages. Also constructed during this period was Battery Dynamite, a pneumatic system which launched 15-inch high-explosive shells. By 1907, the fort held a collection of 16 Endicott Program batteries, mounting a total of 67 guns and mortars. The fort also had a large, controlled submarine mine complex. Fort Scott became the headquarters for the San Francisco harbor defenses and the west-coast Coast Artillery District resulting in a large cantonment area being constructed. In 1937, the completion of the Golden Gate Bridge had a large impact on the old fort and several of the Endicott Program batteries were buried. The only additions to Fort Scott during World War II were four AMTB batteries, a new mine casemate, and a very large Harbor Entrance Control Post. At end of World War II, the fort became the new home of the army's Coast Artillery School which closed in 1949. The army retained the Presidio-Fort Winfield Scott after the end of the coast artillery for use as an administrative and transportation complex from 1950 to the early 1970s. It became home to the 6th U.S. Army and was used for the administration and support of the Nike defenses in the San Francisco Bay area. A Nike launch site (SF-89L) was located on the post. The U.S. Army began phasing out its use of the San Francisco military complex with end of the Nike program in 1970s. A portion of Fort Scott was transferred to National Park Service in 1972 to become part of the Golden Gate National Recreation Area. After a major military base realignment in 1989, the rest of Fort Scott and the Presidio of San Francisco were transferred to the Presidio Trust. In 1996 Congress created the Presidio Trust and transferred jurisdiction of 80 percent of the Presidio to this new federal agency. The Trust was given a mandate to preserve the areas of the Presidio under its jurisdiction and attract non-federal resources to the park to ensure that it would be sustained without direct annual taxpayer support. As a result, the Presidio of San Francisco is managed by two federal agencies in partnership: 300 acres along the coast are managed by the National Park Service, while the rest of the Presidio, 1,191 acres, is managed by the Presidio Trust. Both agencies work in close collaboration with the Golden Gate National Parks Conservancy, a non-profit organization that provides philanthropic and programmatic support. Over the past two decades, the Trust has converted this former military post into a national park site in an urban area. In 2013, the Presidio reached a crucial milestone by becoming financially self-sufficient. The majority of the historic buildings remain at the Presidio and Fort Winfield Scott. The important military defense structures remain along the western and northern edges and can be visited with the exception of the dynamite battery-HECP-Battery Saffold complex and Battery Cranston which are being used as a maintenance yards, and Batteries Stotsenburg and McKinnon, which are being used as storage for wine. Doyle Drive which connects the Golden Gate Bridge with City was replaced with the Presidio Parkway, allowing for access to several Endicott Program batteries that were cutoff or buried by Doyle Drive. Fort Scott is in excellent condition and one of the most important Endicott Program forts to be visited. Of special interest is Battery Chamberlain which displays one of the two surviving 6-inch disappearing guns in the United States.

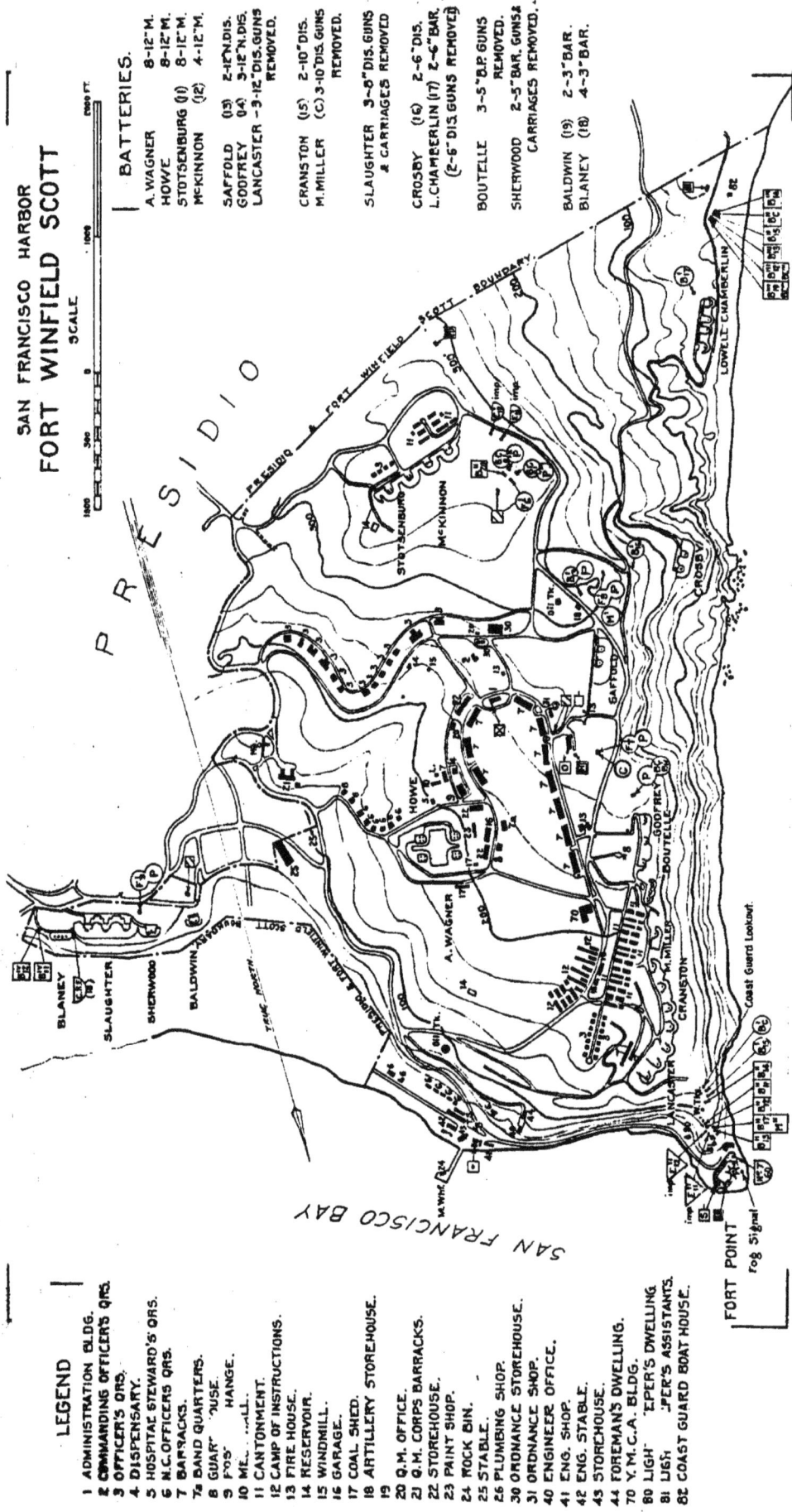

SAN FRANCISCO HARBOR

FORT WINFIELD SCOTT

SCALE

BATTERIES.

A. WAGNER		8-12" M.
HOWE		8-12" M.
STOTSENBURG	(11)	8-12" M.
McKINNON	(12)	4-12" M.
SAFFOLD	(13)	2-12"N.DIS.
GODFREY	(14)	3-12"N.DIS.
LANCASTER		3-12"DIS.GUNS REMOVED.
CRANSTON	(15)	2-10" DIS.
M. MILLER	(C)	3-10"DIS.GUNS REMOVED.
SLAUGHTER		3-5"DIS.GUNS & CARRIAGES REMOVED
CROSBY	(16)	2-6" DIS.
L. CHAMBERLIN	(17)	2-6"BAR. (2-6"DIS.GUNS REMOVED)
BOUTELLE		3-5"D.P. GUNS REMOVED.
SHERWOOD		2-5"BAR. GUNS & CARRIAGES REMOVED.
BALDWIN	(19)	2-3"BAR.
BLANEY	(18)	4-3"BAR.

P R E S I D I O

SAN FRANCISCO BAY

FORT POINT

Fog Signal

Coast Guard Lookout.

LEGEND

1 ADMINISTRATION BLDG.
2 COMMANDING OFFICER'S QRS.
3 OFFICER'S QRS.
4 DISPENSARY.
5 HOSPITAL STEWARD'S QRS.
6 N.C. OFFICERS QRS.
7 BARRACKS.
7a BAND QUARTERS.
8 GUARD HOUSE.
9 POST EXCHANGE.
10 MESS HALL.
11 CANTONMENT.
12 CAMP OF INSTRUCTIONS.
13 FIRE HOUSE.
14 RESERVOIR.
15 WINDMILL.
16 GARAGE.
17 COAL SHED.
18 ARTILLERY STOREHOUSE.
19
20 Q.M. OFFICE.
21 Q.M. CORPS BARRACKS.
22 STOREHOUSE.
23 PAINT SHOP.
24 ROCK BIN.
25 STABLE.
26 PLUMBING SHOP.
30 ORDNANCE STOREHOUSE.
31 ORDNANCE SHOP.
40 ENGINEER OFFICE.
41 ENG. SHOP.
42 ENG. STABLE.
43 STOREHOUSE.
44 FOREMAN'S DWELLING.
70 Y.M.C.A. BLDG.
80 LIGHT KEEPER'S DWELLING
81 LIGHT KEEPER'S ASSISTANTS.
82 COAST GUARD BOAT HOUSE.

(VII-32-I-9)(XI-24-39-1-50P)(2-5000) FORT SCOTT, CALIF. CONFIDENTIAL

Fort Winfield Scott 1939 (NARA)

Fort Winfield Scott 1939 (NARA)

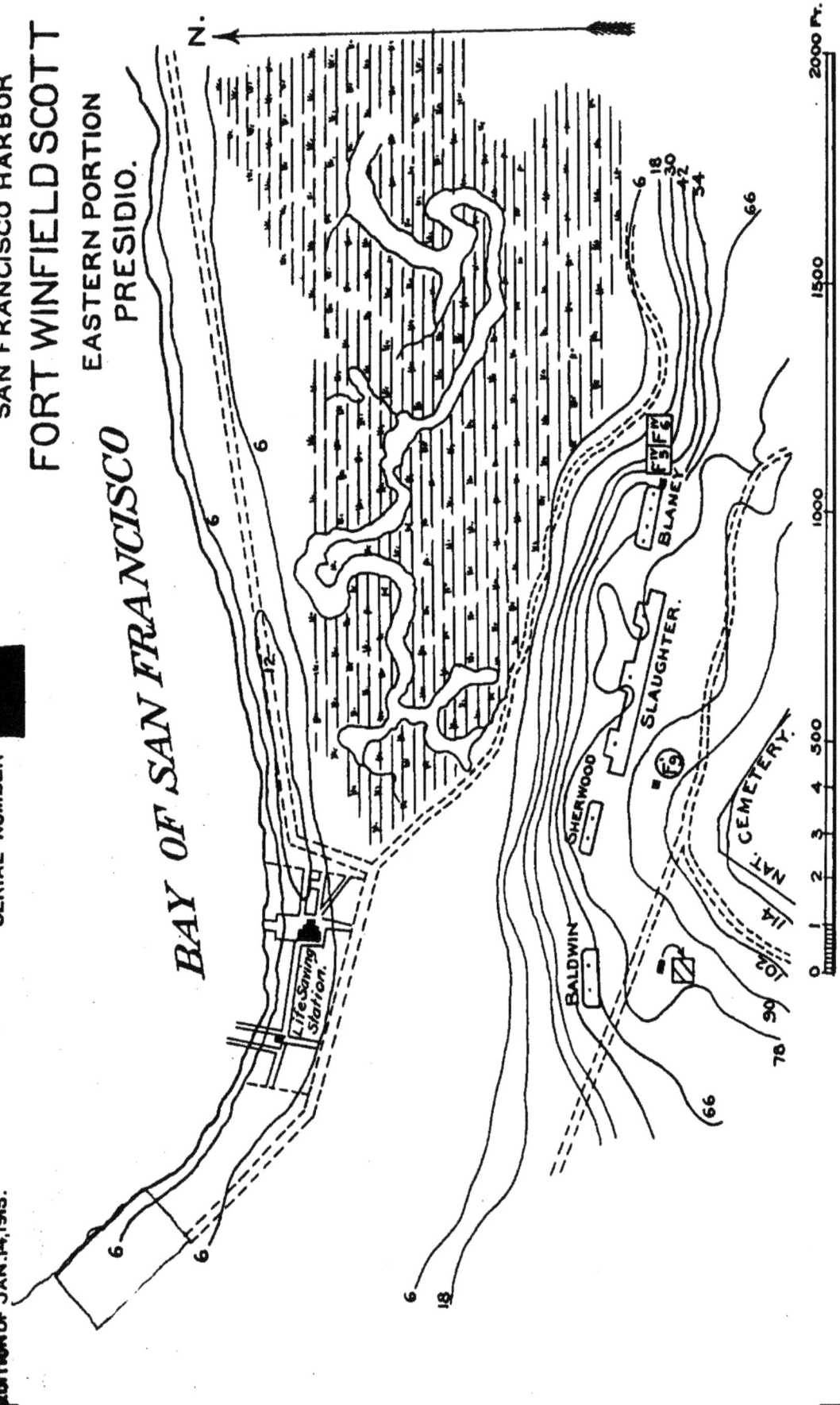

SAN FRANCISCO HARBOR
FORT WINFIELD SCOTT
EASTERN PORTION
PRESIDIO.

BAY OF SAN FRANCISCO

SERIAL NUMBER

EDITION OF JAN.14,1915.

Life Saving Station.

BALDWIN
SHERWOOD
SLAUGHTER.
BLANEY
NAT. CEMETERY.

2000 FT.

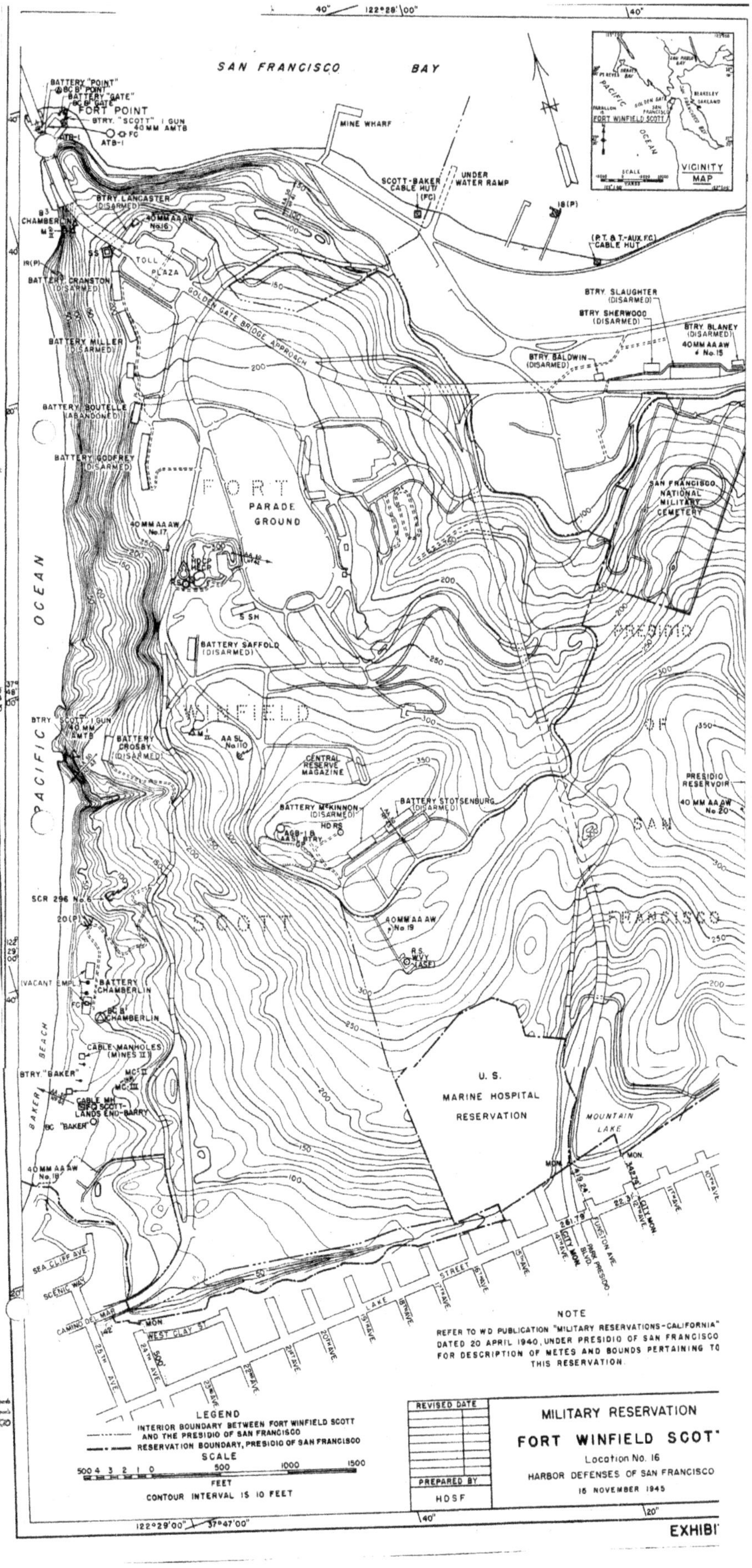
SAN FRANCISCO BAY
PACIFIC OCEAN
BAKER BEACH
FORT POINT
BATTERY "POINT"
BATTERY "GATE"
MINE WHARF
SCOTT-BAKER CABLE HUT (FC)
UNDER WATER RAMP
BTRY LANCASTER (DISARMED)
40MM AAAW No.16
TOLL PLAZA
GOLDEN GATE BRIDGE APPROACH
BATTERY CRANSTON (DISARMED)
BATTERY MILLER (DISARMED)
BATTERY BOUTELLE (ABANDONED)
BATTERY GODFREY (DISARMED)
(P.T.&T.-AUX.F.C.) CABLE HUT
BTRY SLAUGHTER (DISARMED)
BTRY SHERWOOD (DISARMED)
BTRY BLANEY (DISARMED)
40MM AA AW No.15
BTRY BALDWIN (DISARMED)
FORT WINFIELD SCOTT
PARADE GROUND
40MM AA AW No.17
SAN FRANCISCO NATIONAL MILITARY CEMETERY
BATTERY SAFFORD (DISARMED)
PRESIDIO OF SAN FRANCISCO
BTRY "SCOTT" 1 GUN 40 MM AMTB
BATTERY CROSBY (DISARMED)
AA SL No.110
CENTRAL RESERVE MAGAZINE
PRESIDIO RESERVOIR
40 MM AA AW No 20
BATTERY STOTSENBURG (DISARMED)
BATTERY McKINNON (DISARMED)
HD RS
SCR 296 No.6
40MM AA AW No 19
VACANT EMPL.
BATTERY CHAMBERLIN
BC B' CHAMBERLIN
CABLE MANHOLES (MINES II)
BTRY "BAKER"
U. S. MARINE HOSPITAL RESERVATION
MOUNTAIN LAKE
40 MM AA AW No.18
SEA CLIFF AVE
SCENIC WAY
CAMINO DEL MAR
WEST CLAY ST
LAKE STREET
VICINITY MAP
FORT WINFIELD SCOTT
NOTE
REFER TO W.D. PUBLICATION "MILITARY RESERVATIONS-CALIFORNIA" DATED 20 APRIL 1940, UNDER PRESIDIO OF SAN FRANCISCO FOR DESCRIPTION OF METES AND BOUNDS PERTAINING TO THIS RESERVATION.
LEGEND
INTERIOR BOUNDARY BETWEEN FORT WINFIELD SCOTT AND THE PRESIDIO OF SAN FRANCISCO
RESERVATION BOUNDARY, PRESIDIO OF SAN FRANCISCO
SCALE
500 4 3 2 1 0 500 1000 1500
FEET
CONTOUR INTERVAL IS 10 FEET
REVISED DATE
PREPARED BY
HDSF
MILITARY RESERVATION
FORT WINFIELD SCOTT
Location No. 16
HARBOR DEFENSES OF SAN FRANCISCO
15 NOVEMBER 1945
122°29'00"
37°47'00"
122°28'00"
EXHIBI

Fort Winfield Scott Parade, Golden Gate National Recreation Area (Terry McGovern)

Battery Godfrey, Battery Boutelle, and Battery Miller, Golden Gate National Recreation Area (Terry McGovern)

Battery Chamberlin Fort Winfield Scott Golden Gate National Recreation Area (Terry McGovern)

Entrance to Fort Winfield Scott Golden Gate National Recreation Area (Mark Berhow)

Fort Scott Gun Batteries

- **HOWE – WAGNER:** The first of two mortar batteries planned and built at Fort Winfield Scott. It was located about 400-yards inland from the coastal bluff. Engineers submitted their plans for the battery on January 24, 1893. At the time this emplacement was simply known as Mortar Battery No. 1. It followed the type plans for the early quadrangular type emplacement of four pits of four mortars each. It was slightly expanded in plan – the pits being defined by a circle of 325-feet in diameter vs. 300-feet in the type plan and used on other similar batteries. The relatively safe position on this reservation meant that neither a ditch nor complete protective outer parapet would be necessary. Also, the wall thickness was reduced, and the height of the pit side walls decreased to a minimum. Work was done (with early employment of Portland cement) in 1893-1894. The first firing by a mortar mounted here occurred on February 9, 1895—and the emplacement was found most satisfactory. The first pit was armed by September 24, 1894 when a heavy rainstorm caused a slide of the earthen cover, after which some rebuilding had to be done. Transfer to service troops was made on January 18, 1900 for a cost of $144,247.15. The battery was originally named on General Orders No. 16 of February 14, 1902 for Colonel Albion P. Howe of Civil War service. It was armed with sixteen 12-inch cast-iron, Model 1886 mortars on Model 1891 mortar carriages (#43/#22, #29/#26, #38/#21, #39/#23, #47/#51, #46/#48, #44/#45, #35/#47, #40/#71, #42/#72, #36/#46, #53/#67, #8/#34, #6/#32, #7/#33, and #5/#31). In 1897 a casemated firing room was added, and in April of 1898 additional plotting and relocation rooms were made in the wall of a tunnel leading to the northwestern pit. When all the 4 pit mortar batteries were split administratively in 1906, the western two pits of Battery Howe were separated to become Battery Wagner. This was named on General Orders No. 20 of January 25, 1906 for Colonel Arthur Wagner of Spanish American War service. The two eastern pits remained Battery Howe. The battery served successfully with its armament through World War I. The mortars and carriages were finally removed in 1920 and scrapped. After that the battery was not used actively, though its convenient post location facilitated its use for storage of various equipment. Officer's housing was built around the battery in the 1920s, and three of the four pits were eventually filled-in, though the lateral magazine tunnels remain intact underground. As modified, with one relatively intact pit remaining, the emplacement still exists in good condition as part of the Presidio Trust. The battery is not open to the public.

- **Dynamite Battery:** San Francisco received one of the two prototype dynamite gun batteries directly funded by Congress in 1888 for $400,000. Consequently, an emplacement for three pneumatic dynamite guns at the Presidio reservation of Fort Scott was constructed. Even though built by the engineers and tested by the ordnance department, this type of gun never had the support of the army administration. The contract for supplying and emplacing three 15-inch pneumatic guns was given to the Pneumatic Torpedo and Construction Company on September 17, 1889. However, much development time was required, the actual site for the emplacement wasn't selected until October 1894. That position was on the western bluff about 1500-feet southwest of Battery Godfrey. Work on the platforms for each gun and a large, concrete steam and reservoir plant to generate and hold the necessary compressed air was the responsibility of the contractor. That work was apparently completed in 1896. The three 15-inch guns were purchased for $187,500 and received at the post of June 14, 1896. The guns used were Model 1893 guns on carriages #5/#1, #6/#2, and #7/#3. While they were successfully test-fired, the accuracy, rate of fire, and damage caused by detonating a large charge next to the target proved wanting. The battery was never officially transferred to service troops, and in fact was never manned or operated by coast artillery personnel. In 1898 the engineers submitted their own plan and funding request to build protected magazines for the shells and a massive parapet to protect the guns and plant from bombardment. That work was done in 1898-1899 and reported completed on March 26, 1900. Not long after, on February 23, 1901, these weapons were declared obsolete. By December 1904 the guns were gone, sold for scrap. The large work itself was used for many other purposes. In 1912 the power plant was expanded and used as a post power plant. In 1919 the battery was the post telephone and fire control switchboard room. Finally in 1942 it was converted to the San Francisco Harbor Defense Command Post. Even after World War II it was used for a variety of telecommunications functions. While heavily modified for these other uses, it still exists at the former Fort Scott as part of the Golden Gate National Recreation Area/Presidio Trust. The battery is not open to the public.

- **GODFREY**: A battery for three 12-inch guns on barbette carriages emplaced on the western bluff of Fort Winfield Scott. Pending development of satisfactory disappearing carriages, the high bluffs at San Francisco were seen as an ideal location for the early generation of barbette carriages. This battery, as selected on September 28, 1891, was emplaced at positions No. 14, 15, and 16 in the original defensive plan for the Presidio. The adjusted plan was submitted on April 26, 1892, and construction began immediately. The construction entailed the destruction of three old 1870s system magazines and the burying of three others. Work was done by September 1894, except for the foundation platforms that were still subject to change. The engineers were finished with their work on July 20, 1896, but the first gun was mounted a little earlier in June of 1895. It was transferred on August 19, 1896 for a total cost (including Battery Miller) of $299,861.51. The battery was armed with three 12-inch Model 1888 Watervliet guns on Model 1892 barbette carriages (#9/#6, #6/#3, and #4/#2). It was named on General Orders No. 16 of February 14, 1902 for Captain George J. Godfrey killed in action during the Philippine Insurrection. New hoists were added in 1908 and a variety of other modifications made throughout the emplacement's service life. It served as one of the important San Francisco batteries until authorized for removal on January 23, 1943. Actual armament scrapping came later that same year. The emplacement still exists in good condition at the Golden Gate National Recreation Area. The battery site is open, but the interior is closed to the public.

- **SAFFOLD**: A battery for another pair of 12-inch non-disappearing, barbette guns. Plans were submitted on November 10, 1896, funding having originated with the Fortification Act of June 6, 1896. The approved site was a small knoll, near but slightly behind the bluff line on the western side of the Presidio. It was quite a distance from the other batteries already constructed, some 2000-feet south of Battery Godfrey. The field of fire could cover from Lime Point to Baker's Beach. The plan closely followed standard mimeographs, but the magazines and shot rooms were enlarged to hold 200 rounds per gun. The battery was equipped with ammunition lifts. Work was done from 1897-1898. It was transferred on February 14, 1898 for a cost of $107,409 (which also included emplacement No. 3 of 12-inch Battery Lancaster). It was armed with two 12-inch Model 1888 Watervliet guns on Model 1892 barbette carriages (#3/#8 and #19/#7). It was named on General Orders No. 16 of February 14, 1902 for Captain Marion M. Saffold who had been killed in action during the Philippine Insurrection. The emplacement was subsequently modified for new lifts, altered firing platforms and a new battery commander's station. It served through World War I, not being authorized for removal and armament scrapping until February 23, 1943. However, it does not appear that the guns and carriages were actually scrapped until order of January 23, 1946. In relatively intact condition, the emplacement still exists at the Golden Gate National Recreation Area/Presidio Trust. This battery is not open to the public.

- **LANCASTER:** A battery for three 12-inch disappearing guns emplaced at the northwest bluff head at Fort Point, overlooking (and firing over) old Fort Point and the Golden Gate. It fired to the north towards Gravelly Beach with a field on either side. Plans for emplacement No. 8 (the western-most emplacement) were submitted on September 15, 1896. Work was done in 1897-1897. At the site three Rodman guns had to be dismounted and several old platforms and magazines were destroyed of the old West Battery. The final two emplacements (6 and 7) were submitted on August 30, 1898, with an allotment of funds from the Act of June 8, 1898. Work was done on these emplacements from 1898-1900. The first emplacement was a conventional two-story affair using ammunition lifts. It was armed with a Model 1888 M1-1/2 Watervliet gun on Model LF 1896 disappearing carriage (#40/#23). This emplacement was transferred with Battery Saffold on June 15, 1899 for a cost of (both batteries) $107,409. The second two emplacements, designed and built two years later, differed significantly from mimeograph type plans by using adjacent magazines and storing shells under the loading platform. Also, these two emplacements used the later LF Model 1897 disappearing carriage. This part of the battery had two 12-inch Model 1896 Watervliet guns on Model 1897 carriages (#5/#7 and #8/#6). This pair were transferred on April 27, 1900 for $75,000. The three emplacments were named in General Orders No. 16 of February 14, 1902 for Lt. Colonel James M. Lancaster of Civil War service. Changes were made over the next years. In 1899 a new plotting room was added, in 1908 the hoist in emplacement No. 3 was replaced, and in 1911 a new battery commander station was added on the traverse. The gun carriages were not modified for increased elevation in 1916, the short range across the Gate made the change unnecessary and the consequential loss of elevation depression for guns on a cliff top was unacceptable. However, the gun tubes were removed in 1918 for use on railway carriages. The older Model 1888 tube was relined and subsequently sent to Battery Chester at Fort Miley on June 10, 1918. The two Model 1895 tubes were shipped to Watervliet Arsenal on May 15, 1918. The battery was never rearmed but served as local storerooms in the 1920s and 1930s. With the construction of the Golden Gate bridge in the mid-1930s the two western emplacements were covered over, and the eastern emplacment was heavily modified. In that condition remains still exist at the Golden Gate National Recreation Area. A portion of the eastern emplacement of the battery is open to the public.

- **MILLER:** The first new Endicott battery approved for the Presidio, this battery for three 10-inch guns on disappearing carriages was placed on the western bluff, facing west. Following authorization, local engineers submitted a plan on March 2, 1891 for guns emplaced in emplacements 11, 12, and 13. These platforms were among the old 1870s West Battery, and required dismounting ten old rodman guns and destruction of a number of brick magazines. As there was yet no approved carriage type, the actual designation of gun size and carriage model was left open to a future decision. Work quickly got underway; excavation of the site finished by that October. Concrete work was begun on June 2, 1891; in common with all the Endicott batteries on the Pacific Coast, the Engineering Department insisted on using only the superior Portland cement in the construction of gun batteries. The battery was transferred on April 14, 1898 along with Battery Godfrey for a combined total cost of $299,861. The battery was armed with three 10-inch Model 1888 Watervliet guns on LF Model 1894 disappearing carriages (#18/#27, #15/#24, and #5/#34). Initially all five 10-inch emplacements were named together as one battery, on General Orders No. 16 of February 14, 1902 for Lt. Arthur Cranston. Around 1905 the older, three-gun battery was being referred to as "Cranston 2". Then finally the unit was renamed in 1907, on General Orders No. 210 of October 11, 1907 for Brigadier General Marcus Miller of Civil War, Indian War and Philippines service. Being an older design, not surprisingly a number of changes were made from 1896-1916. These included widened loading platforms, a new battery commander's station, latrines and support room additions, and changes in the ammunition hoists. In 1917 the guns were earmarked for removal, which was done in 1918. The carriages, which had been left in place, were scrapped in 1920. In later years the emplacement was used for the Harbor Defense Alternate Transmitter Station. Somewhat modified, the battery emplacement still exists at the Golden Gate National Recreation Area. A portion of the battery is open to the public.

- **CRANSTON**: A second battery for two 10-inch disappearing guns built adjacent, and to the north of Battery Miller and southwest of Battery Lancaster. This work was begun with the Fortification Act funding of March 3, 1897. Work started on June 10, 1897 and was structurally complete by year's end. The armament was received and mounted by March 31, 1898. The battery was transferred on July 11, 1898 for a cost of $55,431.97. It was armed with two 10-inch Model 1888MII Bethlehem guns on Model 1896 disappearing carriages (#7/#29 and #13/#25). All five 10-inch guns were originally designated as Battery Cranston in General Orders No. 16 of February 14, 1902 for Lt. Arthur Cranston killed in action during the Modoc War in 1873. In 1907 the three older southern emplacements were split off and named separately for Brigadier General Marcus Miller. Modifications to the emplacement were made from 1898-1915 to the platforms, stations, and hoists. This was an important defensive element, and with the disarming of Miller and Lancaster in 1918, Battery Cranston continued to serve through World War I and interwar years. Authority for removal did not come until January 13, 1943. The emplacement later served as a dormitory of the nearby Signal Station. It still exists at the Golden Gate National Recreation Area. The battery is a support facility of the Golden Gate Bridge, Highway, and Transportation District's operations. The battery is not open to the public.

- **McKINNON – STOTSENBURG**: The second mortar battery built at Fort Winfield Scott. It was funded from the appropriation of March 3, 1897. The plans were submitted on June 4, 1897. The site was selected to be on Rob Hill, a locally known promontory on the southern part of the fort reservation. It followed the current mimeograph design. By this time, in comparison to Battery Howe, the four pits were all arranged in-line. Powder magazines were in the parapet in front of the pits, with shell rooms on the flanks. Pits were still of the 40-foot, small dimensions. Work on clearing ground started immediately and finished by June 30, 1898. Transfer was made on April 27, 1900 for a cost of $130,188.43. It was armed with sixteen 12-inch Model 1890M1 mortars on Model 1896 carriages (Builders mortars #22/#62, #33/#60, #31/#66, #30/#64, #34/#91, #36/#94, #28/#86, #23/#96, Watervliet mortars #47/#63, #35/#61, #49/#67, #39/#85, #28/#95, #46/#80, Niles mortar #6/#65 and Bethlehem mortar #22/#79). The battery was named in General Orders No. 16 of February 14, 1902 for Captain John M. Stotsenburg who was killed in action during the Philippine Insurrection. In 1906 the two right flank pits were administratively separated as a separate battery named in General Orders No, 20 of January 25, 1906 for Chaplain William D. McKinnon of Spanish American War service. In 1917 various proposals were made to move some of the battery's mortars to other locations—including to Western Washington. This was not implemented, but four mortars and carriages were relocated to arm new Battery Howe at Fort Funston (mortars Bethlehem #22, Builders #23, and Watervliet #28 and #46) in the summer of 1917. The remaining four guns at Battery McKinnon (two per pit) and eight at Stotsenburg (still with four per pit) continued to serve well into World War II. These twelve guns were finally authorized for removal and salvage on January 23, 1943 (actually not implemented until early 1946). The emplacement was subsequently used for several ordnance and private storage uses and still exists under the Presidio Trust. The magazines of the mortar battery are used by the Presidio Wine Bunkers for the storage of wine. The battery is not open to the public.

- **CROSBY:** A battery for two 6-inch disappearing guns as part of the rapid-fire coverage of the offshore mine fields. Plans were submitted on May 25, 1899. The battery was located well to the front and below the heavy gun emplacements on the bluff above. The battery followed type plans of Mimeograph No. 31. Insistence was placed on making sure the guns would be able to use their 5-degree of depression in order to cover the shoreline as close-in as possible. Work was done in 1899-1900. There were problems with the excavation and work. Concrete placement was suspended with the discovery that 2,200 barrels of cement were defective, and further work was delayed when water usage for construction was temporarily suspended in order to conserve it for Philippine-bound troops stationed at Fort Scott in September of 1899. It appears, though, that work was finally completed late in the spring of 1900. The battery was transferred on August 2, 1900 for a cost of $59,038.57. The battery was armed with two 6-inch Model 1897M1 guns on Model 1898 disappearing carriages (#30/#6 and #28/#5). It was named on General Orders No. 16 of February 14, 1902 for 1st Lieutenant Franklin B. Crosby killed in action in 1863 during the Civil War. The battery was retained in active condition for a considerable period of time, being finally removed under authority of August 5, 1943. The emplacement still exists in good condition on Golden Gate National Recreation Area property. The battery is open to the public.

- **BOUTELLE:** A 5-inch rapid-fire gun emplacement authorized for Fort Winfield Scott. Colonel Charles Suter submitted his engineering plans for the emplacement on February 23, 1898. At that time plans called for four such guns to be located in the main line gap between Miller and Godfrey. In the 1870s this area had been occupied by the old mortar battery, most prominent being the brick magazines in the traverses of the mortar platforms. Having rapid-fire guns was deemed indispensable to having adequate coverage of the mine field offshore. Ultimately space for just three such guns was found. Plans followed were the standard 2-gun design with an added flank position. While not possible to use the old magazines for ammunition, effort was made to keep the structures and their entry tunnels for crew shelter and implement storage. Two emplacements were built in early 1899. The third was funded on June 3, 1899 and completed except for armament by May 1900. In plan each emplacement was separate from the others, with its own magazine and platform. Transfer was made on October 1, 1901 for a cost of $27,030.22. The battery was named in General Orders No. 105 of October 9, 1902 for 2nd Lieutenant Henry M. Boutelle killed in action during the Philippine Insurrection. It was armed with three 5-inch Model 1897 Bethlehem guns on Model 1896 balanced pillar mounts (#17/#13, #15/#12, and #3/#14). In 1916 it was proposed to move this armament to San Diego's Fort Rosecrans, but this was not implemented. The gun armament was removed in 1917 for use on field mounts, but the carriages were left in place until scrapped in 1920. The emplacement still exists on the Golden Gate National Recreation Area. The battery is open to the public.

- **CHAMBERLIN:** A battery for four 6-inch disappearing emplaced near the shore on the south-western corner of the Fort Winfield Scott reservation. It was the last Endicott battery approved for the Presidio. Plans were submitted on May 18, 1903 for this site on Baker's Beach. Plans closely followed the mimeograph type—with widely-spaced, separate platforms for Model 1903 disappearing carriages and traverse magazines. Work was done from 1903-1904. It was transferred on December 24, 1904 for a cost of $100,803. It was armed with four 6-inch Model 1903 Watervliet guns guns on Model 1903 disappearing carriages (#27/#27, #28/#28, #30/#26, and #29/#52). A new BC station was soon added to the right, rear flank. The battery was named in General Orders No. 194 on Dec. 27, 1904 for Captain Lowell H. Chamberlin, who served during the Civil War. All four guns were removed on December 18, 1917 and sent to Watervliet for use on field carriages. The old disappearing carriages were scrapped in place in 1920. However, in late 1919 the battery was authorized for modification. Two 6-inch Model 1900 pedestal guns were to be emplaced on new, raised gun blocks in emplacements No. 2 and 3. The Model 1900 guns and pedestals (#31/#41 and #21/#40) were received and mounted on these platforms on January 8, 1920. These particular guns and carriages had been previously used at Battery DeKalb, Fort Taylor, Key West. This new armament served throughout World War II. It became surplus only on May 23, 1948, being the final armed battery at Fort Winfield Scott. The NPS has installed in the No. 4 emplacement a 6-inch Model 1905 gun on a Model 1903 disappearing carriage. This 6-inch DC gun came from the West Point training battery. The emplacement, partly restored for museum display purposes, is open on specified days as part of the Golden Gate National Recreation Area interpretive program.

- **SLAUGHTER:** A battery for three 8-inch disappearing guns emplaced on the eastern side of Fort Scott. The original plan for the San Francisco defenses called for no less than twenty 8-inch guns to cover the inner harbor—the roadstead inside the Golden Gate. Concerns about cover of the inner mine fields and against ships that might run past the main defenses prompted this element. While only eight 8-inch guns were eventually emplaced, this battery was one of the key elements. It was funded from the Act of July 7, 1898. Submission was made on September 27, 1898. It was located on high ground north of the San Francisco National Cemetery. The plans incorporated several non-standard features. There was an increase in spacing between gun centers, allowing room for adjacent, same-level magazines that did not require ammunition hoists. The magazines were built to hold 300 rounds per gun. Originally two guns emplacements were submitted in the plans of September 1898, but a third on the flank was added in December 1898. Work was done and reported ready by January 5, 1900. Transfer was made on January 23, 1900 for $71,062.63. It was named on General Orders No. 16 of February 24, 1902 for 1st Lieutenant William A. Slaughter, killed in action with the White River Indians in Washington Territory in 1855. It appears the gun tubes may have been shipped as early as 1892 to the West Coast, but waited almost ten years before a suitable battery was ready to take them. It was armed with three 8-inch Model 1888 Watervliet guns on LF Model 1896 disappearing carriages (#2/#30, #3/#25, and #8/#2). Most all of the batteries of the inner defenses were disarmed in 1917-1918. Battery Slaughter had its 8-inch gun tubes removed on November 26, 1917 under authority issued the previous month. The carriages were scrapped in place in 1920. In later years the battery was partially filled-in and abandoned with the construction of US Highway 101 Doyle Drive parkway immediately adjacent to it. The outline of Battery Slaughter has been preserved in the newly restored Battery Bluff section of the Golden Gate Recreational Area north of the National Cemetery. The top of the battery is open to the public, but the lower level remains filled with earth.

- **SHERWOOD:** A battery for two 5-inch, rapid-fire, pedestal guns emplaced west of Battery Slaughter as part of the inner harbor defenses. Battery plans were submitted on October 30, 1899. Authorization at this time was for four guns, two here for this location near the National Cemetery and two more for Angel Island. It followed the type plan, with two separate platforms for the guns, and lower-level service magazines. There was an observation station on the right flank. It did not have its own power room, that being supplied from Battery Slaughter. Work was done in 1900. It was transferred on August 2, 1900 for $20,355.07. There was a considerable delay in awaiting the arrival of the armament. It did not finally get installed until early 1908. The battery was armed with two 5-inch Model 1900 Watervliet guns on Model 1903 pedestal mounts (#2/#16 and #13/#17). It was named on General Orders No. 16 of February 14, 1900 for 2nd Lieutenant Walter Sherwood killed in action during Seminole War in 1840. The guns were removed under authority of February 10, 1917 and re-emplaced at Battery Bruff at Fort Funston. The emplacement was subsequently used for storage but still exists at the Golden Gate Recreational Area in the newly restored Battery Bluff section of the Presidio. The battery is open to the public.

- **BALDWIN:** A battery for two 3-inch masking parapet mounts emplaced on the western flank, and somewhat down the hill in elevation of Battery Sherwood with the eastern defenses of Fort Winfield Scott. Plans were submitted on July 2, 1900. It was located 350-feet west of Sherwood, overlooking the shoreline below. Work was done through 1901, and transfer was made on December 3, 1903 at a construction cost of $11,119.89. It was named for 2nd Lieutenant Henry M. Baldwin killed in action in 1864 during the Civil War in General Orders No. 105 of October 9, 1902. The battery was armed with two 3-inch, 15-pounder Model 1898 Driggs-Seabury guns on masking parapet mounts (#73/#73 and #74/#74). Around 1918 these pillars were converted to M1898M1 fixed pedestal mounts. In 1920 the armament was removed and the pillars scrapped. In postwar years the emplacement was buried by the construction of the Highway 101 Doyle Drive overpass. The reconstruction of the highway thrufare resulted in the uncovering of the top of the battery whioch has been restored and marked in the newly restored Battery Bluff section of the Presidio. The top of the battery is open to the public, but lower battery level is filled with earth.

- **BLANEY:** The final 3-inch, masking parapet battery for the eastern side of Fort Scott built as an element of the inner harbor defenses. The plan was submitted on June 3, 1901. It was located on the right flank of the series defined by Batteries Baldwin, Sherwood and Slaughter. It was about 150-feet to the northwest of Battery Slaughter's No. 1 emplacement. Originally designed for three guns, it was completed with four emplacements. It followed the type plans pretty closely. Work was commenced on the first three emplacements in July 1901 and was within a year they were joined by the fourth platform. While completed by the end of 1902, the battery was not transferred until March 22, 1907 for a construction cost of $20,000. It was named on General Orders No. 105 of October 9, 1902 for 2nd Lieutenant Daniel Blaney killed in action during the War of 1812. The armament was slow to arrive. By 1905 only one gun and carriage had been mounted, the other three not going in until 1908. Eventually it was armed with four 3-inch, 15-pounder Driggs-Seabury Model 1898 guns on balanced pillar mounts (#75/#75, #23/#23, #12/#12, and #95/#95). They did not serve long. By 1918 all four mounts had been converted to M1898M1 fixed pedestals. Both the guns and carriages were removed in 1920 under authority of May 26, 1920. Subsequently the emplacement was sometimes used for storage. The battery exterior surfaces and its CRF station have been restored with limited access in the Golden Gate National Recreation Area's newly restored Battery Bluff section of the Presidio.

- **Battery Point:** Under authority of February 21, 1942 four 3-inch Model 1902 guns were removed from Fort Baker's Battery Yates and redistributed to give AMTB capabilities early during the war. Two guns and pedestals (Model 1902 #20/#20 and #21/#21) were emplaced on new concrete platforms atop old Fort Point and designated Battery Point. They were removed again under authority of November 18, 1945 on March 7, 1946. The gun blocks still exist at Fort Point in the Golden Gate National Recreation Area. The battery site is open to the public.

- **Battery Gate:** A second pair of 3-inch Model 1902 guns from Battery Yates were also moved to old Fort Point. Erected on the old ramparts to the west of the Point, they were emplaced under authority of October 21, 1942. It used 3-inch M1902 #22/#22 and #23/#23. These in turn were removed under authority of November 16, 1945 and subsequently moved to Fort Cronkhite. The gun blocks still exist Fort Point in the Golden Gate National Recreation Area. The battery site is open to the public.

- **Battery Baker:** A 1943 Program 90mm AMTB battery for Fort Winfield Scott. Emplaced on Baker Beach at the southern end of the Presidio reservation under authority of June 1, 1943. It consisted of two 90mm fixed and two 90mm mobile mounts. It was removed subsequently post-1945. The blocks were later destroyed or buried; no traces remain today of this battery.

Fort Miley (1900-1948) is located on the top of Point Lobos at the entrance to the Golden Gate, about five miles southwest of Fort Scott. Fort Miley was a small, 12-acre Endicott Program fort with three concrete batteries. It was named in General Orders 43 of 1900 for Lt. Col. John D. Miley who died in Manila, the Philippines on Sept. 19, 1899. A battery of three 12-inch guns (two on disappearing carriages and one on a barbette carriage) was constructed in 1898, while installed in 1900 was two mortar batteries, each mounting eight 12-inch mortars. The garrison area of Fort Miley was turned over the Veterans Administration in the 1930s and was transformed into a hospital complex. The 1940 Program resulted in the addition of a #200 Series battery with two 6-inch shielded barbette guns and a 90mm AMTB battery at Land End. In 1972 a portion of Fort Miley was added to the new Golden Gate Recreation area. Most of the fort's buildings have been destroyed, but the batteries still remain. The is parking access to the Fort Miley battery area and Lands' End. Battery 243 is used as a storage facility and the mortar battery is used by the National Park Service as their horse stable complex.

Fort Miley Gun Batteries

- **CHESTER:** A battery for 12-inch guns authorized for the new Point Lobos reservation of Fort Miley. A plan for two 12-inch guns on Model 1897 disappearing carriages was submitted on August 29, 1899. Originally the Board of Engineers had allocated no less than six 12-inch guns for this bluff-top site facing west from the point. The plan was reduced to two guns and generally followed the type-plan but reduced the thickness of protection due to the extensive front earthen slope (however it also needed an extensive apron in front of the firing pits). Two years later, on June 3, 1901 the plan for a single 12-inch gun on barbette carriage on the left (southeast) flank, and 250-feet distant from the others was added. The availability of an "excess" Model 1892 barbette carriage was a key factor. While the first two emplacements were oriented almost directly to fire to the west, emplacement No. 3 was canted, directed to the southwest. Clearing began on September 23, 1899, and concrete work on the first two emplacements was done by May, 1900. It was transferred on September 26, 1902 for a construction cost of $165,919.39. It was armed with two 12-inch guns Model 1895 on Model 1897 disappearing carriages (#24/#26 and #22/#27). The third emplacement was built in 1901-1902 and included in the transfer and cost mentioned above. It was armed with one 12-inch Model 1888 Bethlehem gun on Model 1892 barbette carriage (#3/#28). This tube had been transferred from a New York battery being taken out of service. Gun platforms were widened in 1910 for a cost of $3600. Ammunition hoists were changed to Taylor-Raymond chain hoists in 1915. A new plotting room was built behind the traverse of emplacements No. 1 and No. 2 in 1914. In 1916 the carriages for the two disappearing guns were altered to allow higher elevation. The battery was named in General Orders No. 194 of December 27, 1904 for Major James Chester of Civil War service. The two Model 1895 guns were removed in 1918 for service on railway mounts, but they were promptly replaced with older Model 1888 tubes (Watervliet #17 from Battery Spencer at Fort Baker, and Model 1888M1-1/2 Watervliet #40 from Battery Lancaster at Fort Winfield Scott). Battery Chester continued to serve until the armament was ordered removed under authority of 1943. The emplacement still exists on property of the Golden Gate National Recreation Area. The battery site is open, but the interior is closed to the public.

GOLDEN GATE STRAIT.

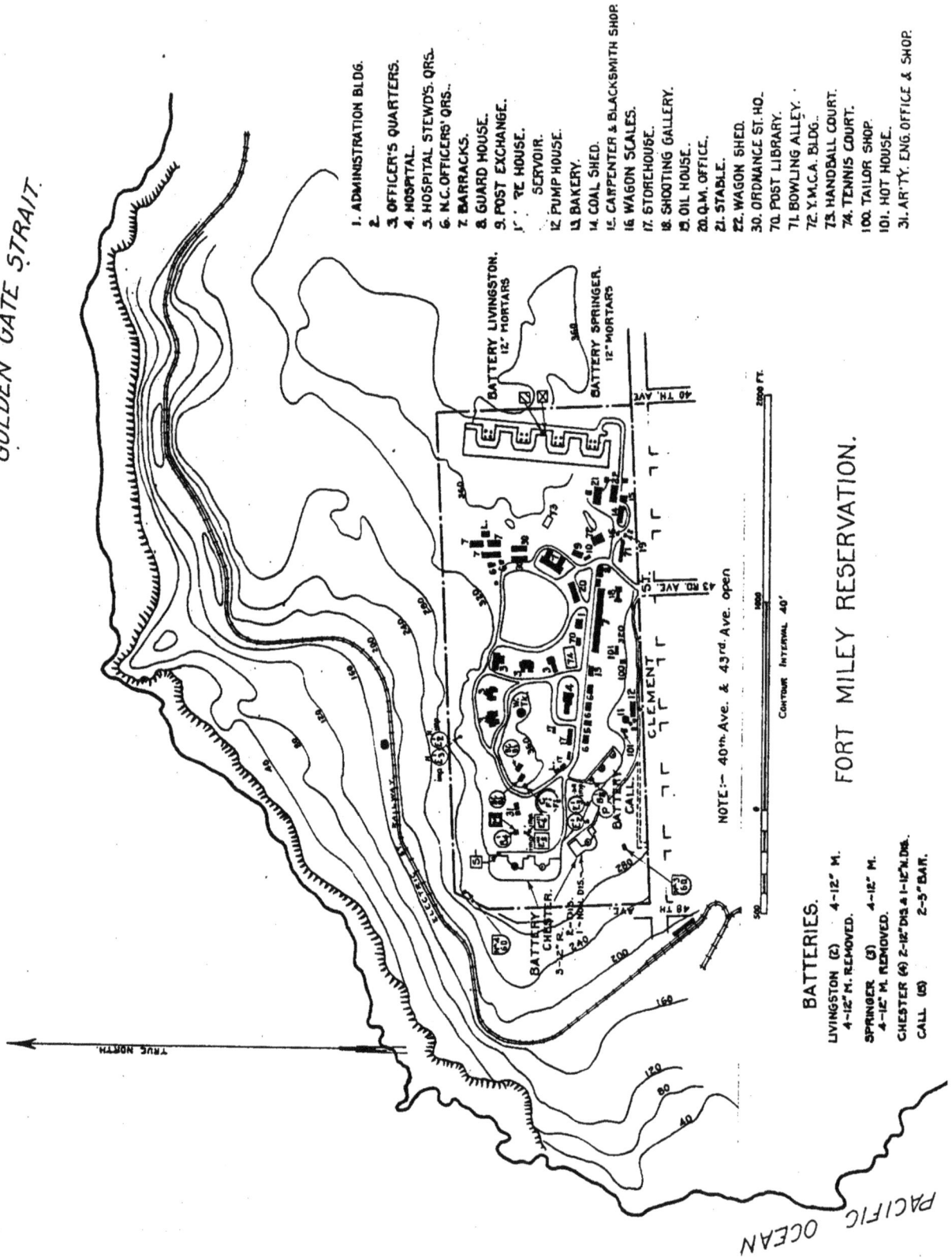

PACIFIC OCEAN

TRUE NORTH

1. ADMINISTRATION BLDG.
2.
3. OFFICER'S QUARTERS.
4. HOSPITAL.
5. HOSPITAL STEWD'S. QRS.
6. N.C. OFFICERS' QRS.
7. BARRACKS.
8. GUARD HOUSE.
9. POST EXCHANGE.
1. P.E. HOUSE.
 SERVOIR.
12. PUMP HOUSE.
13. BAKERY.
14. COAL SHED.
15. CARPENTER & BLACKSMITH SHOP.
16. WAGON SCALES.
17. STOREHOUSE.
18. SHOOTING GALLERY.
19. OIL HOUSE.
20. Q.M. OFFICE.
21. STABLE.
22. WAGON SHED.
30. ORDNANCE ST. HO.
70. POST LIBRARY.
71. BOWLING ALLEY.
72. Y.M.C.A. BLDG.
73. HANDBALL COURT.
74. TENNIS COURT.
100. TAILOR SHOP.
101. HOT HOUSE.
31. ART'Y. ENG. OFFICE & SHOP.

BATTERY LIVINGSTON. 12" MORTARS

BATTERY SPRINGER. 12" MORTARS

BATTERY CHESTER.

NOTE:- 40th AVE. & 43rd AVE. open

CONTOUR INTERVAL 40'

FORT MILEY RESERVATION.

BATTERIES.

LIVINGSTON (2)   4-12" M.
4-12" M. REMOVED.

SPRINGER   (3)   4-12" M.
4-12" M. REMOVED.

CHESTER (4) 2-12" D.R. & 1-12" D.R.

CALL (05)   2-5" BAR.

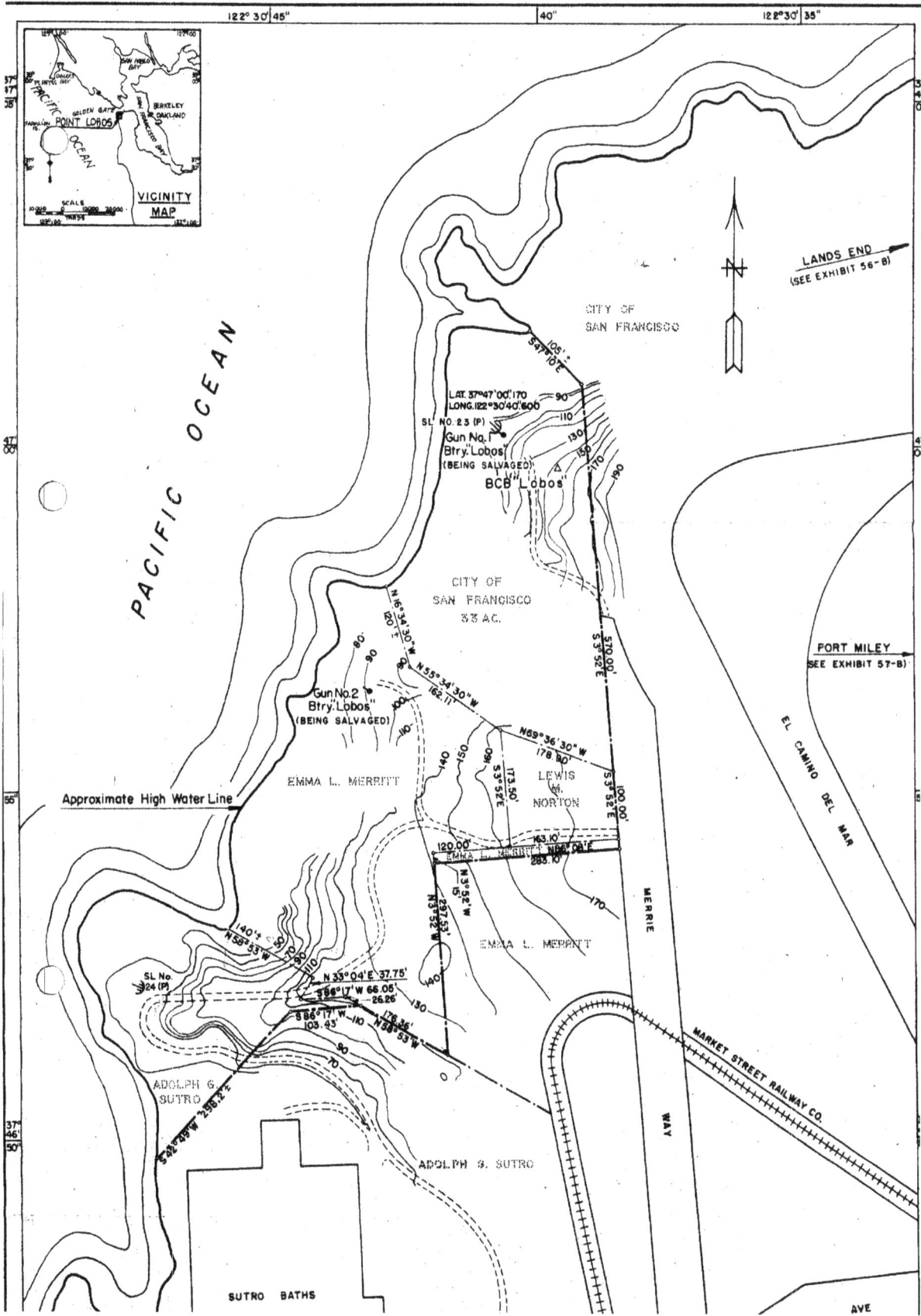

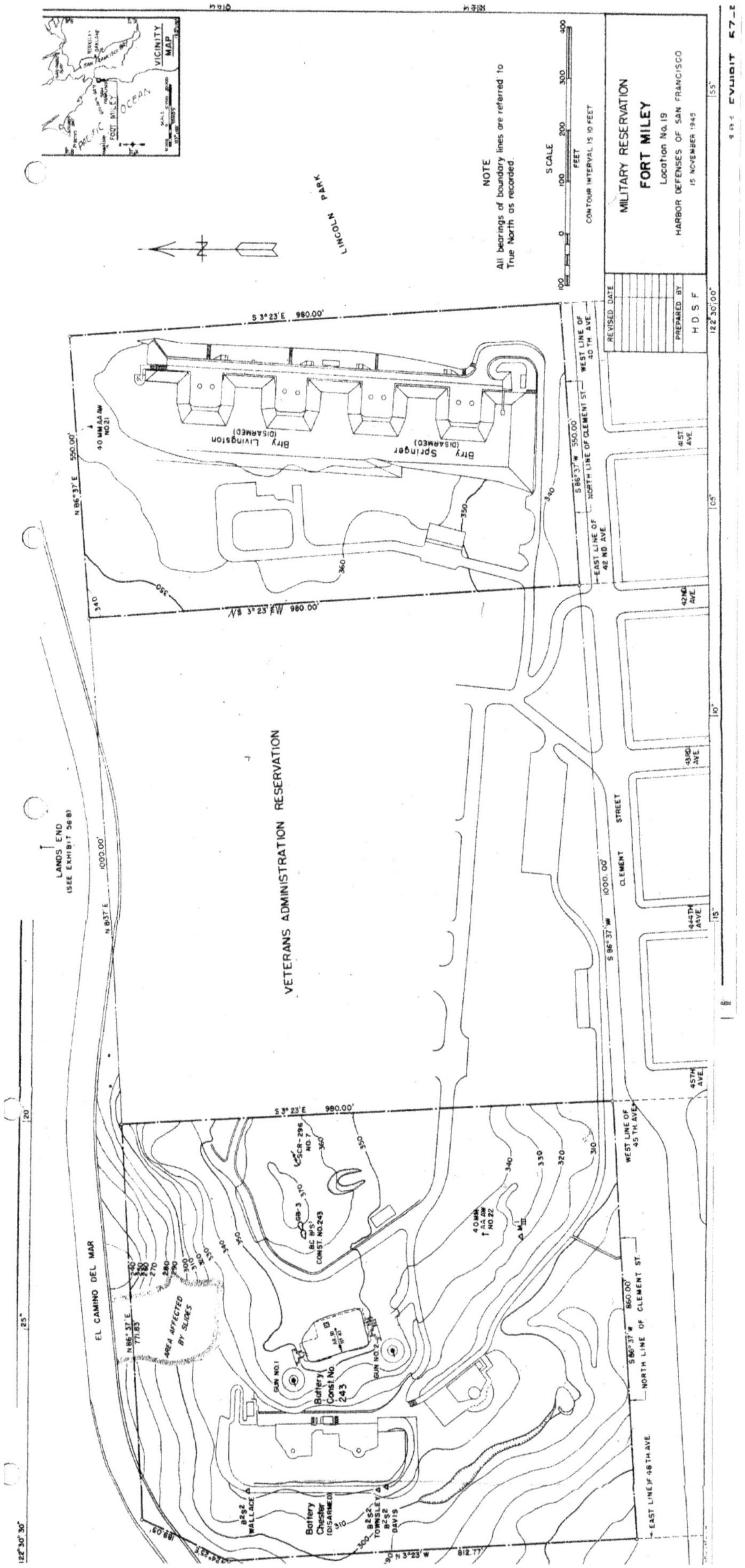

MILITARY RESERVATION

FORT MILEY

Location No.19

HARBOR DEFENSES OF SAN FRANCISCO

15 NOVEMBER 1945

NOTE

All bearings of boundary lines are referred to True North as recorded.

CONTOUR INTERVAL IS 10 FEET

SCALE

FEET

EXHIBIT 57-2

PREPARED BY H.D.S.F.

Fort Miley 1939 (NARA)

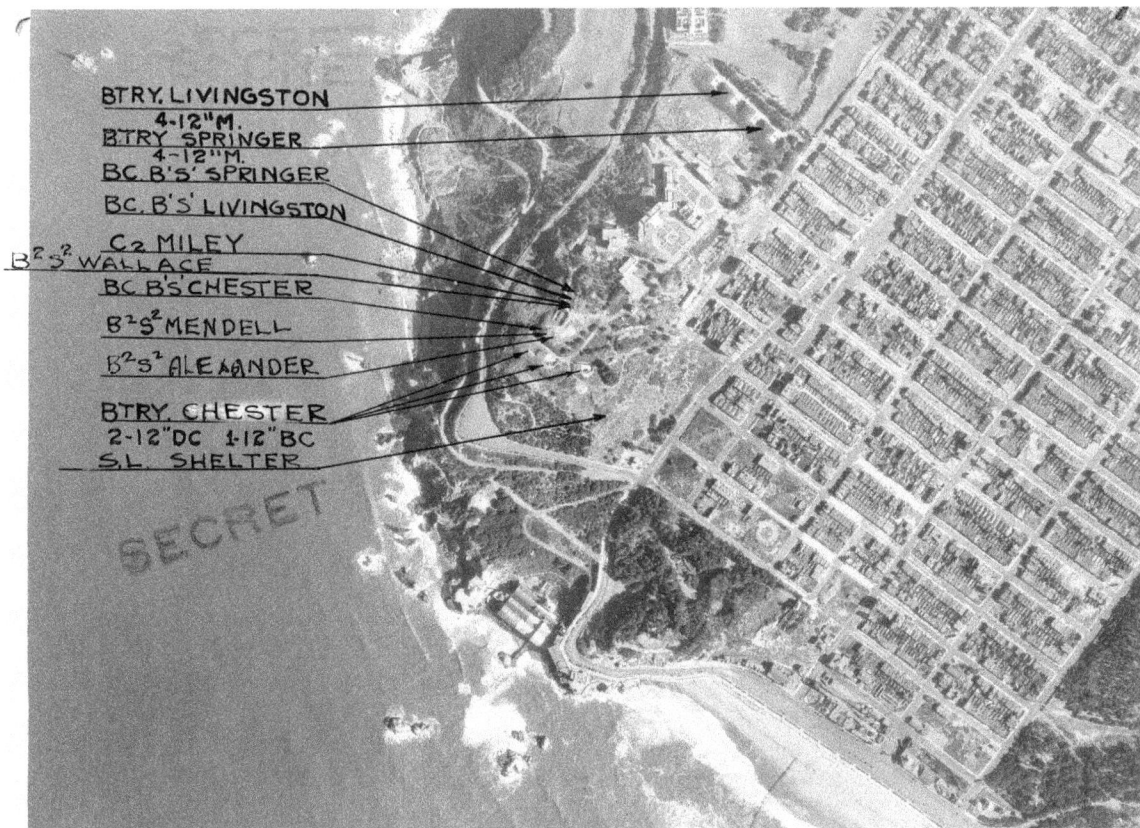

Fort Miley 1938 (NARA)

Battery Chester Fort Miley, Golden Gate National Recreation Area (Terry McGovern)

Battery Livingston-Springer Fort Miley Golden Gate National Recreation Area (Terry McGovern)

- **LIVINGSTON – SPRINGER:** A battery for sixteen seacoast mortars was emplaced near a ridge on the eastern side of the Fort Miley reservation. This was the first modern work for the reservation, being submitted on June 4, 1897. It was designed and planned simultaneously with mortar Battery McKinnon-Stotsenburg at Fort Winfield Scott. Magazines were both in front of the pits and in the flank traverses between the pits. However, before construction on September 19, 1899 the plan was revised to use the new 67-foot wide, enlarged pits vs. the older style of just 40-foot dimensions. Excavation began on November 27, 1899, and concrete work started on March 27, 1900. Work was finished by mid-1901. Transfer was made on September 26, 1902 for a cost of $174,050.89. It was named in General Orders No. 194 on December 27, 1904 for Colonel La Rhett Livingston of Civil War service. It was armed with sixteen 12-inch Model 1890M1 mortars (all of Watervliet manufacture) on Model 1896 carriages (#164/#290, #166/#301, #163/#232, #167/#233, #153/#300, #140/#234, #154/#238, #161/#235, #169/#302, #168/#299, #165/#231, #95/#230, #156/#298, #143/#292, #142/#239, and #149/#236). The slopes began to erode badly and had to be rebuilt in 1905. In 1906, in common with all other large mortar batteries, the unit was tactically split into two 2-pit batteries. The two southern pits (pits C and D) became Battery Springer, named on General Orders No. 20 of January 25, 1906 for Captain Anton Springer who was killed in action during the Philippine Insurrection. On September 5, 1908 the top carriage of Model 1896 carriage #230 was found to be broken after a practice firing, and was taken out of service until a new top carriage was fabricated and supplied. In May of 1918 two mortars were removed from each pit, and the corresponding carriages removed and scrapped in 1920 and 1921. The remaining mortars served until removed under authority of January 23, 1943. From 1955-1968 the old emplacement was occupied and used as an Air Force Reserve Center. The emplacements still exist of Golden Gate Recreational Area property and are used as NPS horse stables. The battery is not open to the public.

- **CALL:** A battery for two 5-inch pedestal guns. An urgent need for rapid-fire guns had long been identified for the southern defenses of San Francisco, but not originally provided for. In 1914 it was recommended to transfer the two guns from Battery Ledyard at Fort McDowell. It was located to the east, behind, and above Battery Chester's No. 3 emplacement. Work began on a new gun block set in December 1914 and was complete by September 15, 1915. Transfer was made on June 15, 1916 for a cost of $3,719.97. It consisted of simple platform blocks, a plotting room and a battery commander's station. It was armed with two 5-inch Model 1900 guns on Model 1903 pedestal carriages with shields (Watervliet guns #1/#14 and #7/#15). The guns were mounted in early 1917. It was named on General Orders No. 23 of April 27, 1915 for 1st Lieutenant Loren H. Call of the Coast Artillery Corps. The armament served only until they were removed in 1921. Around 1936 the emplacement was destroyed by construction at the post, nothing remains of it today.

- **Battery #243:** A 1940 Program dual 6-inch barbette battery emplaced on the bluff directly behind (to the east) of Battery Chester. Its construction necessitated the destruction of various older fire control stations previously located on this hill. It was authorized for construction on September 26, 1940, the Adjutant General issuing commencement orders on March 27, 1941. Structural work was begun in January of 1943 and completed on August 14, 1943. Transfer was made on September 18, 1944 for a total engineering cost of $245,808.25. It was never named, just being known as Battery Construction No. 243. Even after completion, there was considerable delay in mounting the armament. The carriages and shields (Model M4 barbettes #15 and #16) were delivered and mounted by late 1944, however the intended gun tubes were not delivered. After the end of the war, it was decided to complete the two San Francisco 6-inch batteries with tubes from the Columbia River defenses. They were emplaced in the existing carriages in April of 1948. In November 1949 the U.S. Army transferred its San Francisco mine material, along with batteries #243 and #244 (the latter at the Milagra Ridge Military Reservation) to the U.S. Navy. They served only a very short time with that service, being removed for scrap in 1950. The emplacement still exists on Golden Gate National Recreation Area property. The battery is open to the public, but magazine area is closed.

- **Battery Land:** A 1943 Program AMTB battery situated off the Miley reservation at Land's End, firing to the north. It consisted of concrete gun blocks and two fixed 90mm guns and was authorized on June 1, 1943. It was built from September 6, 1943 and transferred to service troops on January 18, 1944 for a cost of $13,050. It served at least through the end of the war in 1945. The blocks still exist on Golden Gate National Recreation Area property, though access along the eroded cliff face is difficult.

Fort Funston (1898-1950) is located on between the Pacific Ocean and State Highway 35 about 15 miles south of the entrance to the Golden Gate. Established as the Laguna Merced Military Reservation due to its location next to Lake Merced, the reservation was renamed in General Orders 76 of 1917 for Maj. Gen. Fredrick Funston, U.S. Army. Fort Funston use as a coast artillery fort began during the Interwar Period when two batteries of relocated guns were established there, Battery Bruff (2-5-inch BC Guns) and Battery Howe (4-12-inch Mortars). Fort Funston was the location of the first casemated 16-inch gun battery built by the U.S. Army in 1939. Battery Davis was the prototype for all the #100 Series batteries that were constructed during the 1940 Program. A portion of the reservation was used as a Nike missile launch site until 1970. In 1972 Fort Funston was incorporated into the new Golden Gate National Recreation Area. Today the area is large mixed-use area with access to the beach. Much of the area is an open access dog park. Two fire control stations remain on the bluffs, but the old Panama mounts have since tumbled into the sea. The Nike Administration area building remain in use by the park and the Nike launch area is now a parking lot. Battery Bruff and Howe have been destroyed, but Battery Davis remains as part of the GGNRA. The site is open to the public.

Fort Funston Gun Batteries

- **HOWE:** A battery for 12-inch seacoast mortars was built to provide better coverage of the southern approaches to San Francisco at this new reservation at Lake Merced in 1917. Plans for additional armament here were discussed as early as 1914-1915. On March 25, 1915, a site in a natural valley near Lake Merced was selected for a mortar battery. The land was immediately purchased from the Spring Valley Water Company. Approval for re-siting four mortars from Battery McKinnon at Fort Winfield Scott was granted under orders of October 12, 1916, and implemented with funding of February 1917. The new battery consisted of simple concrete gun platforms holding base rings arranged in a line just yards from the ocean at the northern extreme of the reservation. An accompanying magazine was built with just light, splinter-proof construction. It was armed with four 12-inch Model 1890M1 mortars on Model 1896 carriages (Bethlehem tube #22/#79, Builders tube #23/#96, Watervliet tube #28/#95 and Watervliet tube #46/#80). The battery was transferred on January 30, 1919 for a cost of $8,356. It was named on General Orders No. 135 of October 25, 1917 for Brigadier General Walter Howe, an Artillery Corps officer. With little modification the battery served throughout the Interwar period, until finally being disarmed under authority of January 19, 1945. The actual emplacement was either destroyed or buried in 1964 for the construction of a modern coastal highway.

SLOAT BOULEVARD

BATTERY BRUFF

PACIFIC OCEAN

TRUE NORTH.

U.S.C.A. LOOKOUT

FORT FUNSTON

Contour Interval 20 Feet.

2600 FEET

S.F.W.Q. DRAINAGE TUNNEL.

BATTERIES

(A) BATTERY BRUFF—2-5" GUNS.
(1) BATTERY WALTER HOWE.
4—12" MORTARS.

2. COMMANDING OFFICERS QUARTERS.
3. OFFICERS QUARTERS.
7. BARRACKS.
8. GUARD HOUSE.
10. BLACKSMITH & ELECTRIC SHOP.
11. COAL BIN.
12. GARAGE.
13. STOREHOUSE.
70. Y.M.C.A.
71. BASKETBALL COURT.

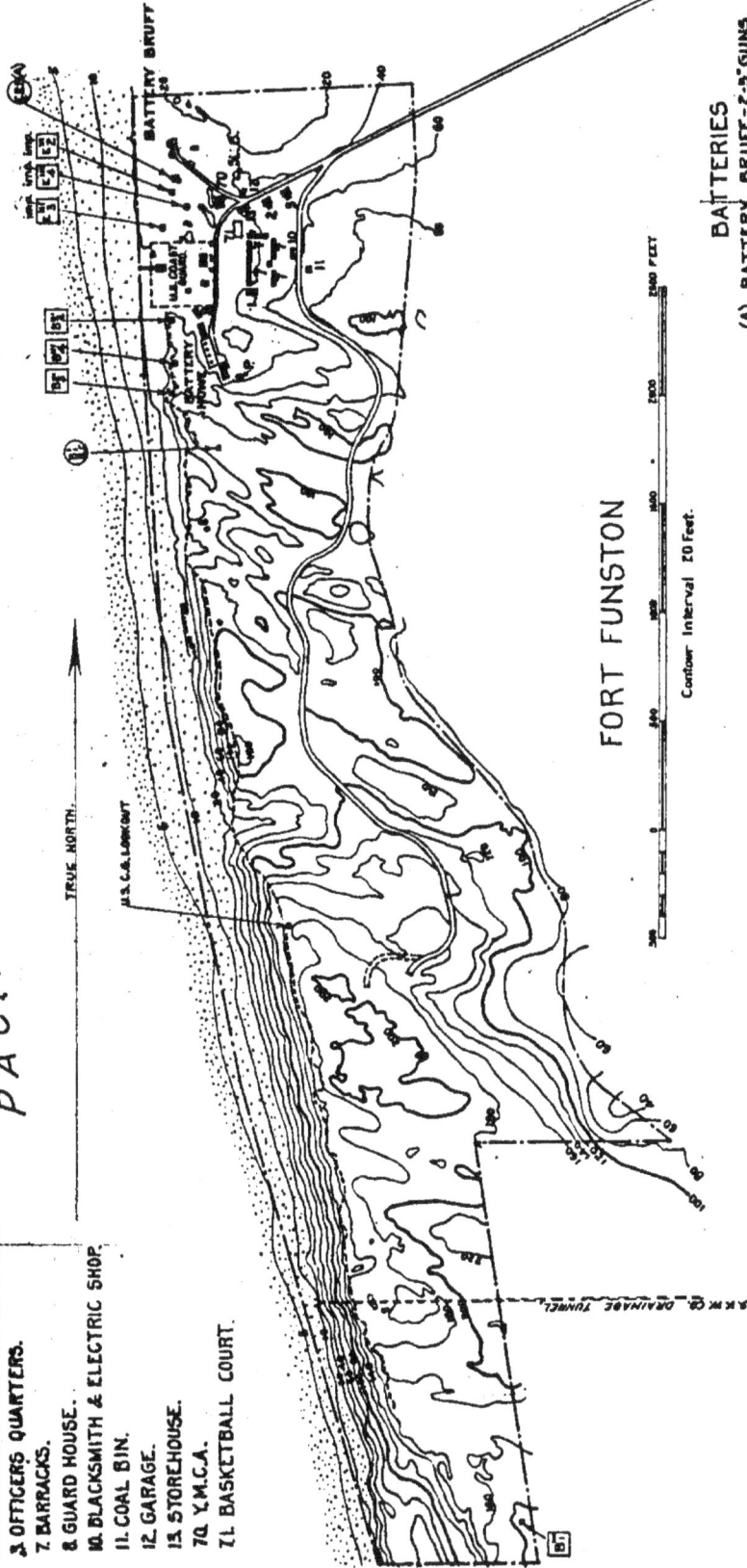

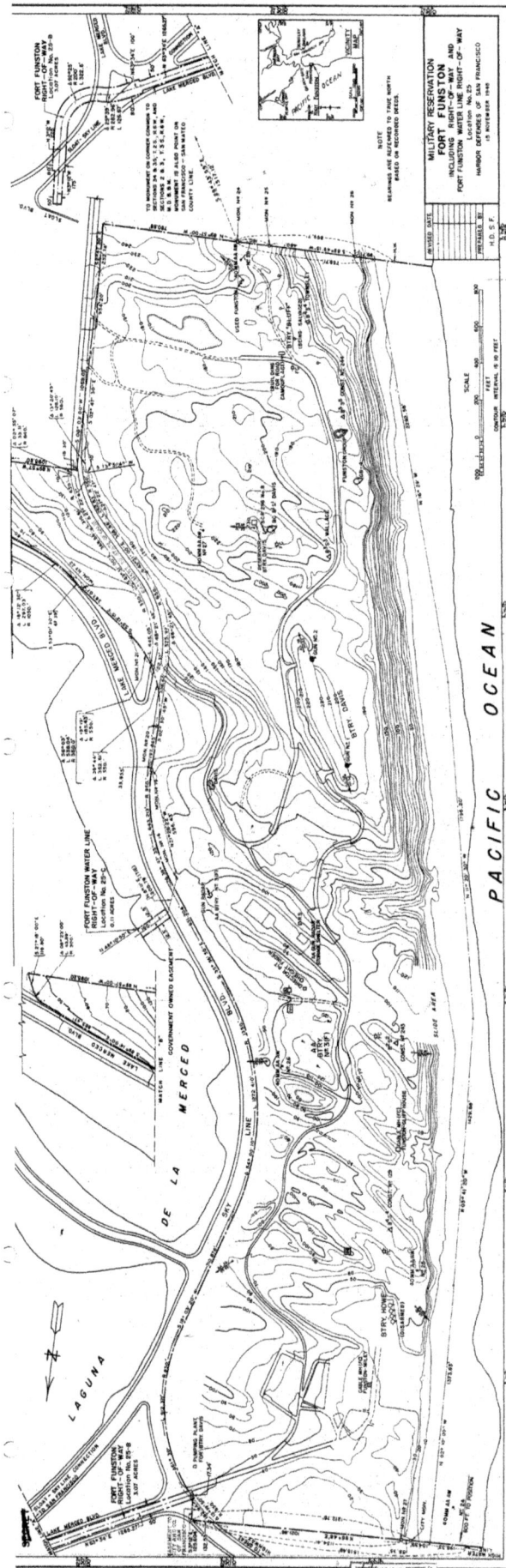

- **BRUFF**: A second temporary battery at the new Lake Merced reservation of Fort Funston. An emplacement for two relocated 5-inch pedestal guns was approved on February 27, 1917, and actually ordered constructed on September 21 of that year. Just simple gun platforms and an adjacent wooden magazine shelter were erected just north of Battery Howe, with a direct field of fire to the west. Transfer was made on January 30, 1919 at a cost of $3,272. It was named in General Orders No. 135 of October 25, 1917 for Colonel Lawrence L. Bruff. The battery was armed with two 5-inch Model 1900 Watervliet guns on Model 1903 pedestal carriages moved from Battery Sherwood at Fort Winfield Scott (#2/#16 and #13/#17). It served only a short time, being declared obsolete under orders of July 22, 1919, being disarmed shortly thereafter. The emplacement was buried or destroyed during World War II.

- **DAVIS**: A battery for two 16-inch barbette guns had long been discussed for a proposed location south of San Francisco. The Board of 1915 had recommended this site at the Lake Merced Military Reservation as early as June 8, 1915 for such a battery. Under significant planning for a number of years, the resultant battery design became the prototype for modern, casemated, long-range batteries. This battery (and Battery Townsley at Fort Cronkhite) were prototype casemates for the standard 100-series batteries of the later 1940 Program. Construction was authorized in 1934. Two-gun platforms were separated by 600-feet (the later standard batteries were 500-feet apart). Each gun was in its own casemated gun room. A protected gallery connected the rooms, with side galleries for powder and shell rooms, a power plant, and support rooms. A separate bomb-proof, protected plotting and switchboard room was also built on the reservation. The appropriation of FY-1937 gave an initial $300,000 to begin construction. Work actually started in October of 1936 and was completed by February 15, 1939. It was located on the southern section of the reservation, parallel to the bluff line with a field of fire to the west. The battery was transferred on September 21, 1940 at a final cost of $860,440.24. It was armed with two 16-inch Navy MkIIM1 guns on Model 1919M2 barbette carriages (#64/#10 and #74/#14). The battery was named in August of 1937 for General Richmond P. Davis of the Coast Artillery Corps. The battery served throughout the war as an important defensive work for San Francisco. The armament was removed and scrapped in 1949. The emplacement still exists at the Golden Gate National Recreation Area. The battery is open to public, though the service galleries and rooms are sealed closed.

Overgrown Battery Davis at the Fort Funston unit of the Golden Gate National Recreation Area.
(Mark Berhow)

Milagra Ridge Military Reservation (1939-1974) is located about seven miles south of San Francisco, between the cities of Pacifica and Daly City, California. In 1943, the U.S. Army started construction of Battery Construction No. 244 on top of Milagra Ridge. Two 6-inch T2/M1 guns mounted on shielded long-range barbette M4 carriages were moved from Fort Columbia to Battery 244 in 1948. The battery was decommissioned in 1950. Plans were drawn up to build a second gun emplacement armed with larger 16-inch guns, Battery Construction No. 130, but World War II ended before construction started on Battery 130. In 1956 Nike missile site SF-51 was built here and converted to the Nike-Hercules system in 1958. Typical of Nike sites, SF-51 was divided into an administrative area (SF-51A), an integrated fire control area (SF-51C), and a launcher area (SF-51L); SF-51A and 51L lie within the area of Milagra Ridge, while SF-51C is in the neighboring Sweeney Ridge open space preserve. In May 1974, the land was turned over to the City of Pacifica. Prior to the land transfer, several areas had already been turned over, including 73 acres in 1962 (eventually becoming residential parcels) and 36 acres in 1972, first to the Department of the Interior and then to the City of Pacifica in 1974. The buildings at SF-51A were demolished in 1983 and a condominium complex was built on that site. In 1987 the National Park Service acquired the remaining 240 acres of Milagra Ridge and made it a part of the Golden Gate National Recreation Area. The reservation is open to public access while the 6-inch battery is closed.

Milagra Ridge M.R. Gun Batteries

- **Battery #244**: A 1940 Program battery built in 1943 as a reinforced concrete coastal artillery battery on Milagra Ridge Military Reservation. The battery was begun 24 Mar 1943, completed 18 Sep 1943 and turned over for service 18 Sep 1944 at a cost of $299,699. The two 6-inch T2(M1) guns mounted on shielded long-range barbette M4 carriages were not mounted until 1948 as the guns were obtained from the Columbia River defenses. It was disarmed in 1950. The emplacement remains on NPS property as part of the Golden Gate National Recreation Area. The battery site is open, but the interior is closed to the public.

- *Battery #130* (planned): Battery Construction Number 130 was to have been a reinforced concrete, World War II 16-inch coastal gun battery on Milagra Ridge Military Reservation, San Mateo County, California. It was cancelled before any construction was begun.

Battery Wallace Fort McDowell Angel Island State Park (Mark Berhow)

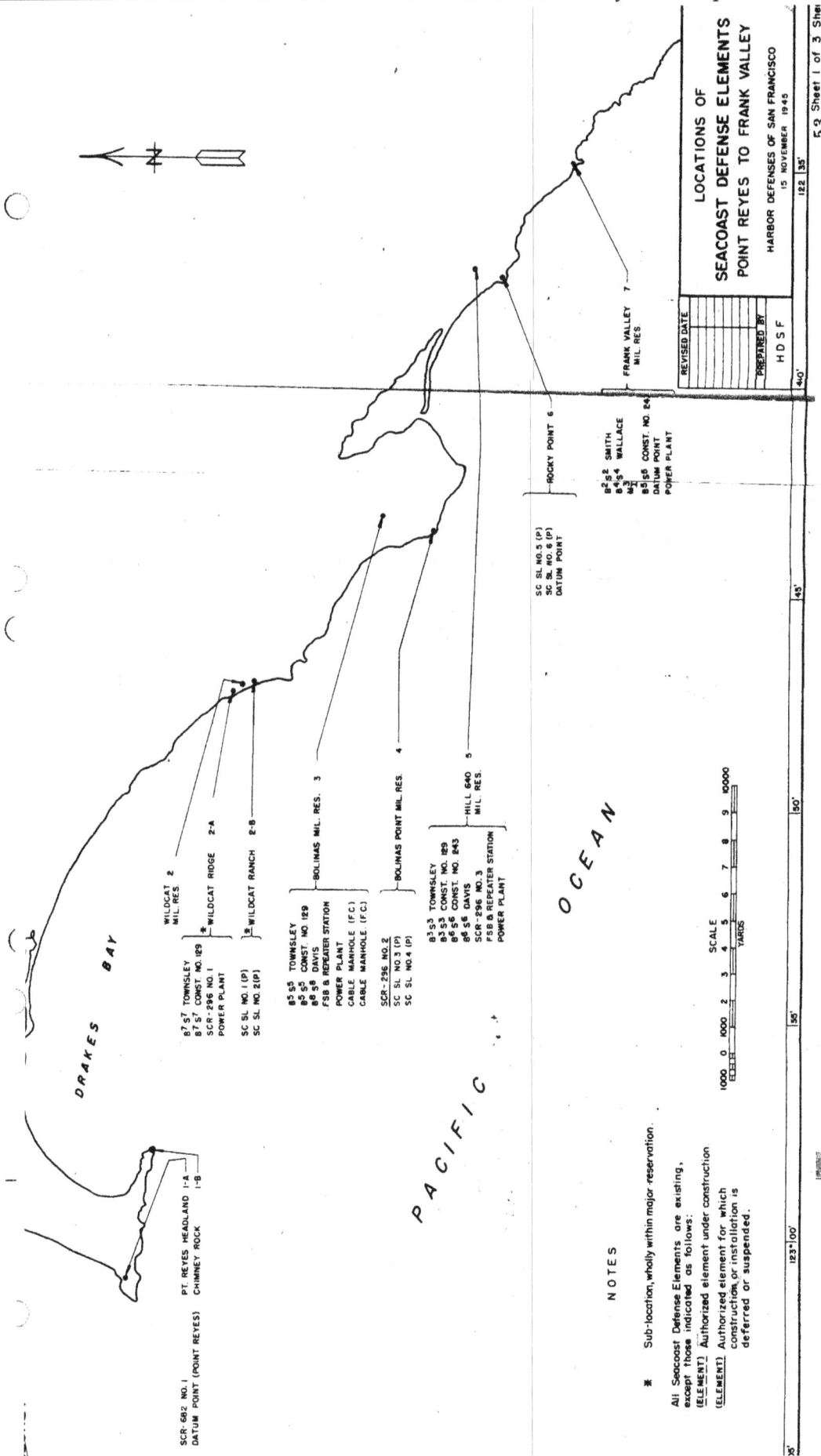

LOCATIONS OF
SEACOAST DEFENSE ELEMENTS
POINT REYES TO FRANK VALLEY

HARBOR DEFENSES OF SAN FRANCISCO
15 NOVEMBER 1945

REVISED DATE	

PREPARED BY

H D S F

53   Sheet 1 of 3 Sheets

122° 35'

FRANK VALLEY 7
MIL. RES.

40'

B²S² SMITH
B⁴S⁴ WALLACE
M²
B⁵S⁵ CONST. NO. 243
DATUM POINT
POWER PLANT

ROCKY POINT 6

SC SL NO. 5 (P)
SC SL NO. 6 (P)
DATUM POINT

45'

SCR-682 NO. 1
DATUM POINT (POINT REYES)

PT. REYES HEADLAND I-A
CHIMNEY ROCK   I-B

DRAKES BAY

WILDCAT 2
MIL. RES.

B⁷S⁷ TOWNSLEY
B⁷S⁷ CONST. NO. 129
SCR-296 NO. 1
POWER PLANT

SC SL NO. 1 (P)
SC SL NO. 2(P)

WILDCAT RIDGE   2-A

WILDCAT RANCH   2-B

B⁵S⁵ TOWNSLEY
B⁵S⁵ CONST. NO. 129
B⁸S⁸ DAVIS
POWER PLANT
FSB & REPEATER STATION
CABLE MANHOLE (F.C.)
CABLE MANHOLE (F.C.)

BOLINAS MIL. RES.   3

SCR-296 NO. 2
SC SL NO. 3 (P)
SC SL NO. 4 (P)

BOLINAS POINT MIL. RES.   4

B³S³ TOWNSLEY
B³S³ CONST. NO. 129
B⁶S⁶ CONST. NO. 243
B⁸S⁸ DAVIS
SCR-296 NO. 3
FSB & REPEATER STATION
POWER PLANT

HILL 640   5
MIL. RES.

PACIFIC

OCEAN

SCALE

1000   0   1000   2   3   4   5   6   7   8   9   10000
YARDS

NOTES

*   Sub-location, wholly within major reservation.

All Seacoast Defense Elements are existing,
except those indicated as follows:

(ELEMENT)   Authorized element under construction

(ELEMENT)   Authorized element for which
construction or installation is
deferred or suspended.

123° 00'   55'   50'   45'

LOCATIONS OF
SEACOAST DEFENSE ELEMENTS
FORT CRONKHITE TO FORT SCOTT
HARBOR DEFENSES OF SAN FRANCISCO
15 NOVEMBER 1945

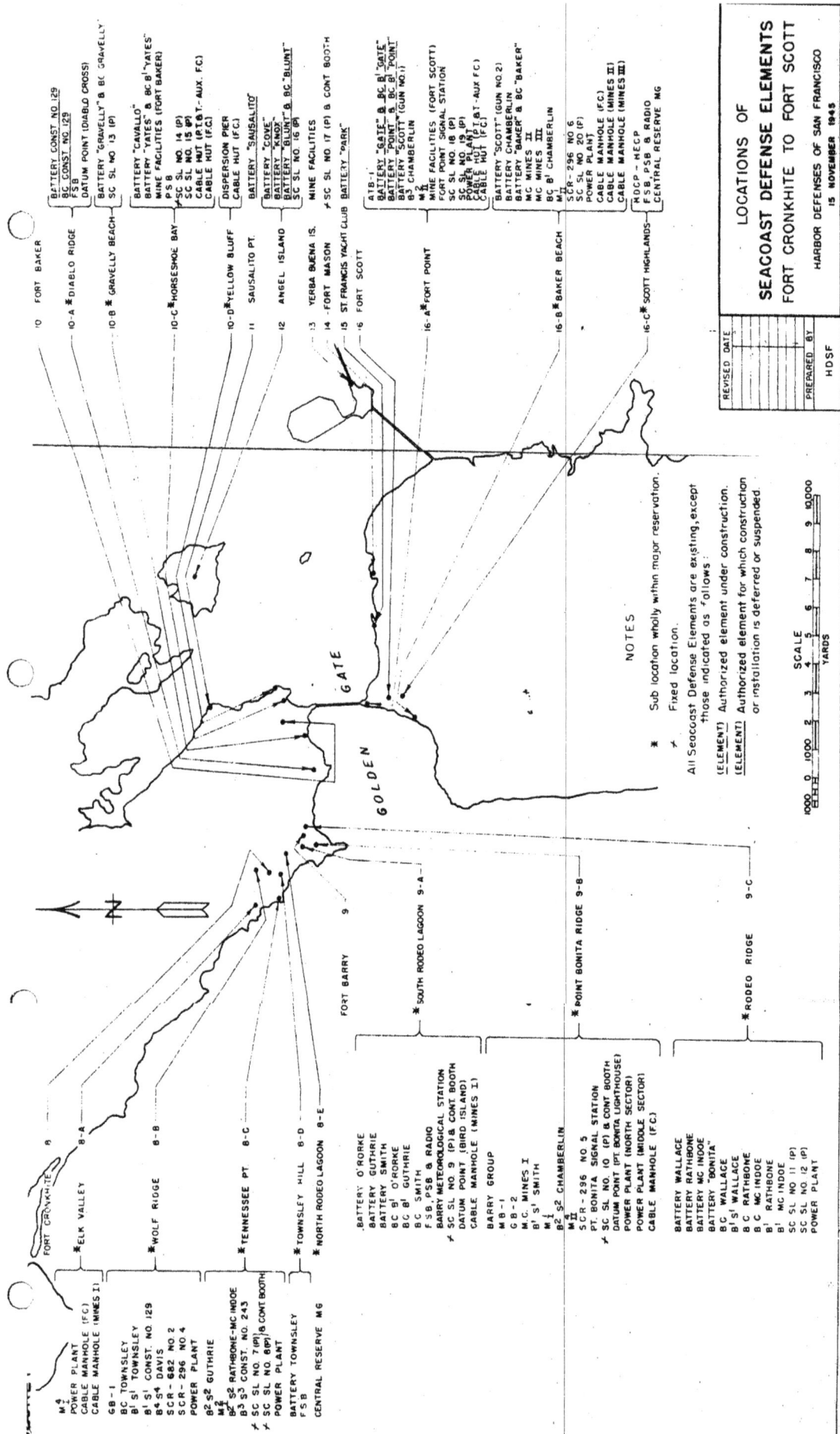

REVISED DATE		
PREPARED BY	HDSF	

54   Sheet 2 of 3 Sheet   EXHIBIT 2.

GOLDEN GATE

10 FORT BAKER

10-A *DIABLO RIDGE*
BATTERY CONST NO 129
BC CONST NO 129
FSB
DATUM POINT (DIABLO CROSS)

10-B *GRAVELLY BEACH*
BATTERY "GRAVELLY" & BC B' GRAVELLY
SC SL NO 13 (P)

10-C *HORSESHOE BAY*
BATTERY "CAVALLO"
BATTERY "YATES" & BC B' "YATES"
MINE FACILITIES (FORT BAKER)
PSB
SC SL NO 14 (P)
SC SL NO 15 (P)
CABLE HUT (PI.B'T.- AUX. FC)
CABLE HUT (FC)

10-D *YELLOW BLUFF*
DISPERSION PIER
CABLE HUT (FC)

11 SAUSALITO PT.
BATTERY "SAUSALITO"

12 ANGEL ISLAND
BATTERY "COVE"
BATTERY "KNOX"
BATTERY "BLUNT" & BC "BLUNT"
SC SL NO 16 (P)

13 YERBA BUENA IS.
MINE FACILITIES

14 FORT MASON
SC SL NO 17 (P) & CONT BOOTH

15 ST FRANCIS YACHT CLUB
BATTERY "PARK"

16 FORT SCOTT

16-A *FORT POINT*
ATB-1
BATTERY "GATE" & BC B' "GATE"
BATTERY "POINT" & BC B' "POINT"
BATTERY "SCOTT" (GUN NO.1)
B3 CHAMBERLIN
M2
MINE FACILITIES (FORT SCOTT)
FORT POINT SIGNAL STATION
SC SL NO 18 (P)
SC SL NO 19 (P)
POWER PLANT
CABLE HUT (PI.B'T-AUX FC)
CABLE HUT (FC)

16-B *BAKER BEACH*
BATTERY "SCOTT" (GUN NO 2)
BATTERY CHAMBERLIN
BATTERY "BAKER" & BC "BAKER"
MC MINES II
MC MINES III
BC B' CHAMBERLIN
M1

16-C *SCOTT HIGHLANDS*
SCR-296 NO 6
SC SL NO 20 (P)
POWER PLANT
CABLE MANHOLE (FC)
CABLE MANHOLE (MINES II)
CABLE MANHOLE (MINES III)
HDCP - HECP
FSB, PSB & RADIO
CENTRAL RESERVE MG

8 FORT CRONKHITE
M4
POWER PLANT
CABLE MANHOLE (FC)
CABLE MANHOLE (MINES I)

8-A *ELK VALLEY*
GB-1
BC TOWNSLEY
B' S' TOWNSLEY
B4 S4 DAVIS
SCR - 682 NO 2
SCR - 296 NO 4
POWER PLANT
B2 S2 GUTHRIE

8-B *WOLF RIDGE*
M2
B2 S2 RATHBONE-MC INDOE
B3 S3 CONST. NO 129
SC SL NO 7 (P) & CONT BOOTH
SC SL NO 8 (P)
POWER PLANT

8-C *TENNESSEE PT*
BATTERY TOWNSLEY
FSB

8-D *TOWNSLEY HILL*
CENTRAL RESERVE MG

8-E *NORTH RODEO LAGOON*

9 FORT BARRY

9-A *SOUTH RODEO LAGOON*
BATTERY O'RORKE
BATTERY GUTHRIE
BATTERY SMITH
BC B' O'RORKE
BC B' GUTHRIE
BC SMITH
FSB, PSB & RADIO
BARRY METEOROLOGICAL STATION
SC SL NO 9 (P)B & CONT BOOTH
DATUM POINT (BIRD ISLAND)
CABLE MANHOLE (MINES I)

9 *BARRY GROUP*
M B-1
G B-2
M.C. MINES I
B' S' SMITH
M I
B2 S2 CHAMBERLIN
M II
SCR-296 NO 5
PT. BONITA SIGNAL STATION
SC SL NO 10 (P) & CONT BOOTH
DATUM POINT (PT BONITA LIGHTHOUSE)
POWER PLANT (NORTH SECTOR)
POWER PLANT (MIDDLE SECTOR)
CABLE MANHOLE (FC)

9-B *POINT BONITA RIDGE*
BATTERY WALLACE
BATTERY RATHBONE
BATTERY MC INDOE
BATTERY "BONITA"
BC WALLACE
B' S' WALLACE
B C RATHBONE
B C MC INDOE
B' RATHBONE
B' MC INDOE
SC SL NO 11 (P)
SC SL NO 12 (P)
POWER PLANT

9-C *RODEO RIDGE*

NOTES

* Sub location wholly within major reservation.
* Fixed location.

All Seacoast Defense Elements are existing, except those indicated as follows:

(ELEMENT) Authorized element under construction.

(ELEMENT) Authorized element for which construction or installation is deferred or suspended.

SCALE
1000   0   1000   2   3   4   5   6   7   8   9   10,000
YARDS

122°10'   20'   25'   30'   35'   40'

40'   35'   30'   122° 25'

GOLDEN GATE

SC SL NO. 21 (P) — CHINA BEACH 17
BATTERY "LAND" & BC "LAND"
BATTERY "BUCK"
SC SL NO. 22 (P) — LANDS END 18
DATUM POINT (MILE ROCK LIGHT)
CABLE MANHOLE (F.C.)

GB-3
BATTERY CONST. NO. 243
BC B¹ S¹ CONST. NO. 243
B² S² TOWNSLEY
B² S² WALLACE — FORT MILEY 19
B² S² DAVIS
M³⅗
SCR-296 NO. 7
FSB

SC SL NO. 23 (P) — POINT LOBOS 20
SC SL NO. 24 (P)

B² S² CONST. NO. 129
M³⅔ — SUTRO HEIGHTS 21
SC SL NO. 25 (P)
POWER PLANT

DATUM POINT (LURLINE BATHS PIER)
CABLE MANHOLE (F.C.) — CLIFF HOUSE 22
CABLE MANHOLE (MINES II & III)

SC SL NO. 26 (P) — SOUTH WINDMILL 23
M²⅔ — GREAT HIGHWAY MIL. RES. 24
FORT FUNSTON 25

SC SL NO. 27 (P) — NORTH OF FORT FUNSTON 25-A
B⁴ S⁴ CONST. NO. 129
B² S² CONST. NO. 243
FUNSTON METEOROLOGICAL STATION
POWER PLANT (MIDDLE SECTOR) — ✱ NORTH FUNSTON 25-B
POWER PLANT (SOUTH SECTOR)
CABLE MANHOLE (F.C.)
CABLE MANHOLE (F.C.)

FUNSTON GROUP
GB-4
BATTERY DAVIS
BC B¹ S¹ DAVIS
B⁴ S⁴ TOWNSLEY
B³ S³ WALLACE — ✱ SOUTH FUNSTON 25-C
B⁵ S⁵ CONST. NO. 244
SCR-296 NO. 8
FSB
POWER PLANT (MIDDLE SECTOR)
POWER PLANT (SOUTH SECTOR)

SC SL NO. 28 (P) — SOUTH OF FORT FUNSTON 25-D
B⁴ S⁴ CONST. NO. 243
B³ S³ CONST. NO. 244 — MUSSEL ROCK MIL. RES. 26
POWER PLANT

SC SL NO. 29 (P)
SC SL NO. 30 (P) — MUSSEL ROCK 27
DATUM POINT

MILAGRA MIL. RES. 28
B⁶ S⁶ TOWNSLEY
B³ S³ DAVIS — ✱ MILAGRA KNOB 28-A
POWER PLANT

BATTERY CONST. NO. 244
BC B¹ S¹ CONST. NO. 244
B⁶ S⁶ CONST. NO. 129 — ✱ MILAGRA RIDGE 28-B
SCR-296 NO. 9
FSB

SC SL NO. 31 (P)
DATUM POINT — SAN PEDRO POINT 29

B⁸ S⁸ TOWNSLEY
B⁸ S⁸ CONST. NO. 129
B⁵ S⁵ DAVIS
SCR-296 NO. 10 — DEVILS SLIDE MIL. RES. 30
FSB & REPEATER STATION
POWER PLANT
CABLE MANHOLE (F.C.)

FUNSTON GROUP FLANK CP & OP
B² S² CONST. NO. 244 — LITTLE DEVILS SLIDE MIL. RES. 31
POWER PLANT

SC SL NO. 32 (P) — SOUTH LITTLE DEVILS SLIDE 32

SC SL NO. 33 (P)
DATUM POINT (MONTARA LIGHT) — MONTARA POINT 33

DATUM POINT — SEAL COVE 34

B⁹ S⁹ CONST. NO. 129
B⁷ S⁷ DAVIS
B⁴ S⁴ CONST. NO. 244
SCR-296 NO. 11
SC SL NO. 34 (P) — PILLAR POINT 35
SC SL NO. 35 (P)
DATUM POINT
POWER PLANT
CABLE MANHOLE (F.C.)

DATUM POINT — PURISIMA 36

SOUTH EAST FARALLON IS.   DATUM POINT (FARALLON LIGHT)

NOTES

1. Sub-location, wholly within major reservation.

Seacoast Defense Elements are existing, except those indicated as follows:

---- Authorized element under construction.

===== Authorized element for which construction or installation is deferred or suspended.

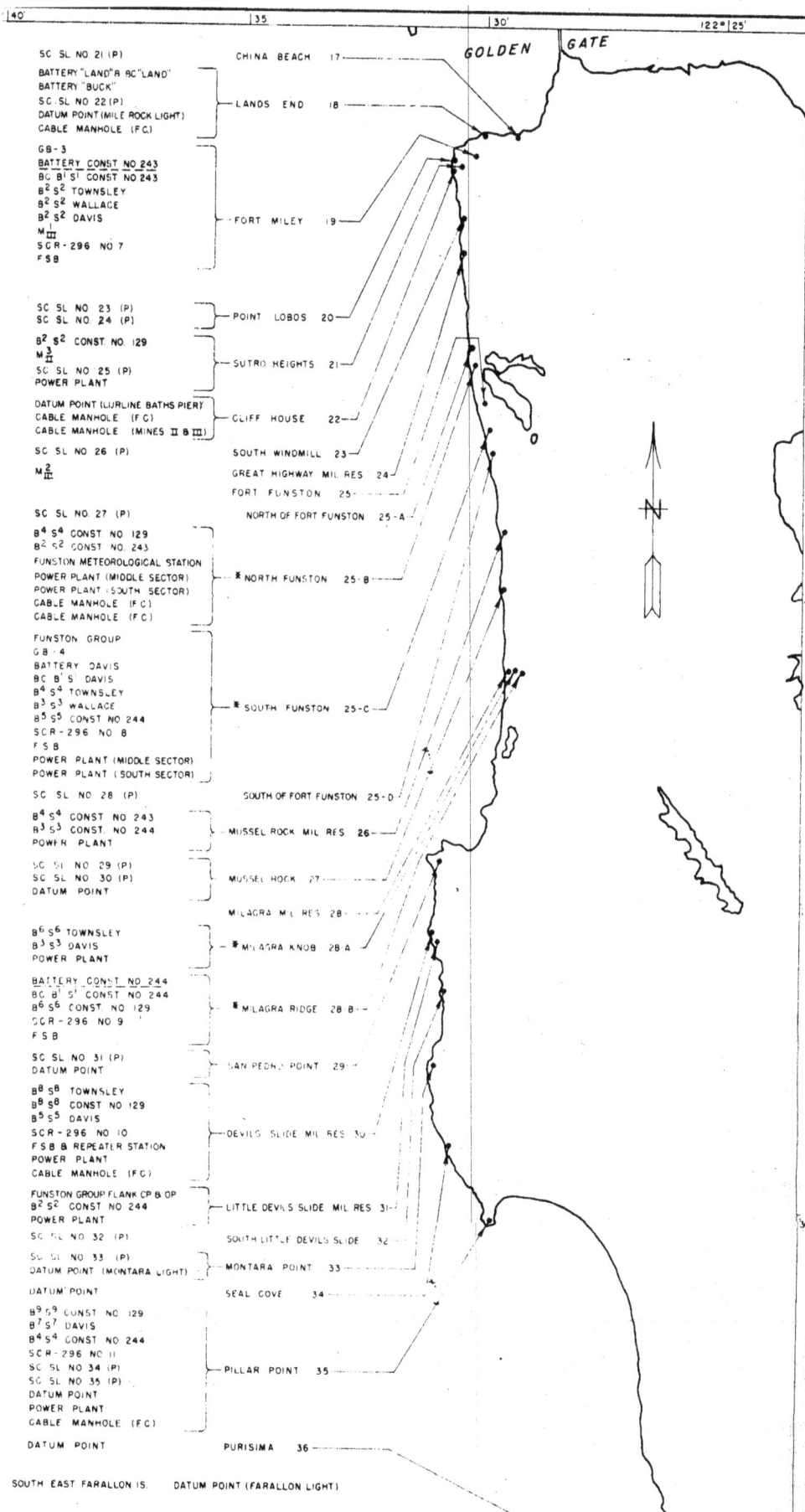

LOCATIONS OF
SEACOAST DEFENSE ELEMENTS
CHINA BEACH TO PURISIMA
HARBOR DEFENSES OF SAN FRANCISCO
15 NOVEMBER 1945

REVISED DATE

PREPARED BY   HDSF

SCALE
1000 2 3 4 5 6 7 8 9 10,000
YARDS

San Francisco World War II-era Site Locations. Stations housed in a single structure are connected by dashes (-)

location	Loc#	Purpose
Point Reyes	1	SCR 682
Wildcat	2	BS7/Townsley, BS7/129, SCR296-1, SL 1,2
Bolinas	3	BS5/Townsley, BS5 129, BS5/Davis, SCR296-2, SL 3,4
Hill 640	5	BS3/Townsley, BS3/129, BS6/Davis, SCR296-3
Rocky Point	6	SL 5,6
Frank Valley	7	BS2/Smith, BS4/Wallace, M3/7, BS5/243
Elk Ridge	8	M4/7
Wolf Ridge	8	Gn1, BC-BS1/Townsley, BS1/129, BS4/Davis, SCR682, SCR296-4
Tennesse Point 7,8	8	Batt. Tact. #2 Townsley, BS2/Guthrie, M2/7, BS2/Rathbone-McIndoe, SL
Fort Cronkhite	8	Reserve Magazine
Fort Barry South/ Rodeo Lagoon	9	Batt. Tact. #8 Mines I, Batt. Tact. #12 Guthrie, Batt. Tact. #14, Smith, Batt. Tact. #15 O'Rorke, BC-B1/O'Rorke, BC-B1/Guthrie, BC/Smith, Met, SL9
Point Bonita Ridge	9	MB-1, G2, MC-2, BS1 Smith, BS2/Chamberlin, M4/II, SCR296-5, SS, SL 10
Rodeo Ridge	9	Batt. Tact. #4 Wallace, Batt. Tact. #10 Rathbone, Batt. Tact. #13 McIndoe, BC/Wallace, BS1/Wallace, BC/Rathbone, BCMcIndoe, B1/Rathbone, B1/McIndoe, SL 11,12"
Ft Baker Diablo Ridge	10	Batt. Tact. #1 BCN 129, BC/129
Gravelly Beach	10	BC/Gravelly
Horseshoe Bay	10	BC-B1/Yates, PSR/129, SL 14,15, Mine
Angel Island/ Fort McDowell	12	BC/Blunt, SL 16
Yerba Buena Island	13	mine
Fort Winfield Scott/ Fort Point	16	BC-B1/Point, BC-B1/Gate, B3/Chamberlin, M2/II, Mines, SL 18,19
Baker Beach	16	Batt. Tact. #7 Mines II, Batt. Tact. #9 Mines III, Batt. Tact. #11 Chamberlin, BC/Baker, BC-B1/Chamberlin, MC II, MC III, M1/II, SCR-296-6
Scott Highlands	16	HDCP-HECP, Reserve magazine
Lands End	18	SL 21,22, BC/Land
Fort Miley BS2/Davis, M1/III,	19	Batt. Tact. #5 BCN 243, G3, BC-BS1/243, BS2/Townsley, BS2/Wallace, SCR296-7
Point Lobos	20	SL 23,24
Sutro Heights	21	BS2/129, M3/II, SL 25
South Windmill	23	SL 26
Great Highway	24	M2/III
Fort Funston North	25	SL 27, BS4/129, BS2/243, Met
Fort Funston South	25	Batt. Tact. #3 Davis, G4, BC-B1/Davis, BS4/Townsley, BS3/Davis, BS3/244, SCR296-8, FSB, SL 28
Mussel Rock	26	BS4/243, BS3/244
Mussle Rock	27	SL 29,30

Milagra Ridge	28	Batt. Tact. #6 BCN 244, BS6/Townsley, BS3/Davis, BC-B1/244, BS6/129, SCR296-9, FSB
San Pedro Point	29	SL 31
Devils Slide	30	BS8/Townsley, BS8/129, BS5/Davis, FSB, SCR296-10
Little Devils Slide	31	Funston Gp OP, BS2/244, SL 32
Montara Point	33	SL 33
Pillar Point	35	BS9/129, BS7/Davis, BS4/244, SCR296-11, SL34,35

Chin, Brian B. *Artillery at the Golden Gate, the Harbor Defenses of San Francisco in World War II.* Hole in the Head Press, Bodega Bay, CA 2015

Clauss, Francis J. *Angel Island, Jewell of San Francisco Bay.* Angel Island Association. Tiberon, CA, 1982.

Baron, Kristin L. and John A. Martini. *Fort Baker through the Years, the Post, the Park, the Lodge.* Hole in the Head Press. Bodega Bay, CA, 2011.

Kent, Matthew. *Harbor Defenses of San Francisco – A Field Guide 1890-1950,* 2nd Ed. www.blurb.com/bookstore/detail/2421994, 2011.

Thompson, Erwin N. *Historical Resource Study, Seacoast Fortifications of San Francisco Harbor,* Golden Gate National Recreation Area, California. Denver Service Center, Historic Preservation Division, US Dept. Interior, NPS. Denver, CO, 1979.

Battery Chamberlin, Fort Winfield Scott Golden Gate NRA (Mark Berhow)

THE HARBOR DEFENSES OF THE COLUMBIA RIVER —
OREGON AND WASHINGTON STATE

Explored by a number of British and American expeditions in the late 1700s and early 1800s, the area was claimed by both the British and Americans. The boundary settlement of 1846 put the Columbia River in American hands along with a network of goods and fur trade stations. Earthwork defenses were established during the Civil War by the U.S. Army to protect the commercial and military assets along the river navigable up to Portland. The defenses were upgraded during the Endicott Program with gun and mine defenses at three sites. The defenses were again upgraded with new defenses during the 1940s then closed down and declared surplus in 1947.

EDITION OF APR. 23, 1915.
REVISIONS NOV. 8, 1916; DEC. 15, 1919;
APR. 26, 1921; APR. 23, 1925;
OCT. 13, 1929; OCT. 1934

SERIAL NUMBER

MOUTH OF THE
COLUMBIA RIVER
OREGON AND WASHINGTON

ILWACO

NORTH HEAD L.H.
FORT CANBY

DATUM PT.
CANBY LT.

SAND ISLAND
DATUM PT.
DATUM PT.

FORT COLUMBIA

DATUM PT.

OCEAN

COLUMBIA RIVER

PACIFIC

DATUM PT.
DATUM PT.
DESDEMONA LT.
DATUM PT.
DATUM PT.

CLATSOP SPIT

DATUM PT.
PT. ADAMS

DESDEMONA SANDS

FORT STEVENS

ASTORIA

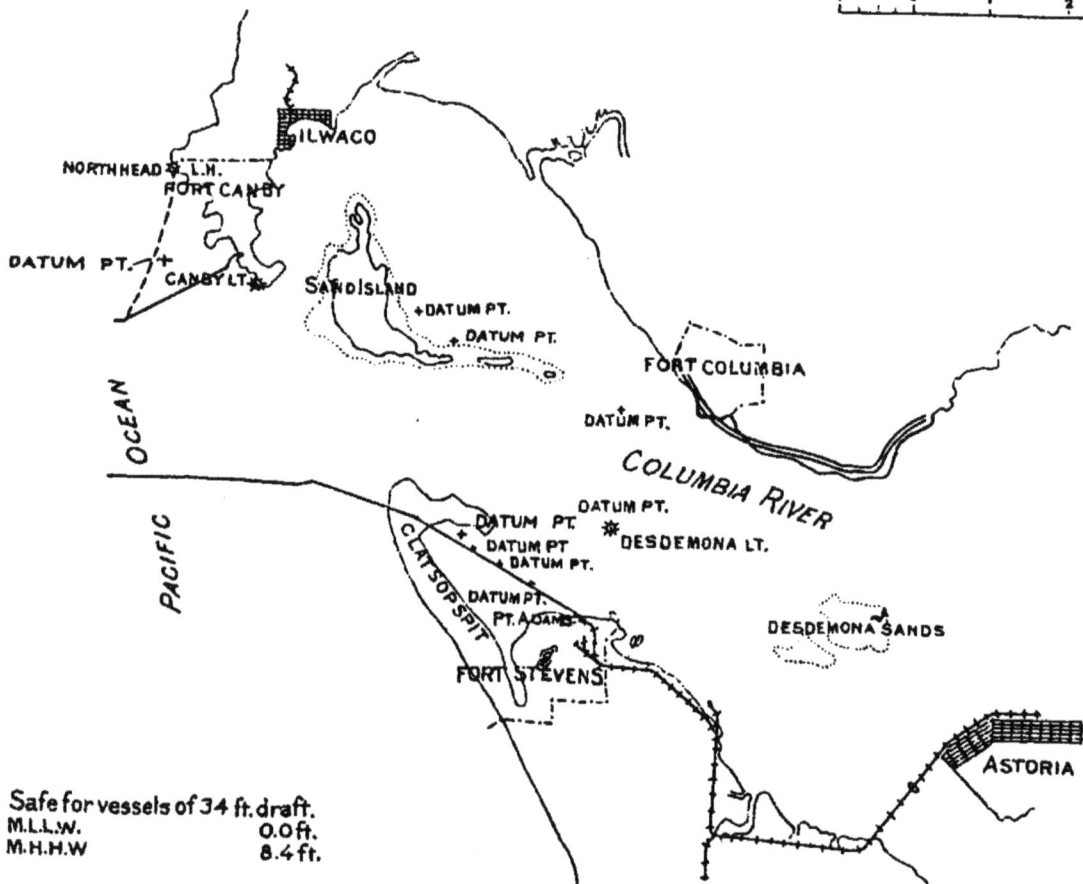

Safe for vessels of 34 ft. draft.
M.L.L.W. 0.0 ft.
M.H.H.W 8.4 ft.

Caretaking Status

Fort Canby (1864-1947) is located at Cape Disappointment, on the north side of the Columbia River at its mouth, about two miles southwest of Ilwaco, Washington State. Fort Canby mounted several muzzle-loading cannons in earthen batteries during the Civil War. The post was named in General Orders 5 of 1875 for Maj. Gen. Edward R.S. Canby, U.S. Army killed by Modoc Indians on April 11, 1873. The military reservation received two concrete batteries during the Endicott Program. In 1921, Battery Guenther was constructed to hold four 12-inch mortars from Battery Clark at Fort Stevens. This relocation was to allow for the mortars to provide better coverage of the mouth of the river. The battery marked the construction of the last permanent mortar battery in the United States. The 1940 Program saw the addition of a #200 Series battery with two 6-inch barbette guns (Battery Construction #247) at North Head and a 90mm AMTB battery on the Columbia River North Jetty. Part of reservation transferred to the Coast Guard and the rest to the State in 1950s. The State Park was developed in the 1960s. All garrison structures have been destroyed. State built the Lewis & Clark Interpretive Center on the parapet of Battery Allen in the 1980s. The park was renamed Cape Disappointment State Park in the 1990s. Battery O'Flyng and Battery Guenther are on Coast Guard property. Battery Allen and Battery 247 are open to the public. All Washington State parks require a daily parking pass.

Fort Canby Gun Batteries

- **ALLEN**: One of two 6-inch disappearing gun batteries built at the Cape Disappointment reservation of Fort Canby. The two batteries, one for three guns and one for two were submitted together on July 16, 1904. This battery on the western side of the reservation was for three guns. It was several hundred yards northeast of the lighthouse, on the site of an older Civil War earthwork. It had a crest of 205-feet, being high on the cape's bluff. It fired to the south, southwest. The plan followed conventional mimeograph type, with a wider, shared magazine in the traverse between emplacements No. 1 and 2, and a smaller, single magazine between No. 2 and No. 3. Gun centers were 125-feet between the first pair, and just 103-feet between the second set. It was built as one interior, and two flank emplacements. Work was done from August 1904 to December 1905. Transfer was made on February 27, 1906 at a cost of $58,935. It was named on General Orders No. 194 of December 27, 1904 for Lt. Colonel Harvey L. Allen, 2nd U.S. Artillery, who served in the Mexican and Civil wars. It was armed in 1908 with three 6-inch Model 1905 Watervliet guns in Model 1903 disappearing carriages (#13/#86, #17/#87 and #16/#88). In October 1917 it was recommended to disarm the battery to release gun tubes for wheeled carriages, which was implemented early in 1918. After the war in 1920, the gun tubes from Battery O'Flyng were moved to Battery Allen to replace the guns in emplacements No. 1 and No. 2 (6-inch Model 1905 guns on Model 1903 carriages #1/#86 and #11/#87). Leftover carriage #88 in emplacement No. 3 was soon scrapped. This armament served until dismounted by authority of November 1944, and the battery was deactivated subsequently on March 9, 1945. The emplacement still exists at the Fort Canby State Park, immediately adjacent to the Lewis and Clark visitor center. The battery is open to the public.

- **O'FLYNG**: The second 6-inch disappearing gun battery built at Fort Canby during the Endicott Program. The two batteries, one for three guns and one for two were submitted together on July 16, 1904. This battery of two guns was emplaced to the west of Battery Allen at the northern end of the old Civil War earthen Center Battery. The plan was of the standard mimeograph type, with two flank emplacements and a shared magazine in the center traverse. It had a crest height of 225-feet. On the right flank it had a substantial, straight retaining wall where the battery abutted a knoll. Work was done from August 1904 into mid-1905. Transfer was made on February 27, 1906 at a construction cost of $39,290. It was named in General Orders No. 194 of December 27, 1904 for

Mouth of the Columbia River
FORT CANBY
General Map
Scale of Feet

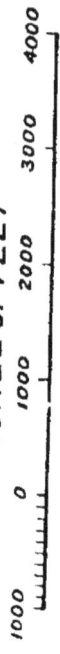

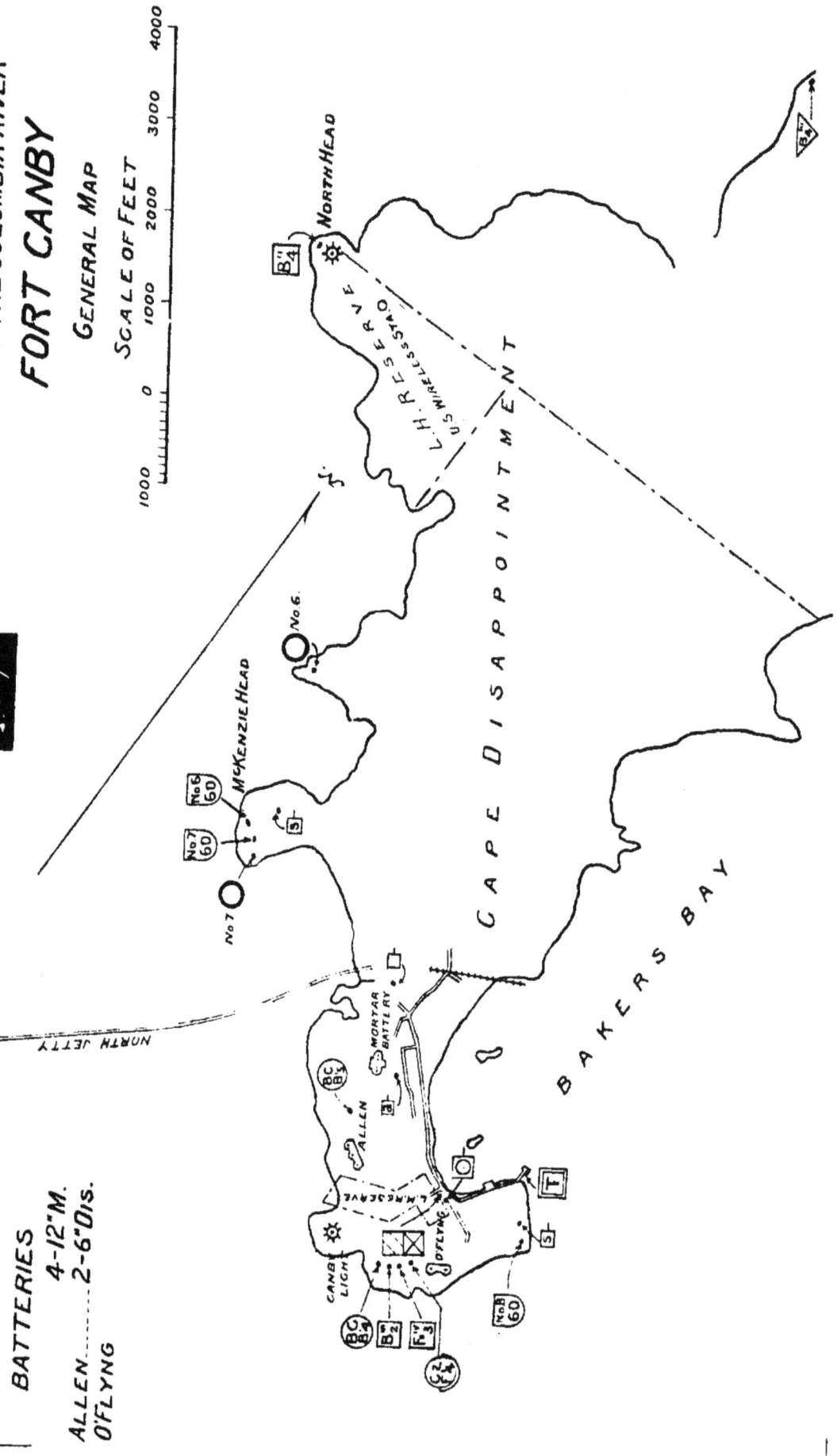

Serial Number 124

Edition of Apr. 26, 1921.

Batteries
4 - 12" M.
Allen ...... 2 - 6" Dis.
O'Flyng

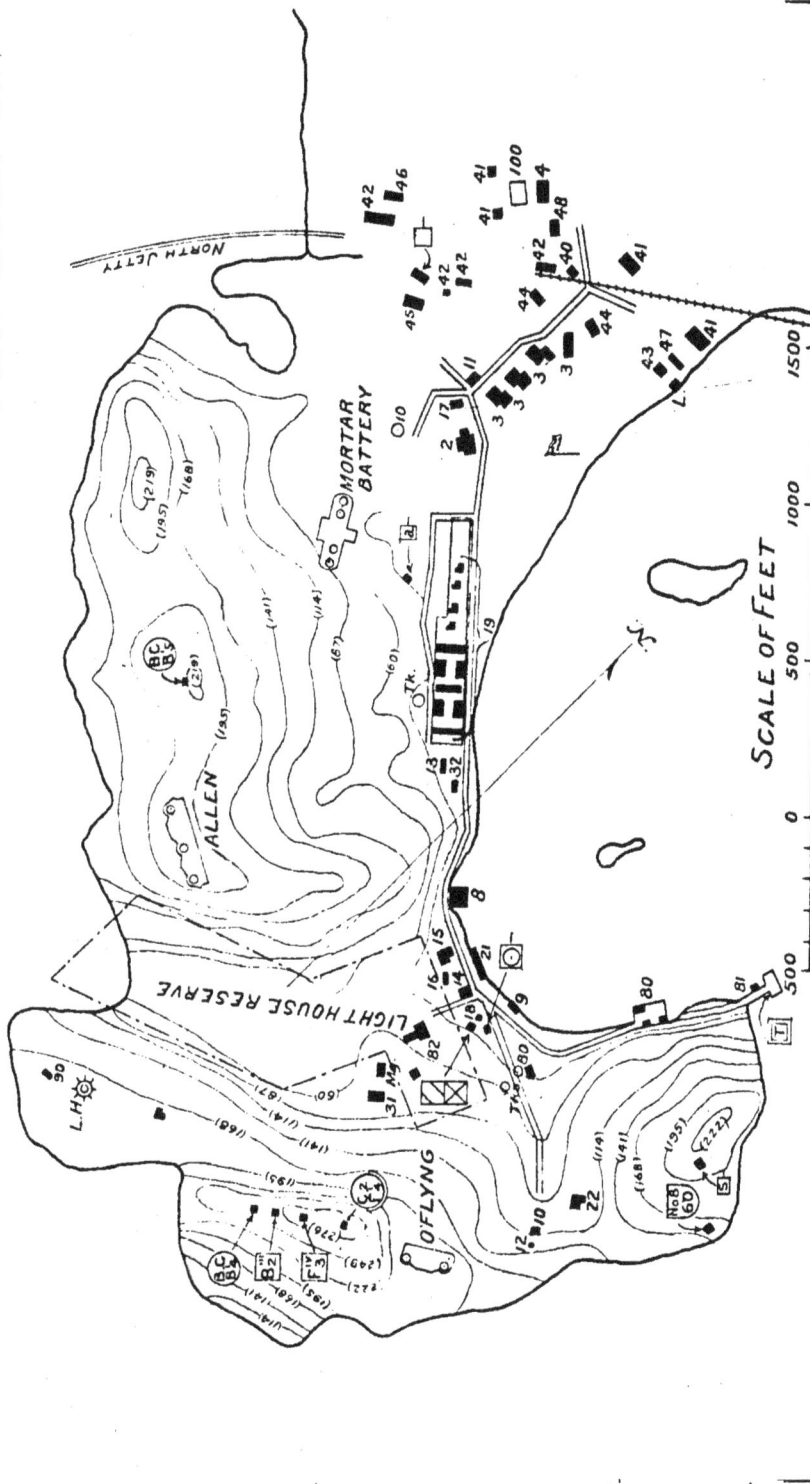

MOUTH OF THE COLUMBIA RIVER
FORT CANBY-DI.

SERIAL NUMBER 124

EDITION OF APR. 26,1921.

LEGEND

1.
2. COMDG OFFICERS QRS.
3. OFFICERS QRS.
4. HOSPITAL.
5.
6.
7.
8 GUARD HOUSE.
9 POST EXCHANGE.
10. RESERVOIR.
11. WINDMILL.
12 SPRING.
13. BAKERY.
14. WOOD SHED.
15. STABLE.
16. WAGON SHED.
17. SHED.
18. FORMER SW. BD. ROOM.
19. ELEVEN CANTMT. BLDGS.
100. FOUR CANTMT. BLDGS.
21. Q.M. & COM. ST. HO.
22. O. M. EMPLOYEES QRS.
31. ORDN. STOREHOUSE.
32. ORDN. SERGEANTS QRS.
40. ENGR. OFFICE.
41. ENGR. QUARTERS.
42. ENGR. STOREHOUSE.
43. ENGR. WASH ROOM.
44. ENGR. FOREMAN QRS.
45. ENGR. CARPENTER SHOP.
46. ENGR. STABLE.
47. ENGR. BUNK HOUSE.
48. ENGR. HOSPITAL.
80. COAST GUARD STATION.
81. COAST GUARD BOAT HOUSE.
82. L.H. KEEPERS QRS.
90. LOOKOUT STATION.

BATTERIES

ALLEN .... 4-12"M.
O'FLYNG ..... 2-6" DIS.

Fort Canby 1938 (NARA)

Fort Canby 1938 (NARA)

Cape Disappointment Washington (Terry McGovern)

Fort Columbia State Park (Terry McGovern)

Ensign Elijah O'Flyng, 23rd U.S. Infantry who died on September 18, 1814 at Fort Erie during the War of 1812. It was armed with two 6-inch gun Model 1905 on Model 1903 disappearing carriages (Watervliet tubes #1/#84 and #11/#85). While initially retained during the World War I years, the gun tubes were removed in 1920 and remounted in two of the recently disarmed emplacements of nearby Battery Allen. Subsequently the two empty carriages at Battery O'Flyng were scrapped. The emplacement was later used as a troop shelter during World War II. The emplacement still exists on the property of the Coast Guard Station, Cape Disappointment. The battery is not open to the public except with permission from the Coast Guard.

- **GUENTHER:** A late mortar battery built at the Cape Disappointment fort for four mortars relocated from Battery Clark at Fort Stevens. This project was initially proposed in early June of 1917. A more complete field of fire, particularly to the beaches to the north, was obtained with this move. At first the design was just a simple temporary emplacement for four-gun blocks in a line with light, temporary magazines were proposed, but soon the plan emerged as a substantial, much more permanent emplacement. The site was located in a ravine to the rear of the quartermaster quarters on the Baker Bay side of Cape Disappointment, and some 1300-feet north of Battery Allen. The new plan was submitted on March 7, 1918 for an estimated cost of $73,670. There were two pits each with two mortars (but all four still in a line, with a large traverse between the pairs). Each pit had its own magazines, and power and plotting rooms in the traverse. While incorporating many of the design features of the later Taft mortars, it was truly a unique design used only this one time. Work was started in June 1918, but completion was delayed for a variety of reasons to mid-1922. It was armed with four Model 1890M1 mortars and Model 1896 carriages removed from Battery Clark (Builders tubes #29/carriage #124 and #35/carriage #123, Watervliet tubes #48/carriage #119 and #52/carriage #120). It was named on General Orders No. 13 of March 27, 1922 for Brigadier General Francis L. Guenther, U.S. Army who died on March 25, 1920. The battery continued to serve with this armament through World War II. It was finally authorized for armament removal on August 12, 1942, and this was implemented in early 1943. The emplacement still exists but of limited access at the Coast Guard Station, Cape Disappointment. The battery is not open to the public except with the permission from the Coast Guard.

- **Battery #247:** A 1940 Program dual 6-inch barbette battery built to the northwest of the reservation on a cliff site known as MacKenzie Head. It was of typical 200-series type plan. Originally it was given a low national priority of #23 (on a list of September 11, 1940) and was not funded until the FY-1943 Budget. Work was done from February 9, 1943 to August 31, 1944 for final transfer on October 28, 1944 at a construction cost of $243,397. It was completed and armed with two 6-inch Model M1(T2) guns on Model M4 barbette carriages (#15/#3 and #17/#4). The battery was never named, just being known as Battery Construction No. 247. The battery served until the end of the war, but the armament was taken out and used in 1946 to arm a San Francisco battery just being completed. The emplacement was abandoned by June 30, 1947, but still exists as part of the Fort Canby State Park. The battery is open to the public.

- **AMTB Cape Disappointment:** A 1943 AMTB Program battery built on the sand near the southeastern tip of the Fort Canby reservation. It served after completion as local Tactical Battery No. 1. Work was done from June 10, 1943 to the end of that month. It was transferred on April 5, 1944 for a cost of $10,031.45. The battery was armed on December 24, 1943 with two 90mm M1 fixed guns mounted on two concrete blocks. No trace of the blocks remains on Coast Guard property.

Fort Columbia (1896-1947) is located at Chinook Point, on the north side of the Columbia River, about six miles inside its mouth, between the old Megler ferry landing and Chinook, Washington State. The 600-acre military reservation had three concrete batteries, and a mine casemate built during the Endicott Program. It was named in General Orders 134 of 1899 for its location on the Columbia River. The 1940 Program added a 200-Series battery with two 6-inch barbette guns. Property was transferred to State of Washington in 1950s. The historical day-use park features a nearly complete set of garrison buildings, three Endicott Program batteries, and a 6-inch shielded barbette battery from the 1940 Program. In 1994, Washington State Parks was able to arrange the relocation of two 6-inch shielded barbette guns from Fort McAndrews in Argentia, Newfoundland, Canada to Fort Columbia's Battery 246. Only six of these guns remain from the 1940 Program. The park and grounds are open for day use; all Washington State parks require a daily parking pass. The Fort Columbia Interpretive Center, a restored coast artillery barracks, is open daily from July to September

Fort Columbia Gun Batteries

- **ORD**: A battery of 8-inch disappearing guns emplaced as the primary heavy armament of the Chinook Point reservation of Fort Columbia. A plan of submission for two guns using the Model 1896 disappearing carriage was made on January 22, 1897. It was of type plan, with adjacent, internal type platforms and a shared magazine in the lower traverse between them. It was sited centrally on the reservation, firing to the southwest. Work began on July 7, 1897 and completed by that same October. The guns arrived on July 24, 1898, and both carriages by the end of April 1898. Transfer was made on July 16, 1898 for a cost of $137,298.79. It was armed with two 8-inch Model 1888M1 Watervliet guns on Model 1896 LF disappearing carriages (#20/#1 and #23/#2 later #10). It was named on General Orders No. 194 of December 27, 1904 for 1st Lieutenant Jules G. Ord, U.S. Infantry who was killed at San Juan Hill, Cuba in 1898. On mounting, the carriage in emplacement No. 2 overturned, breaking the elevating arm and damaging the chassis. The carriage (Pond Machinery #2) had to be returned to the manufacturer for repair and was replaced with carriage #10. Shortly after this construction was completed, a third single-emplacement for another 8-inch disappearing gun was built a short distance away on the left flank. When completed, it was commissioned as a separate battery as noted below. However, by 1906 it was tactically combined as the No. 3 emplacement of Battery Ord. The battery received the later common modifications made to many disappearing batteries, including in 1900 a large new battery commander's station with plotting room immediately behind the central traverse (costing $1,666 and transferred on November 29, 1900). In 1917 all three guns of the battery were ordered dismounted for use on railway carriages. This was accomplished by mid-1918, the carriages being left in place until scrapped in 1920. The emplacement still exists at the Fort Columbia State Park. The battery is open to the public.

- **NEARY**: An emplacement for a single 8-inch disappearing gun at Fort Columbia. Plans were submitted for the emplacement on July 25, 1897, using funds made available in the March 3, 1897 Fortification Act. Specifically, the chief of engineers directed that the emplacement utilize the original type Model 1894 experimental disappearing carriage. The 8-inch experimental carriage was directed to be used in an active emplacement once the evaluation trials had been performed, mostly in a false sense economy. The site was roughly in line with the two guns of Battery Ord and fired in the same direction. With the exception of having a unique base ring and racer to suit the experimental carriage, the emplacement was generally of the type design, with a single, lower-level magazine on the right flank. Concrete work was done from January to March 1898. It was transferred with Battery Ord on July 16, 1898 and included in the costs for that battery. It originally was armed with one

MOUTH OF THE COLUMBIA RIVER.

FORT COLUMBIA.

CHINOOK POINT, WASHINGTON.

GENERAL MAP.

BATTERIES

ORD --------
MURPHY -------- 2-6"
CRENSHAW --------

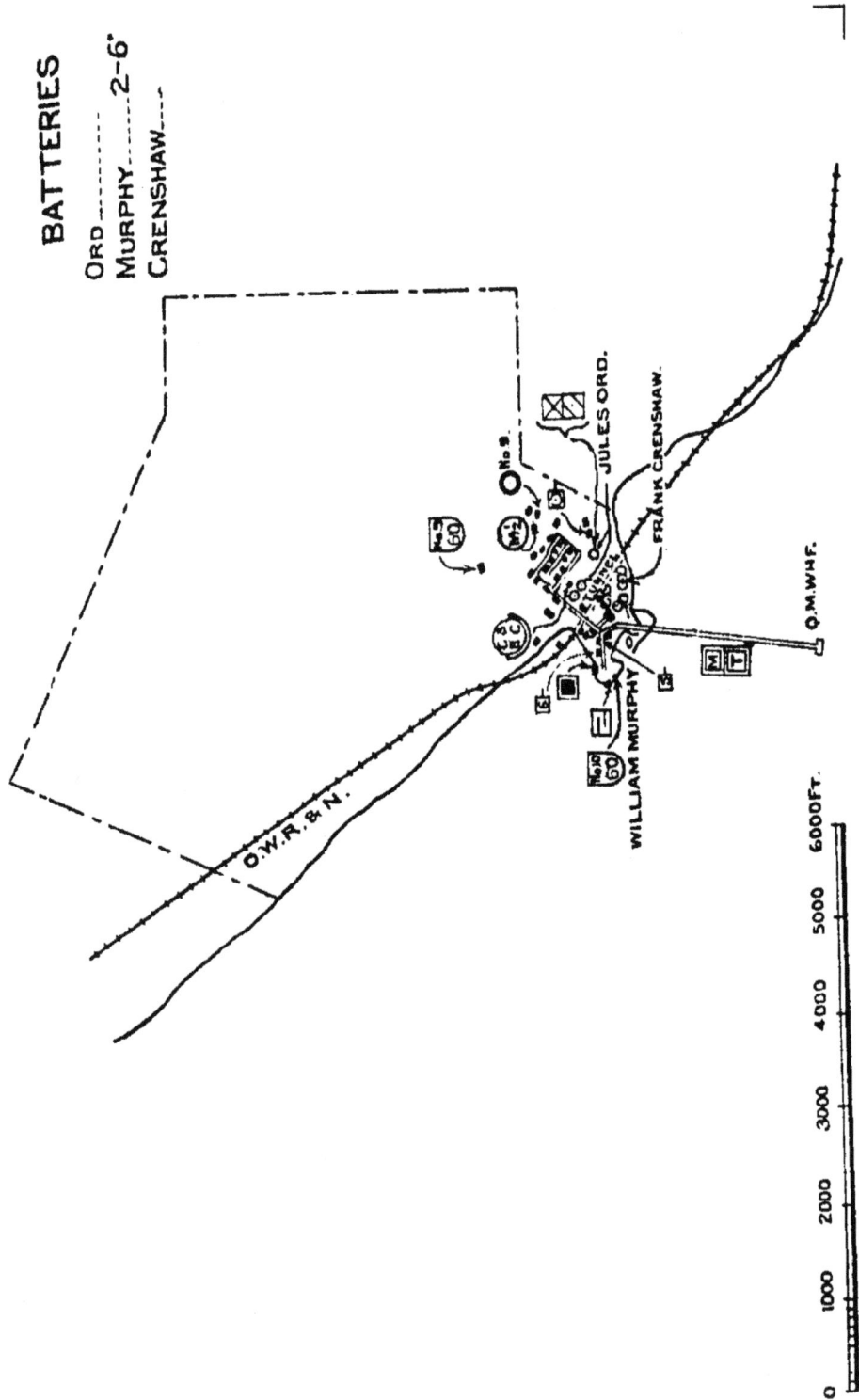

SERIAL NUMBER 124

EDITION OF APR. 23, 1915.
REVISIONS: DEC. 7, 1915;
NOV. 8, 1916; DEC. 15, 1919
APR. 26, 1921.

O.W.R. & N.

No. 9

JULES ORD.

FRANK CRENSHAW.

WILLIAM MURPHY

Q.M. WHF.

0   1000   2000   3000   4000   5000   6000 FT.

MOUTH OF THE COLUMBIA RIVER
FORT COLUMBIA-D-1.
CHINOOK POINT, WASHINGTON.

SERIAL NUMBER 124

EDITION OF APR. 26, 1921.

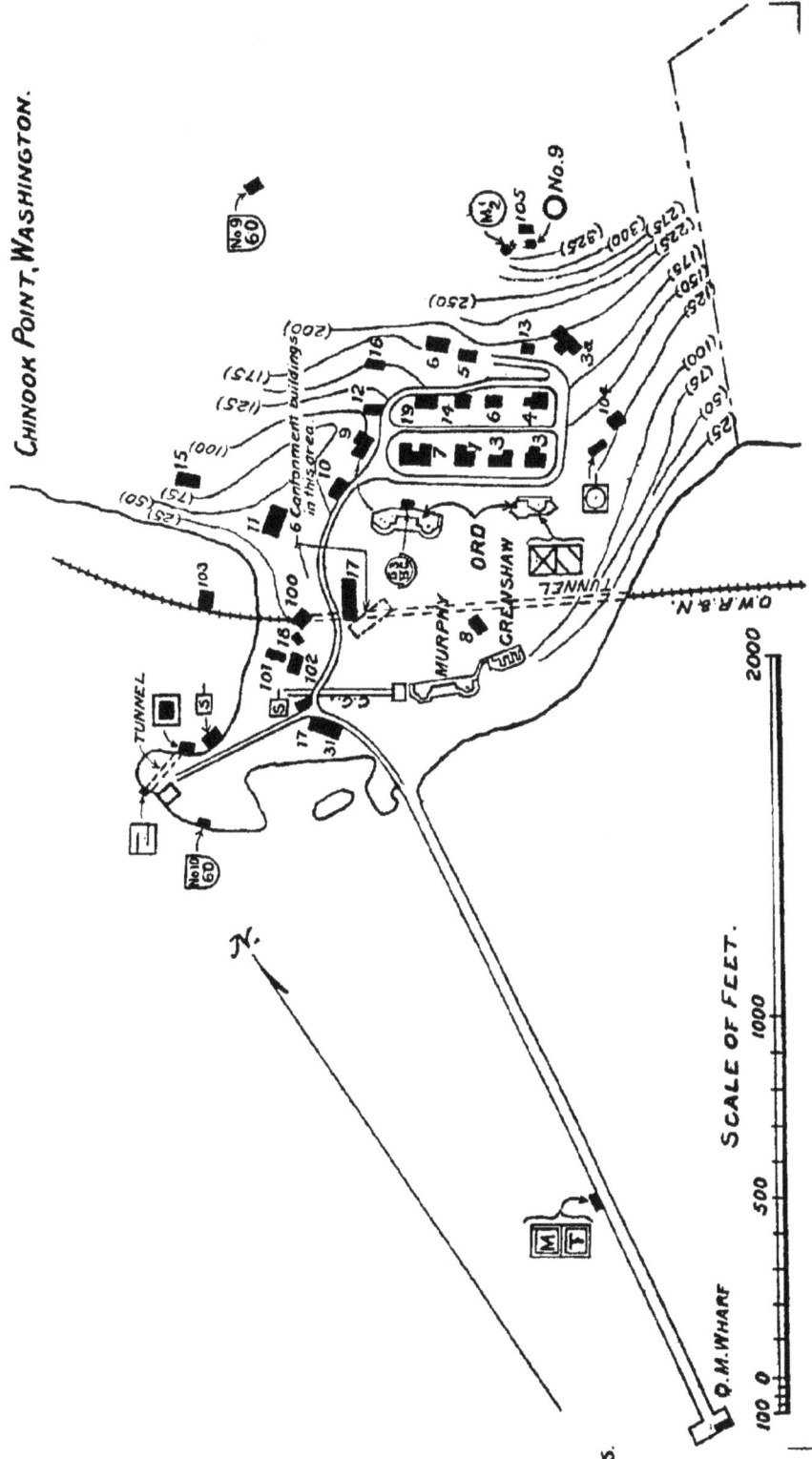

LEGEND

1. ADMINISTRATION BLDG.
2.
3. OFFICER'S QUARTERS.
3a. SURGEON'S "
4. HOSPITAL.
5. " STWD'S. QRS.
6. N.C. OFFICERS QRS.
7. BARRACKS.
8. GUARD HOUSE.
9. POST EXCHANGE.
10. OLD POST EXCHANGE.
11. GYMNASIUM.
12. FIRE APPARATUS HO.
13. RESERVOIR.
14. BAKERY.
15. LAUNDRY.
16. OIL HOUSE.
17. COAL SHED.
18. SHOP.
19. STOREHOUSE.
100. TEAMSTER'S HOUSE.
101. STABLE.
102. WAGON SHED.
103. DEPOT.
104. FORMER SW. BD. ROOM.
105. DORMITORY.
31. ORDNANCE ST. HO.

BATTERIES

ORD. ――
MURPHY. 2-6"Dis.
CRENSHAW. ――

N.

SCALE OF FEET.

100 0   500   1000   2000

Q.M. WHARF

Fort Columbia 1938 (NARA)

Fort Columbia 1942 (NARA)

8-inch Model 1888M1 Watervliet gun #44 on the experimental Model 1894 disappearing carriage which had no serial number. The battery was named in General Orders No. 194 of December 27, 1904 for 1st Lieutenant William C. Neary, 4th U.S. Infantry who was killed in Cuba in 1898. Apparently, the Coast Artillery service was never happy with the re-purposed experimental carriage. Plans were suggested in early 1898 to modify the carriage to M1894M1 standard, but this was not implemented. Instead in September 1908 it was replaced with a standard Model 1896 DC carriage. Modifications of the emplacement, block and racer were estimated at $5,064. This work was accomplished and by 1909 the emplacement had gun tube #44 mounted on LF 1896 disappearing carriage #32, which had previously been use at Battery Burnham at Fort Mason, San Francisco. Also, around 1906 the separate status of the battery was eliminated, and the gun became the No. 3 emplacement of Battery Ord. The gun was removed in 1918. The emplacement magazine was then modified to serve as the post switchboard room, and the original gun pit was filled-in for additional bombproof protection. Almost completely buried, only parts of the original gun emplacement still exist at the Fort Columbia State Park. The battery site is open to the public.

- **MURPHY**: A battery for two 6-inch disappearing guns authorized for Fort Columbia. The plans for the battery were submitted on April 10, 1899. It was located on the slope, in front of and below the crest of 8-inch Battery Ord. It was sited to have a crest of just 65-feet, so that Battery Ord with its 105-foot crest could safely fire over the 6-inch battery. It was of standard type plan for the Model 1897 disappearing gun type, but with its own included latrine and duplicate oil engine plant. Excavation was complete by September 1899; concrete work being done from October to January 1900. By the middle of 1900 it was complete with carriages, awaiting gun tube delivery. The battery was transferred on June 28, 1900 for $58,623.82. It was named in General Orders No. 194 of December 27, 1904 for Captain William L. Murphy, 39th Infantry, who was killed during the Philippine Insurrection in 1900. It was armed with two 6-inch Model 1897M1 guns on Model 1898 disappearing carriages (#15/#9 and #24/#10). It had a long service life, not being deactivated until January 26, 1944. It was disarmed by authority of July 20, 1945. The emplacement still exists at the Fort Columbia State Park. The battery is open to the public.

- **CRENSHAW**: An emplacement for three 3-inch, rapid-fire gun at Fort Columbia to cover the minefields. Originally it was intended for the rocky point on the southwest tip of the reservation. However local engineers argued that the site was too inaccessible and suffered erosion and switched the location to the left flank of 6-inch Battery Murphy. It was built on a site with a crest of 60-feet, still below the firing crest of the 8-inch battery. It also fired to the southwest. It was of typical type plan. Plans for the first two guns were submitted on April 10, 1899 using funds from the Act of July 7, 1898. The final gun on the left flank was submitted on August 5, 1900. Work was done from August 1899 to the fall of 1900. Transfer was made on June 28, 1900 for a cost of $15,462.61, the final gun was transferred on October 29, 1900. It was named in General Orders No. 194 of December 27, 1903 for Captain Frank F. Crenshaw, 28th Infantry, who was killed in 1900 during the Philippine Insurrection. It was armed with three 3-inch, 15-pounder Model 1898 Driggs-Seabury guns on masking parapet mounts (#97/#97, #98/#98, and #104/#104). The pillars were converted to Model 1898M1 pedestals in 1916. The battery was disarmed in June of 1920, and not subsequently used for armament. The emplacement still exists at the Fort Columbia State Park. The battery is open to the public.

- **Battery #246:** A 1940 Program dual 6-inch barbette battery emplaced between 8-inch Battery Ord and 6-inch Battery Murphy on the slope of Fort Columbia below the cantonment area. It is of generally standard design, but as one of the side-wrapped gallery type, with no rear entry. It was assigned a lower priority than the similar generation 200-series emplacements at Fort Stevens and Canby. Work was finally begun on October 28, 1942, and was turned over incomplete on February 2, 1945 for a to-date cost of $220,168.76. Two M4 barbette carriages were supplied (#17 and #18), but no gun tubes were ever supplied or mounted. The battery still incomplete, was abandoned in 1946. It was never named, simply being known as Battery Construction No. 246 during work. The emplacement was supplied with two 6-inch guns and shielded barbette carriages for public display purposes moved here from Construction No. 282 at Argentia, Newfoundland. The battery is open to the public at Fort Columbia State Park.

Fort Stevens (1852-1947) is located at Point Adams, on the south side of the Columbia River at its mouth, about nine miles west of Astoria, Oregon. An enclosed earthwork completed with several muzzle-loading cannons in 1865 as part of the Civil War defenses. It was named after Brevet Major General Isaac Ingalls Stevens, USV, killed during the Battle of Chantilly in 1862 (officially recognized in General Orders 2 of 1938). Eight concrete batteries were built during the Endicott Program on the 800-acre military reservation, including a unique 360 degree all around fire disappearing gun battery. Battery Mishler was design to fire on targets both in Pacific Ocean and in the Columbia River. The emplacement completely surrounded each of the two disappearing guns so from the surface only two large circular opening were visible. When the battery was converted to use as a HECP in 1941, the disappearing guns were removed, and these "pits" were given cement roofs and false floors. The fort also supported a controlled submarine mine complex and a large cantonment area. During World War II, Fort Stevens came under shellfire from a Japanese submarine's 5.5-inch deck gun on the night of June 21, 1942. While the forts' batteries did not fire back, the shelling did little or no damage to the fort. The primary significance of this shelling was the first to occur on continual American shores since the War of 1812. The 1940 Program saw the addition of a #200 Series battery with two 6-inch barbette guns (Battery Construction #245) and a 90mm AMTB battery at Clatsop Spit. The garrison buildings were sold to private owners after the post was closed. Fort Stevens State Park was created in the 1950s and eventually obtained the area around Battery Russell and the main gun line. This large state park is popular destination for camping, lake boating, swimming, beach access, and features an extensive bike and hiking trail system. The historical area features a museum, an active interpretation program, a replica earthwork, and a full-scale replica of one of Battery Pratt's 6-inch guns on a disappearing carriage. The park is open year-round, but the historical area is closed during the week during the winter season. A day pass is required for parking in the park.

Fort Stevens Gun Batteries

- **LEWIS – WALKER:** Part of the six-gun "West Battery" located at the Point Adams reservation of Fort Stevens. The four eastern guns of this battery were designed for 10-inch, limited-fire, disappearing carriages all firing to the north across the mouth of the Columbia River. The site was an open clearing about 1500-feet west of the old Civil War Fort Stevens. On July 22, 1896 the plans for the first three guns were submitted (emplacement No. 2, 3, and 4) for Model 1894 disappearing carriages. They generally followed the early style of type plans, with 124-foot gun centers, all magazines at the same lower level, and use of platform lifts and cranes for ammunition. On January 2, 1897 authority was granted to submit the final emplacement No. 1. Located on the right flank, it was designed for the later Model 1896 disappearing carriage and was oriented for greater field

MOUTH OF THE COLUMBIA RIVER.

FORT STEVENS

POINT ADAMS, OREGON.

GENERAL MAP.

Caretaking Status.

SERIAL NUMBER

EDITION OF APR. 23, 1915.
REVISIONS: DEC. 7, 1915;
NOV. 8, 1916; DEC. 15, 1919;
APR. 26, 1921; APR. 23, 1923; APR. 23, 1925;
OCT. 15, 1928; OCT. 1934

BATTERIES

```
CLARK------4-12'
RUSSELL----2-10'
† LEWIS------
† MISHLER-----2-10" "
† WALKER-----
† FREEMAN----
  PRATT------2-6"Dis
† SMUR ------
```

† *Armament removed*

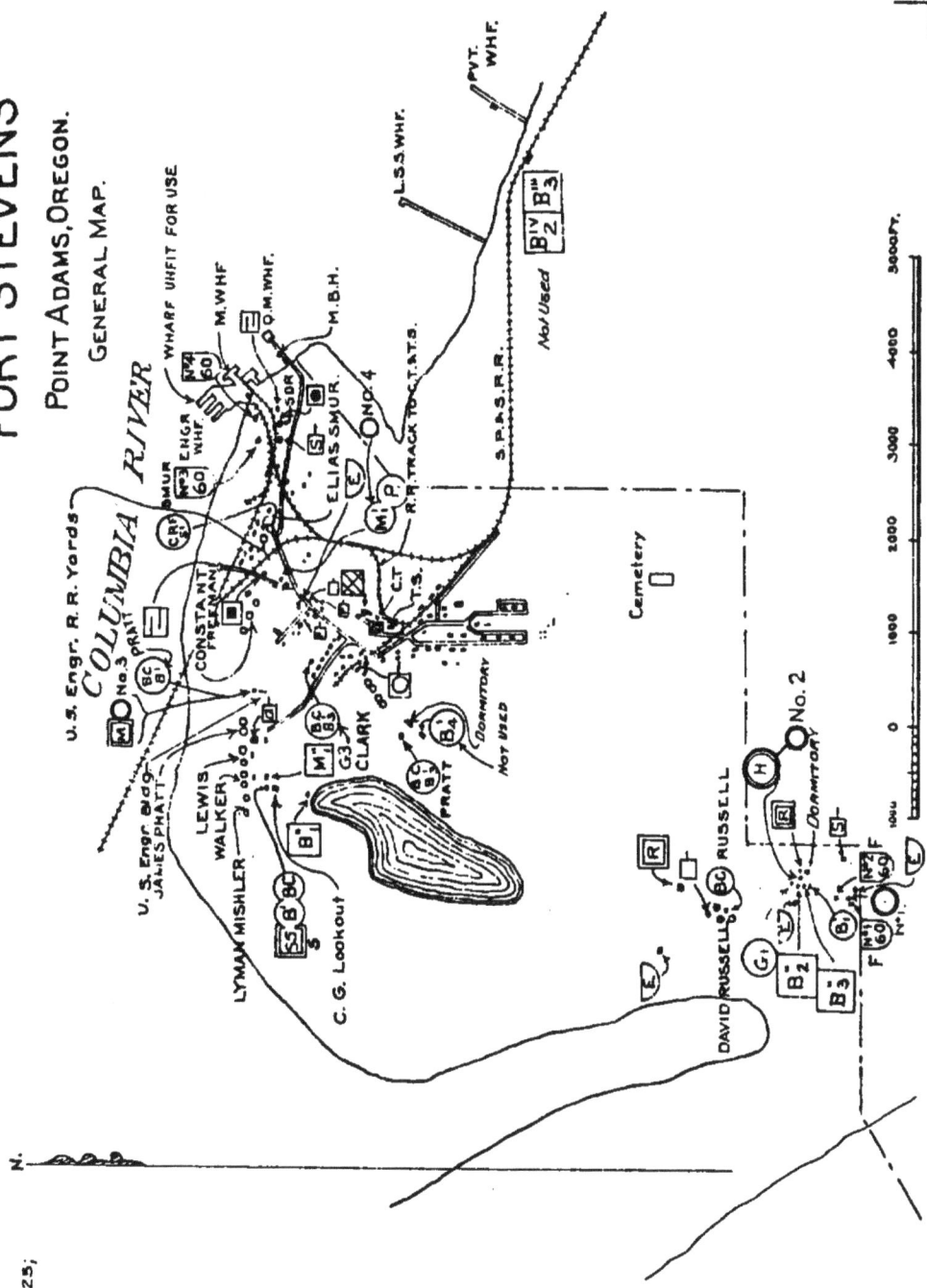

COLUMBIA RIVER

U.S. Engr. R.R. Yards

WHARF UNFIT FOR USE

M. WHF.

PVT. WHF.

L.S.S. WHF.

Q.M. WHF.

M.B.H.

Not Used

S.P.& S.R.R.

R.R. TRACK TO C.A.T.S.

Cemetery

DORMITORY

Not Used

No. 2

No. 1

Dormitory

DAVID RUSSELL

C.G. Lookout

CONSTANT FREEMAN

LEWIS WALKER

LYMAN MISHLER

U.S. Engr. Bldg.
JAMES PRATT

CLARK

PRATT

RUSSELL

MOUTH OF THE COLUMBIA RIVER

FORT STEVENS-D 1.

POINT ADAMS, OREGON.

SERIAL NUMBER

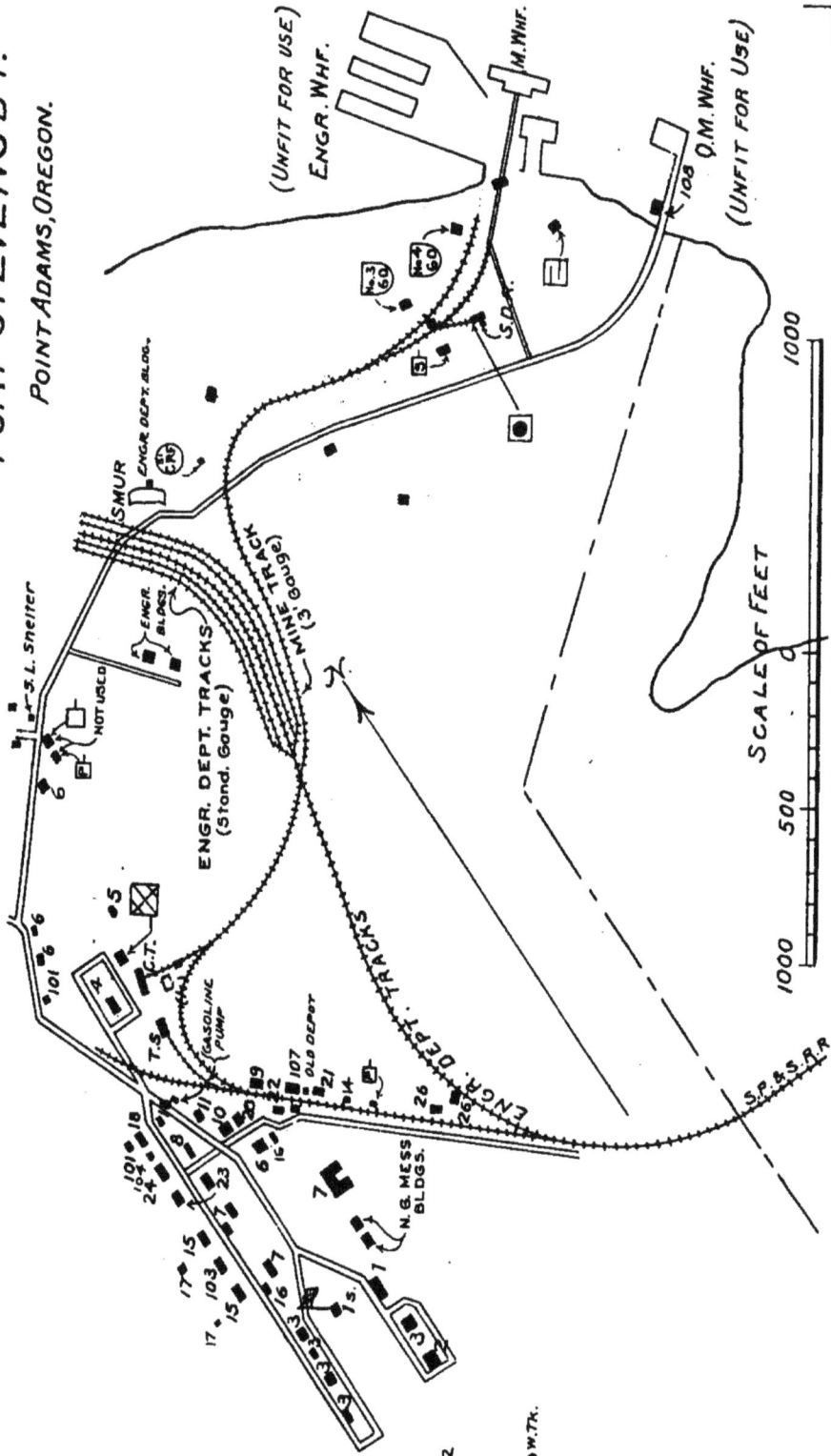

EDITION OF APR. 26, 1921.
REVISIONS: APR. 23, 1925.
OCT. 15, 1928; OCT. 1934.

ENGR. DEPT. TRACKS (Stand. Gauge)

MINE TRACK (3' Gauge)

ENGR. DEPT. TRACKS

S.P. & S.R.R.

SCALE OF FEET

1000   500   0   1000

Caretaking Status.

(UNFIT FOR USE)
ENGR. WHF.

M. WHF.

Q.M. WHF.
(UNFIT FOR USE)

ENGR. DEPT. BLDG.

S.L. Shelter

SMUR

NOT USED.

GASOLINE PUMP

OLD DEPOT

N.B. MESS BLDGS.

W.TK.

LEGEND

1. ADMINISTRATION BLDG.
2. COMMANDING OFF. QRS.
3. OFFICERS QRS.
4. HOSPITAL.
5. HOSPITAL STWD'S. QRS.
6. N.C. OFFICERS QRS.
7. BARRACKS.
8. GUARD HOUSE.
9. POST EXCHANGE.
10. GYMNASIUM.
11. FIRE HOUSE.
14. PLUMBER SHOP.
15. LAVATORIES.
16. FUEL SHED.
17. COAL SHED.
18. WOOD SHED.
101. SHOP.
103. MESS HALL.
104. WAGON SHED.
107. DEPOT & POST OFFICE.
108. BOAT HOUSE.
20. Q.M. OFFICE.
21. Q.M. MESS. (NOT USED).
22. Q.M. & COMMISSARY ST. HO.
23. Q.M. EMPLOYEES QRS.
24. Q.M. STABLE.
25.
26. Q.M. STOREHOUSE.
72. SCHOOL HOUSE.

BATTERIES.

† SMUR - - - - - -

† Armament removed

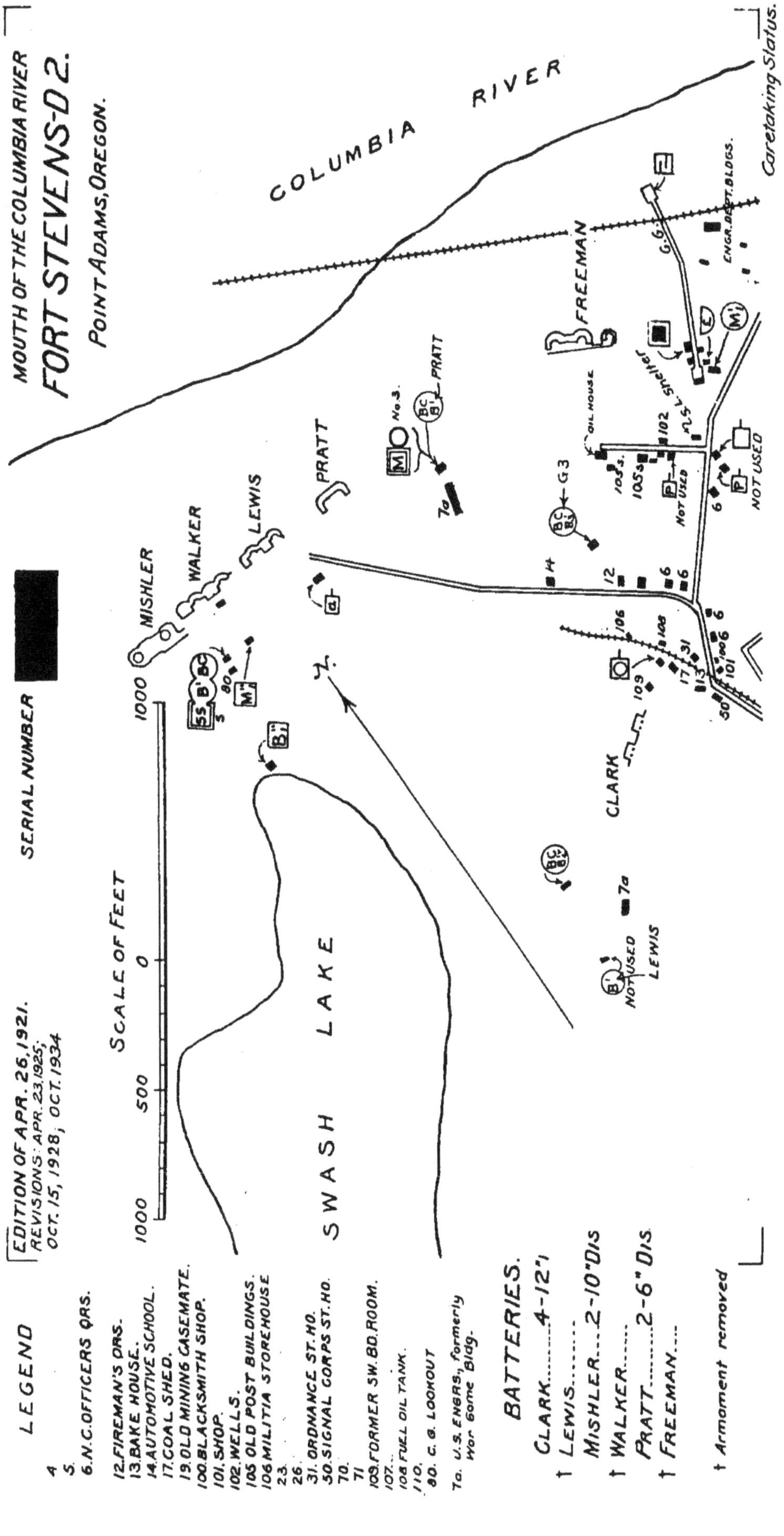

MOUTH OF THE COLUMBIA RIVER
FORT STEVENS-D 2.
POINT ADAMS, OREGON.

SERIAL NUMBER

EDITION OF APR. 26, 1921.
REVISIONS: APR. 23, 1925;
OCT. 15, 1928; OCT. 1934

SCALE OF FEET
1000 500 0 500 1000

COLUMBIA RIVER

SWASH LAKE

FREEMAN
PRATT
MISHLER
WALKER
LEWIS
CLARK

OIL HOUSE
U.S.L. shelter
NOT USED
ENGR. DEPT. BLDGS.
Caretaking Status.
No.3

LEGEND
4.
5.
6. N.C.OFFICERS QRS.
12.FIREMAN'S QRS.
13.BAKE HOUSE.
14.AUTOMOTIVE SCHOOL.
17.COAL SHED.
19.OLD MINING CASEMATE.
100.BLACKSMITH SHOP.
101.SHOP.
102.WELLS.
105 OLD POST BUILDINGS.
106 MILITIA STOREHOUSE
23.
26.
31. ORDNANCE ST. HO.
50.SIGNAL CORPS ST. HO.
70.
71.
103.FORMER SW. BD. ROOM.
107...
108 FUEL OIL TANK
110.
80. C.G. LOOKOUT
7a. U.S. ENGRS., formerly War Game Bldg.

BATTERIES.
CLARK.....4-12"I
† LEWIS.........
MISHLER....2-10" DIS
† WALKER......
† PRATT.......2-6" DIS
† FREEMAN......
† Armament removed

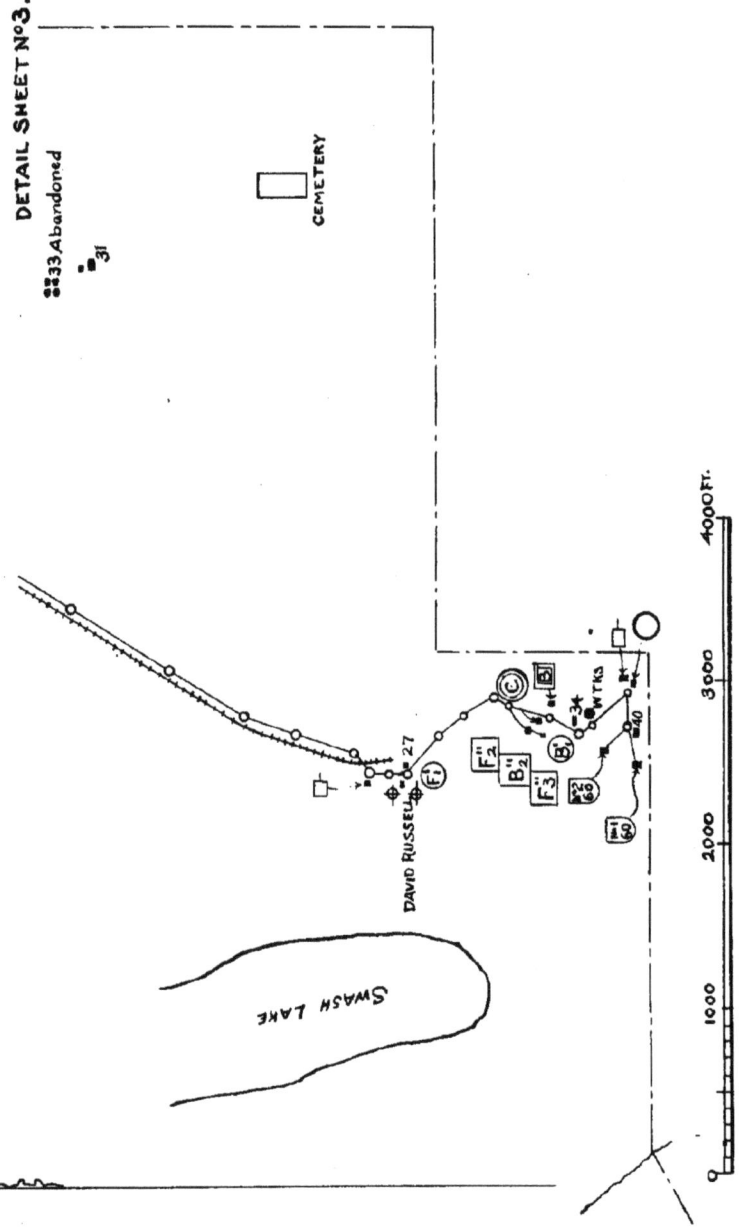

MOUTH OF THE COLUMBIA RIVER
FORT STEVENS-D3.
POINT ADAMS OREGON.
DETAIL SHEET N°3.

SERIAL NUMBER **128**

EDITION OF APR.23,1915.
REVISIONS: DEC.7,1915
NOV. 8,1916.

LEGEND
27. WOOD SHED
31. SCHOOL HOUSE
33. WELLS (ABANDONED)
34. DORMITORY
40. OUTLOOK BOOTH FOR No.1 S.L.

BATTERIES.
RUSSELL.....2-10"DIS.

33 Abandoned
31

CEMETERY

DAVID RUSSELL

Swash Lake

4000 FT.
3000
2000
1000

Fort Stevens 1938 (NARA)

Fort Stevens 1938 (NARA)

Batteries Lewis, Walker, Mishler and #245, Fort Stevens State Park (Terry McGovern)

Battery Russell, Fort Stevens State Park (Terry McGovern)

of fire on this flank. Work was done in 1896-1897. As soon as 1899 modifications were made to allow better communications between platforms. Loading platforms were widened, new battery commander stations added, and in 1905 the battery was equipped with chain hoists. The three early guns were mounted in July 1897, the fourth in April 1898. Transfer came on April 3, 1898 for a cost of $148,014.67 ($118,609 for No. 2 through No.4, and $29,406 for No. 1). In General Orders No. 43 of April 4, 1900, the four emplacements were named Battery Lewis for Captain Meriwether Lewis of the Lewis and Clark Expedition. Then on January 28, 1909 in General Orders No. 15 Emplacements No. 3 and 4 were split off and named Battery Walker for Colonel Leverett H. Walker, an early Coast Artillery officer. The batteries were armed with four 10-inch Model 1888 guns on three Model 1894 and one Model 1896 disappearing carriage (Bethlehem tube #11/carriage M1896 #4, Watervliet tube #49/carriage M1894 #3, Watervliet tube #52/carriage M1894 #4, and Watervliet tube #48/carriage M1894 #5). Apparently, an accident in mounting damaged one carriage, but it was repaired on site and placed into service. All four guns were listed for removal in 1917, the tubes being sent away in early 1918. They were never replaced and the carriages scrapped in place in 1920. The deteriorating emplacements still exist at Fort Stevens State Park. The battery is closed to the public due to safety concerns.

- **MISHLER**: The final two western guns of the new West Battery were designed with unusual Model 1896 all-round firing disappearing carriages. The exposed southern flank of the Point Adams spit could possibly be occupied by enemy warships, and guns capable of firing to the south were needed. The special Model 1896 ARF carriages would fulfill that requirement. Original letters from the Chief of Engineers called for two emplacements of this type here on June 4, 1898. Plans were submitted on August 4, 1898. To provide the required 225-degree field of fire, West Battery emplacements No. 5 and 6 received two of only three such carriages fabricated, and was the only battery truly designed for full utilization of this mount. Two equivalent emplacements featured the centrally placed carriage, capable of being fully traversed 360 degrees. There was a rear entrance with covered postern, and passages and stairways to each gun loading platform and adjacent magazine. The batteries were "sunk," appearing only as open pits from the ground surface. They were substantially more expensive to build. Concrete work was begun on October 28, 1898 and finished by May 27, 1899. Guns and carriages were received in November and the battery finished with various details in mid-1900. Transfer was made on June 28, 1900 for a cost of $130,000. It was armed with two 10-inch Model 1888 Watervliet guns on Model 1896 ARF carriages (#33/#2 and #32/#3). The battery was named in General Orders No. 20 of January 25, 1906 for Brevet Captain Lyman Mishler who was killed in action at the Battle of Valverde, NM, in 1862. Unlike its sister batteries of the gunline, Mishler continued to serve throughout the 1920s and 1930s. It was finally disarmed by authority granted on September 18, 1942. At that time the guns pits were roofed over for conversion of the battery to the Harbor Defense Command Post. The guns and carriages were in fact originally left inside but finally removed in 1945. As modified the emplacement still exists at the Fort Stevens State Park. The battery site is open, but the interior is closed to the public except for special tours.

- **RUSSELL:** A late Endicott 10-inch disappearing battery placed on the south and western side of the reservation, firing to the west. This was an extension to the fort, requiring new land purchase. It was located near the beach and had a 67-foot crest. Plans were submitted on December 26, 1902. At first a new all-round battery type was considered, but it was decided to use the model 1900 guns and Model 1901 carriages in a conventional emplacement. It generally complied with type plans, but the gun centers were increased to 138-feet (vs. 124-feet in the type plans) to allow a more generous allocation of rooms for tools, supplies and an internal plotting room. Work was done from June 1903 to August 1904 for transfer on August 12, 1904 for a cost of $125,000. A battery commander station was added and transfer occurred on September 13, 1904 for another $7,832.20. The battery was armed with two 10-inch Model 1900 Watervliet guns on Model LF 1901 disappearing carriages (#11/#4 and #4/#5). It was named on General Orders No. 194 of December 27, 1904 for Brevet Major General David A. Russell who was killed in action at the Battle of Opequon in, VA, 1864. There was a considerable delay in mounting the armament, the gun tubes not arriving until September 1907 and were not mounted until November. Russell remained a part of the local defenses for years. It was involved in the famous engagement with a Japanese submarine on June 21, 1942. The battery was not abandoned until completion of more modern works, authority coming on December 23, 1944, the guns not being taken out until early 1945. The emplacement still exists at the Fort Stevens State Park. The battery is open to the public.

- **CLARK:** The mortar battery for Fort Stevens. Plans were submitted on May 25, 1897 utilizing funds from the Act of March 3, 1897. The site selected was a tract of heavily forested land in the rear of the old quarters at Fort Stevens; a few hundred yards south of the main 10-inch gun line. It was composed of two four-mortar pits of the earlier small width type. Magazines were located under the front parapet and in the center traverse. Provision was made for 360 rounds/mortar in the magazines. Work was done in 1897-1898, the final platform being completed in November 1898. Transfer was made on January 17, 1899 for a construction cost of $71,546.63. It was named in General Orders No. 43 of April 4, 1900 for Captain William Clark of the Lewis and Clark Expedition. The original armament was eight 12-inch Model 1890M1 mortars on Model 1896 carriages (Builders tube #32/#121, Watervliet tube #48/#119, Watervliet tube #50/#122, Watervliet tube #52/#120, Watervliet tube #51/#125, Builders tube #35/#123, Watervliet tube #34/#133, and Builders tube #29/#124). Shortly after installation carriage #125 was wrecked with a broken racer and was immediately replaced with new carriage #237. In 1905 the battery got new telautograph booths. Plans were actively discussed about rebuilding the battery in 1912-1914, but ultimately that was never funded. Instead, a new mortar battery was built at Fort Canby and four of the mortars from Clark (#48/#119, #52/#120, #35/#123, and #29/#124) were authorized for transferred on July 18, 1918. The remaining four mortars continued to serve at Battery Clark until removed in 1942. The emplacement still exists at the Fort Stevens State Park. The battery is open to the public.

- **PRATT:** An Endicott battery for two 6-inch disappearing guns emplaced to the eastern flank of the 10-inch series. The plan was submitted on March 27, 1899. It followed type plans in design, but with an added storage battery room and latrine. The designed allowed a field of fire of 140-degrees. Work was done in 1899-1900. Transfer was made on June 28, 1900 for a cost of $58,886.52. It was named in General Orders No. 20 of January 25, 1906 for Brevet Captain James P. Pratt, who was killed in action at the Battle of Bethesda Church, VA, in 1864. The battery was armed with two 6-inch Model 1897M1 Watervliet guns on disappearing carriages Model LF 1898 (#4/#7 and #11/#8). It seems that the battery originally received gun #1 in error, for it was promptly returned to Sandy Hook in return for #4. The battery served with its armament until authorized for removal in November 1943 (tubes were soon removed, but the carriages apparently lingered on for some time). Recently a full scale reproduction gun and carriage have been fabricated and emplaced for display purposes at the Fort Stevens State Park. The battery is open to the public.

- **FREEMAN:** A mixed battery of 6-inch and 3-inch guns emplaced at the old Civil War earthwork of Fort Stevens. The project started by letter of May 4, 1899 for a 6-inch disappearing gun battery here but was suspended in June pending a decision on the type of mount to use. Initial funding occurred on December 2, 1899, but again instructions were issued to not start work. Finally, authority was given on June 3, 1901 along with $29,000 from the Act of March 1, 1901. Meantime, the first submission was for a battery of a single 3-inch masking parapet gun was made on June 7, 1900. It was intended to be on the right flank of the proposed 6-inch battery. The plan was typical for a type platform with a 140-degree field of fire. Work was done in 1900 and transfer made on November 12, 1900 for $5,398.05. It was armed with a single Model 1898 Driggs-Seabury gun on a masking parapet mount (#115/#115). Separated from this mount by 139-feet was the 6-inch battery. Its plan was submitted on May 23, 1901. Initial construction occurred late in 1901, being reported as complete on December 20, 1901. In common with other designs for the Model 1900 pedestal at that time, it had a single platform holding two adjacent guns. Ammunition was stored below and raised to the platform by hoists. However, in 1902, before the emplacement was armed, specification for emplacements changed. The battery was partially rebuilt according to plans issued on December 29, 1902 for a cost of $4,377. Two platforms were now used with one gun each, and the parapets adjusted for the new shield design. Transfer of the two sections was made on January 17, 1903 for a cost of $34,538.37. Both units were located within the confines of the Civil War work and entailed removal of much of the earthen wall and platforms to clear fields of fire. The 6-inch battery was armed with two 6-inch Model 1900 guns and pedestal mounts (#8/#24 and #4/#25). The battery was named in General Orders No. 194 of December 27, 1904 for Brevet Colonel Constant Freeman who served during the War of 1812. The 3-inch pillar mount was converted to a fixed pedestal Model 1898M1 in 1915-1916. The battery served for only a short while, the two 6-inch guns and carriages were moved in early 1917 to arm a new battery at Willapa Bay. The single 3-inch gun was dismounted in June of 1920 and scrapped. The emplacements were never rearmed. The battery was destroyed shortly before World War II in the late 1930s, no traces remain today.

- **SMUR:** A battery for two 3-inch masking parapet guns erected on the eastern side of the reservation covering the mine wharf and field on that flank. Plans were submitted on March 22, 1899. The design followed type plans for masking parapet batteries. Work was quickly done and completed in April 1900. Transfer was made on June 28, 1900 at a cost of $11,954.57. It was named in General Orders No. 194 of December 27, 1904 for Lieutenant Elias Smur of War of 1812 service. It was armed with two 3-inch, 15-pounder Model 1898 guns and balanced pillar mounts (#30/#30 and #35/#35). These were modified to M1898M1 pedestals shortly before World War I. The armament was removed in July of 1920 with the removal of all the balanced pillar ordnance items. The emplacement still exists at the Fort Stevens State Park. A M1898 gun barrel has been mounted in a display carriage in the right emplacement, another replica gun and carriage is located in the left emplacement. The battery is open to the public.

- **Battery #245**: A 1940 Program dual 6-inch barbette battery for Fort Stevens. It was to be emplaced to the west of Mishler, firing to the west. Building was begun under the FY-1943 Budget, but then work was pressed with some urgency. Actual concrete construction was done from October 28, 1942 to August 31, 1944. Transfer was made on October 28, 1944 for a cost of $248,578. It was armed with two 6-inch gun M1(T2) on Model M4 barbette carriages and shields (#26/#8 and #27/#7). The battery followed standard plans for 1940 Program 200-series emplacements. The armament was shipped here in June and then mounted in July 1944. The battery was never named, simply being known as Battery Construction No. 245. It served until abandonment by authority of June 30, 1947. The gun tubes were removed and sent away in January 1947 for installation in a San Francisco gun battery. The emplacement still exists today at the Fort Stevens State Park and has two ex-navy 5-inch/38 shielded guns emplaced for display as substitutes for the wartime coast artillery mounts. The battery is open to the public.

- **AMTB Clatsop Spit**: A 1943 Program AMTB battery of two 90mm fixed and two 90mm mobile guns. It was to be emplaced west of Battery #245 on the jetty sand spit (Clatsop Spit). Work on the gun blocks was done from June 10 to June 30, 1943 for arming on December 24, 1943. It was transferred on April 5, 1944 for a cost of $9,856.82. It consisted of the two fixed blocks and temporary ammunition storage with a wooden battery commander's station. It served until war's end. The blocks still exist at the Fort Stevens State Park. The battery is open to the public.

Full scale model of a 6-inch disappearing gun in Battery Pratt, Fort Stevens State Park (Mark Berhow)

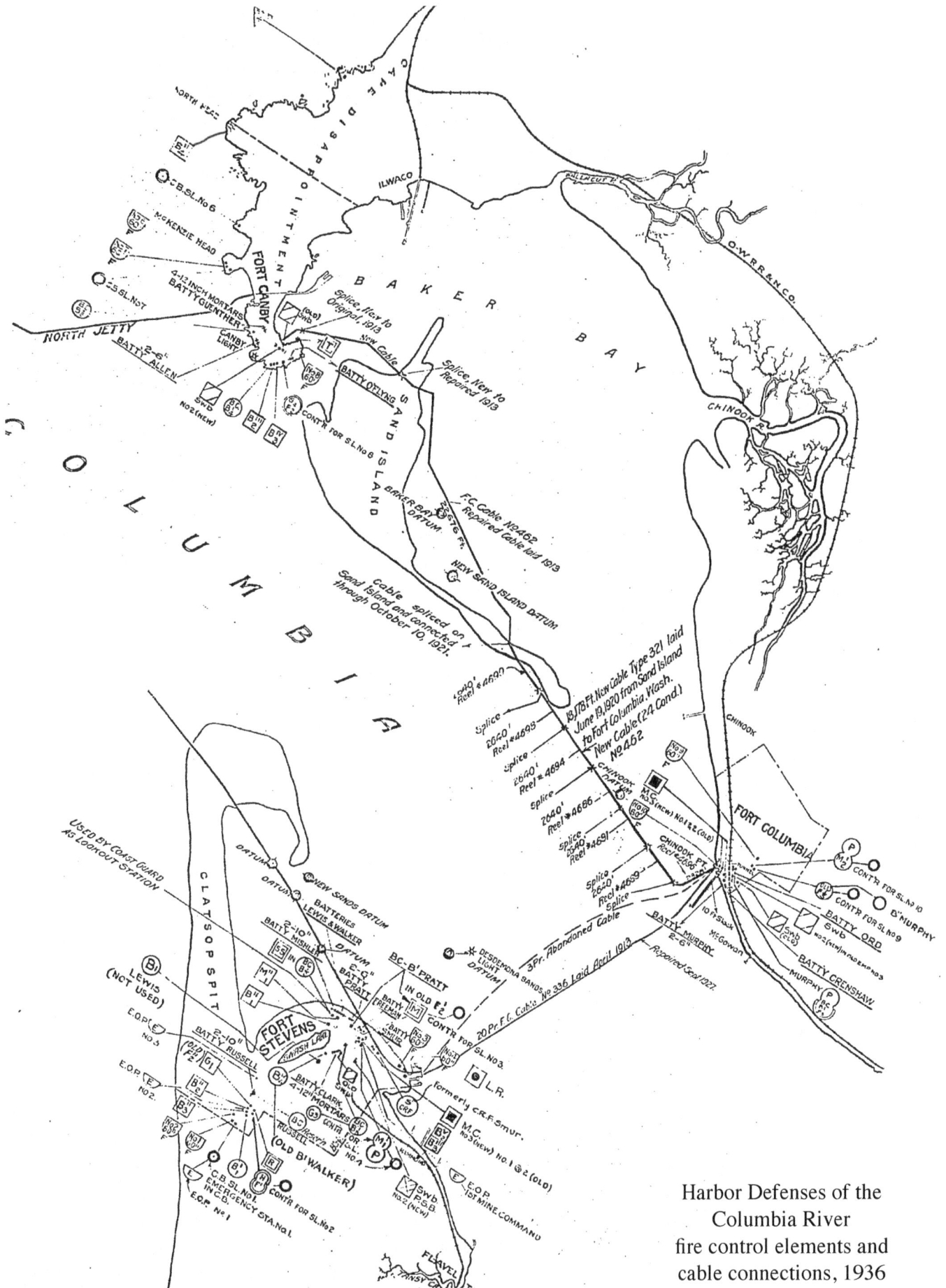

Harbor Defenses of the
Columbia River
fire control elements and
cable connections, 1936

RECLASSED UNCLASSIFIED
DOD DIR 5200 30

Ocean Park

Nahcotta

Klipson Beach

WILLAPA BAY

| Portable Mobile S.L. No.1 | E | 1 |
| Portable Mobile S.L. No.2 | E | A |

B$\frac{6}{5}$ S$\frac{7}{5}$ Btry. 247	E	
Portable Mobile S.L. No.3	E	1
Portable Mobile S.L. No.4	E	

SCR-296 Btry 246	E	
SCR-682		
B$\frac{7}{5}$ S$\frac{7}{5}$ Btry 245	E	
B$\frac{6}{5}$ S$\frac{7}{5}$ Btry 246	E	2
B$\frac{7}{5}$ S$\frac{7}{5}$ Btry 247	E	
Fixed S.L. No.5	E	
Fixed S.L. No.6	E	

Tioga

Long Beach

W A S H I N G T O N

| Btry No.1 - 2-6" BC (247) | E | |
| BCB S$\frac{1}{5}$ Btry 247 | E | 3 |

| SCR-296 Btry 247 | E | 3 A |

| Ft. Canby SWB Rm | E | 4 |

G.II Station	E	
B$\frac{5}{4}$ S$\frac{5}{4}$ Btry 245	E	
B$\frac{5}{4}$ S$\frac{4}{4}$ Btry 246	E	
Portable Mobile S.L. No.7	E	5
Portable Mobile S.L. No.8	E	
Fixed S.L. No.9	E	
HECP-HDCP Sig. Sta.	E	
Portable Mobile S.L. No.15	E	
Portable Mobile S.L. No.16	E	9
Portable Mobile S.L. No.17	E	A
Btry No.6 AMTB No.2	E	

HDCP - HECP	E	
Btry No.2 2-6" BC (245)	U	
CC B$\frac{5}{4}$ S$\frac{5}{4}$ Btry 245	E	9
B$\frac{5}{4}$ S$\frac{4}{4}$ Btry 246	E	
H.D. Amm. Storage Mag.	E	
S.L. Shelter	E	

| M$\frac{3}{4}$ Mines | E | 10 |
| HD Radio Trans. Sta. | E | |

Tide Station	E	
Mine Wharf	E	
Boat House	E	
Mine Loading Room	E	11
Mooring Basin	E	
Mine Store Room No 2	E	
Fixed S.L. No 13	E	
Fixed S.L. No 14	E	

| TNT Magazine No 1 | E | |
| TNT Magazine No 2 | E | |

Meteorological Station	E	
Mine Store Room No 1	E	13
Cable Tank House	E	
Ft. Stevens SWB Rm	E	

G.I. Station	E	
B$\frac{5}{4}$ S$\frac{5}{4}$ Btry 245	E	
B$\frac{5}{4}$ S$\frac{4}{4}$ Btry 246	E	
B$\frac{7}{5}$ S$\frac{5}{4}$ Btry 247	E	14
Fixed S.L. No 18	E	
Fixed S.L. No 19	E	
SCR-296 Btry 245	E	

CC S$\frac{5}{4}$ Btry 245	E	
B$\frac{5}{4}$ S$\frac{5}{4}$ Btry 246	E	15
B$\frac{7}{5}$ S$\frac{5}{4}$ Btry 247	E	

P A C I F I C   O C E A N

Seaview

Holman

Ilwaco

B A K E R   B A Y

SAND ISLAND

FORT CANBY

| M$\frac{3}{4}$ Mines | E | 6 |

| Fixed S.L. No.11 | E | 6 |
| Fixed S.L. No.12 | E | A |

BCB B$\frac{5}{4}$ S$\frac{5}{4}$ Btry 246	E	
B$\frac{6}{5}$ S$\frac{7}{5}$ Btry 245	E	7
Btry. No.4 MINES		
MCM$\frac{1}{4}$ & Plotting Rm.	E	

Btry No.3 2-6" BC 246	U	
Mine Casemate	E	8
Ft Columbia SWB Rm	E	

FORT COLUMBIA

Btry No.__ (AMTB No.1)	E	5
Portable Mobile S.L. No.10	E	A
SCR-547		

M$\frac{4}{4}$ Mines	E	
M$\frac{4}{4}$ Mines	E	5
Sand Island Military Res.	E	B
Hammond Military Res.	E	B A

C O L U M B I A   R I V E R

FORT STEVENS

Horn head

Warrenton

O R E G O N

Camp Clatsop

L E G E N D
U - UNDER CONSTRUCTION
E - EXISTING

YARDS
2000   0   2000   4000   6000

Revised

| Portable Mobile S.L. No.20 | E | |
| Portable Mobile S.L. No.21 | E | 16 |

B$\frac{5}{4}$ S$\frac{4}{4}$ Btry 245	E	
B$\frac{5}{4}$ S$\frac{5}{4}$ Btry 246	E	17
C$\frac{4}{4}$ S$\frac{4}{4}$ Btry 247	E	

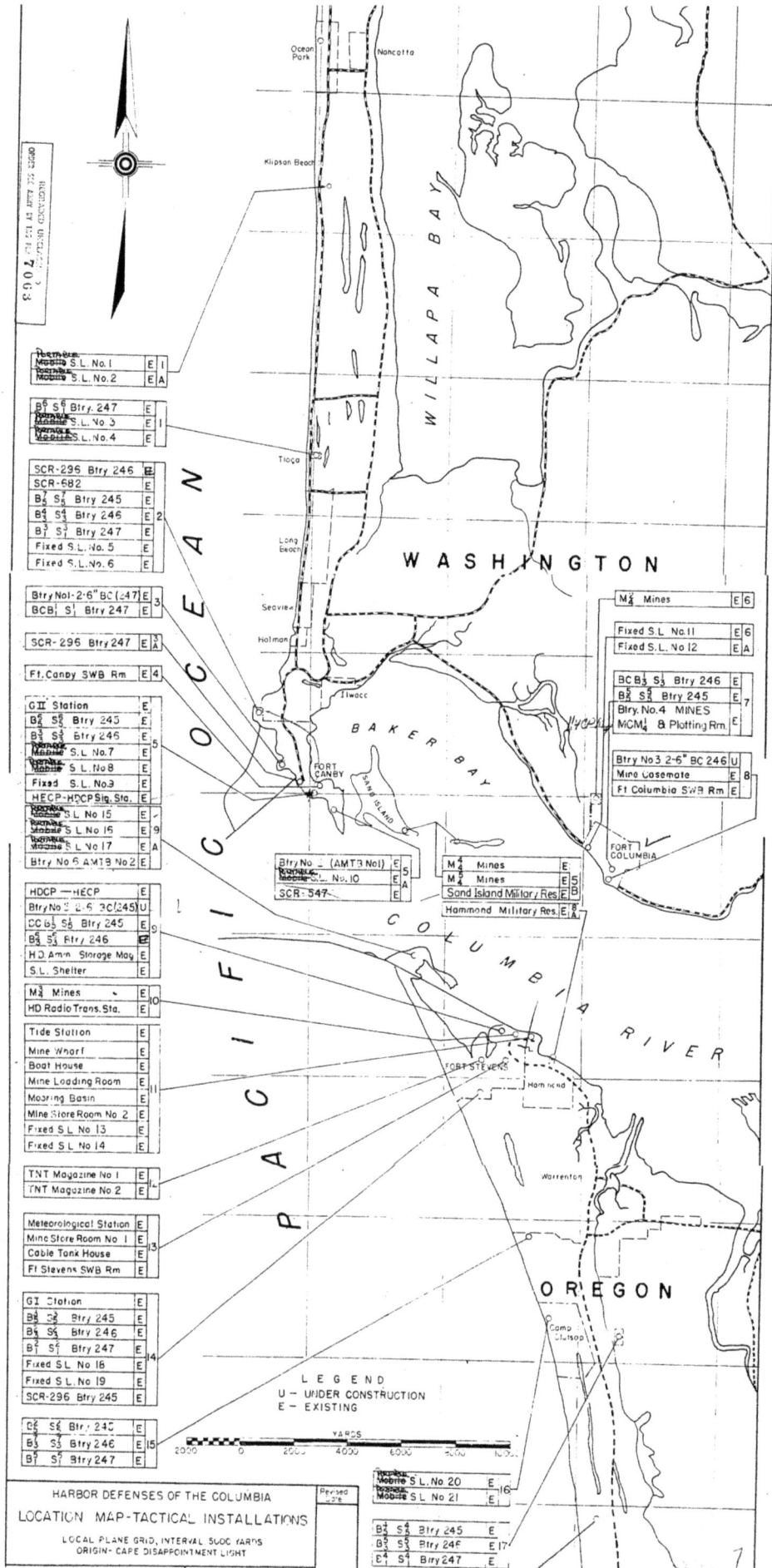

HARBOR DEFENSES OF THE COLUMBIA
LOCATION MAP-TACTICAL INSTALLATIONS
LOCAL PLANE GRID, INTERVAL 5000 YARDS
ORIGIN - CAPE DISAPPOINTMENT LIGHT

Columbia River World War II-era Site Locations. Stations housed in a single structure are connected by dashes (-)

location	Loc#	Purpose
Kilpsan Beach	1A	SL 1,2
Tioga	1	BS5/247, SL 3,4
North Head	2	BS7/246, BS4/245, BS3/247, SCR682, SCR296-246, SL 5,6
Mackenzie Head	3	Batt Tact. #1 BCN 247, BC-BS1/247
Fort Canby	3A	SCR296-247
Cape Disappointment	4	SWB
Cape Disappointment	5	BS6/245, B3/245, G2
Cape Disappointment	5A	Batt Tact. #2 AMTB 2, SL 7.8, AMTB 2-SCR547
Sand Island	5B	M5/4, M1/4
Chinook	6	M2/4
Fort Columbia	6A	SL 11,12
Fort Columbia	7	Batt Tact. #3 BCN 246, BC-BS1/246, MC3, M1/4, B5/245
Fort Columbia	8	Batt Tact. #4 Mines
Hammond MR	8A	not used
Fort Stevens	9	BS2/246, BS2/247, BS3/245, G1, SL 18,19
Fort Stevens	9A	AMTB1 BC, SL 15, 16, 17
Fort Stevens	10	M3/4
Fort Stevens	11	SL 20,21
Fort Stevens	12	military reservation
Fort Stevens	13	military reservation
Fort Stevens	14	Batt Tact. #5 BCN 245, Batt Tact. #6 AMTB 1, HDCP HECP, HDOP, B6/246, SCR296-245
Fort Stevens	15	BS2/245, BS5/247, BS7/246
Camp Clatsop	16	SL 20, 21
Columbia Beach	17	BS5/246, BS4/245, BS4/247
Gearhart	18	SL 22

Hanft, Marshall, *The Cape Forts: Guardians of the Columbia*. Oregon Historical Society. Portland, OR, 1973.

Hanft, Marshall. *Fort Stevens, Oregon's Defender at the River of the West*. Oregon State Parks and Recreation Branch. Salem, OR, 1980.

Hussey, John. *Chinook Point and the Story of Fort Columbia*. Washington State Parks and Recreation Commission. Olympia, WA, 1967.

Lucero, Donella J., and Hobbs, Nancy L. *Columbia River Forts and History of Fort Columbia*, privately published,

Lucero, Donella J., and Hobbs, Nancy L. *Columbia River Forts and History of Fort Canby*, privately published

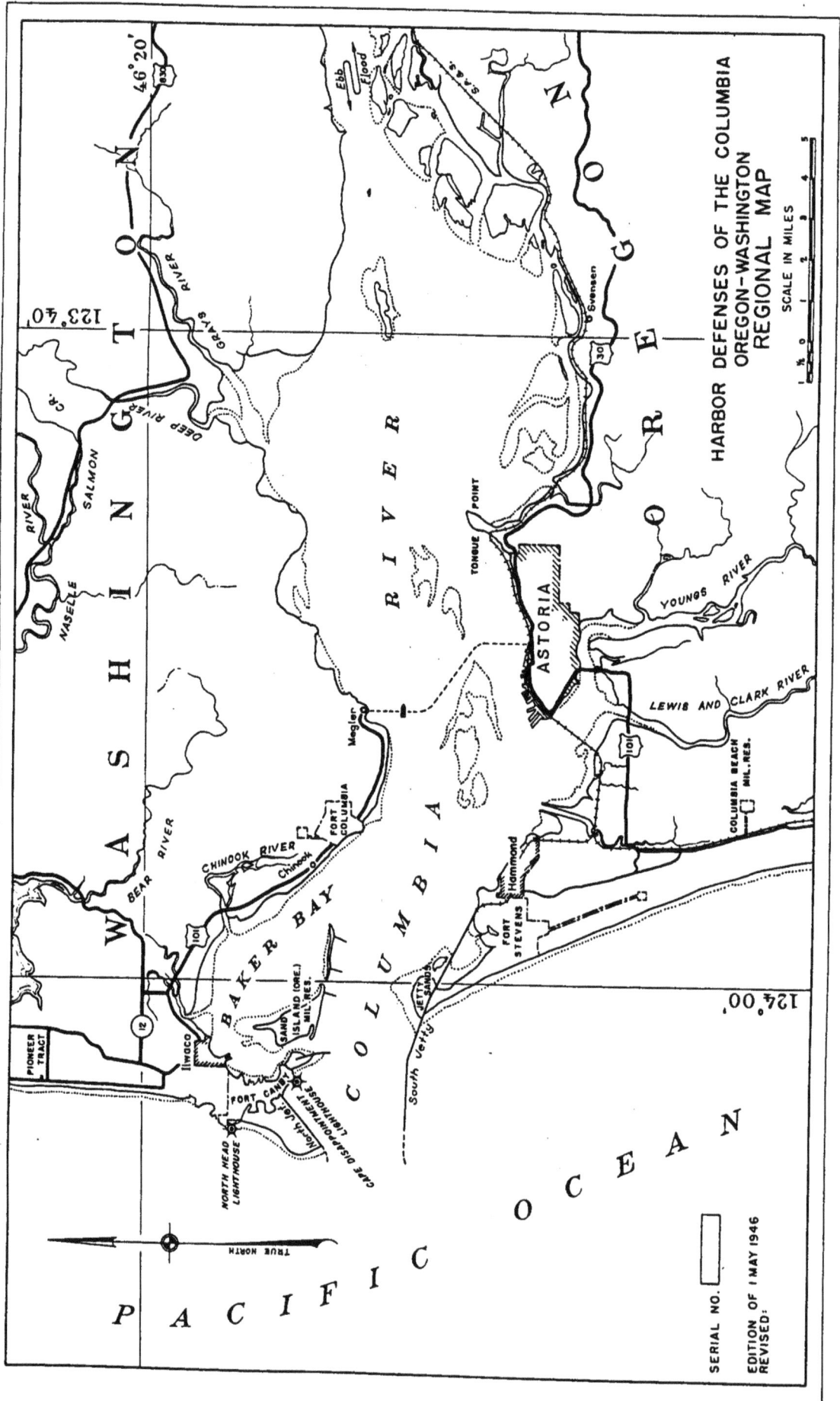

HARBOR DEFENSES OF THE COLUMBIA
OREGON-WASHINGTON
REGIONAL MAP

SCALE IN MILES

PIONEER TRACT

PACIFIC OCEAN

WASHINGTON

OREGON

COLUMBIA RIVER

BAKER BAY

Ilwaco

FORT COLUMBIA

Chinook

CHINOOK RIVER

BEAR RIVER

NASELLE RIVER

SALMON RIVER

DEEP RIVER

GRAYS RIVER

46°20'

123°40'

124°00'

ASTORIA

TONGUE POINT

YOUNGS RIVER

LEWIS AND CLARK RIVER

COLUMBIA BEACH MIL. RES.

Hammond

FORT STEVENS

SANDY ISLAND (ORE.) MIL. RES.

South Jetty

SAND ISLAND MIL. RES.

FORT CANBY

CAPE DISAPPOINTMENT LIGHTHOUSE

NORTH HEAD LIGHTHOUSE

TRUE NORTH

SERIAL NO.

EDITION OF 1 MAY 1946
REVISED:

1. Post Administration Bldg., IA. 249ᵗʰ Admin. Bldg.
1B. Q.M. Admin. Bldg., IC. Court Martial Admin. Bldg.
1D. Misc. Admin. Bldg., IE. Admin. and Supply.
1F. Military & Civ. Admin., IG. Storehouse & Admin.
1H. Ordn. & Post Engr. Admin., IJ. Ant. Engr. Administration.
2. Commanding Officer's Quarters.
3. Officer's Qtrs. 3A. Officer's Mess. 3B. Orderly Room
3C. Bachr. Officer's Qtrs, Mess.
4. Hospital.
5. Nurses Quarters.
6. Non-Commissioned Officer's Quarters.
7. Barracks, 7A. Isolation Barracks.
8. Guard House.
9. Post Exchange, 9A. P.X. Office, 9B. P.X. Beer Parlor.
10. Fire Station.
11. Oil House.
12. Storehouse.
13. Garage.
14. Recreation (Day Room), 14A. Officer's Recreation.
15. Latrine, 15A. Isolation Latrine
16. Kitchen-Mess.
17. Theater.
18. Artillery Shop.
19. Repair Shop.
20. Gasoline Station.
21. Gate House (Guard).
22. Wood Shed.
23. Tool House.
24. Civilian Personnel.
25. Bakery.
26. Dispensary.
27. Mess.
28. Office (Btry."B").
29. Motor Dispatch Office.
30. Civilian Bachelor Quarters.
31. Cistern.
32. Hose Tower and Shop.
33. Tool and Blacksmith Shop.
34. Machine Shop.
35. Drill Hall and Gym.
36. Signal Corps Repairs and Storage
37. Fuel Shed.
38. Non-Com. Officers Club.
39. Water Tank.
40. Post Office.
41. Motor Transp. Repair Shop.
42. Central Heating Plant.
43. Paint Shop.
44. Grease Rack.
45. Motor Pool Shop.
46. Gasoline Storage Tank.
47. Hydrogen Generator.
48. Waiting Room.
49. Dock., 49A. Ferry Slip.

50. Battery Mishler   (Disarmed)
51. Battery Walker
52. Battery Lewis   "   "
53. Battery Pratt   "   "
54. Battery Smur   "   "
55. Battery Clark   "   "
56. Battery Russell   "   "
57. Battery Ord   "   "
58. Battery Ord, Gun #3 "   "
59. Battery Crenshaw   "   "

70. Chapel

80. Navy Compass Station

100. Boat Storage.
101. Boat House.
102. Cold Storage Room.
103. Garbage and Can Washing.
104. Water Clarifier.
105. Wash Room.
106. Shoe Repair Shop.
107. Typewriter Repair Shop.
108. Gas Instruction.
109. Laundry.
110. Sales Commissary.
111. Scales.
112. Indoor Rifle Range.
113. Camera Tower.

500. Battery Murphy   (Disarmed)
501. Battery O'Flyng
502. Battery Guenther   "   "
503. Battery Allen   "   "
504. 90mm A.M.T.B. Bty.
505. 90mm A.M.T.B. Bty.

⊚	HDCP	Harbor defense command post.
○	HECP	Harbor entrance command post.
○	BC ≡	Battery command post.
○	EBC	Emergency battery command post.
◁	CB	Searchlight controller booth.
▣	B ≡, S ≡	Battery observation post, Spotting post.
■	M ≡	Mine observation post.
◪	M	Meteorological station.
⊠	T	Tide station.
⊡	SS	Signal station.
		Mine casemate.
⊕		Fire-control switchboard room.
		Combined F-C and post switchboard room.
		Cable terminal.
▭	P	Power house.
	S	Pumping plant.
	P-G	Searchlight power house.
		Pump-Generator house.
	CT	Cable tank.
	LR	Mine loading room.
	R	Radio.
⋔	T	Transmitter.
	S Sh	Searchlight shelter.
	SL	Searchlight, F, fixed, P, portable, M, mobile.
⌃	SCR	Surface craft detection station.

SERIAL NO. ▭

EDITION OF 1 MAY 1946
REVISED

HARBOR DEFENSES OF THE COLUMBIA
OREGON-WASHINGTON
LEGEND

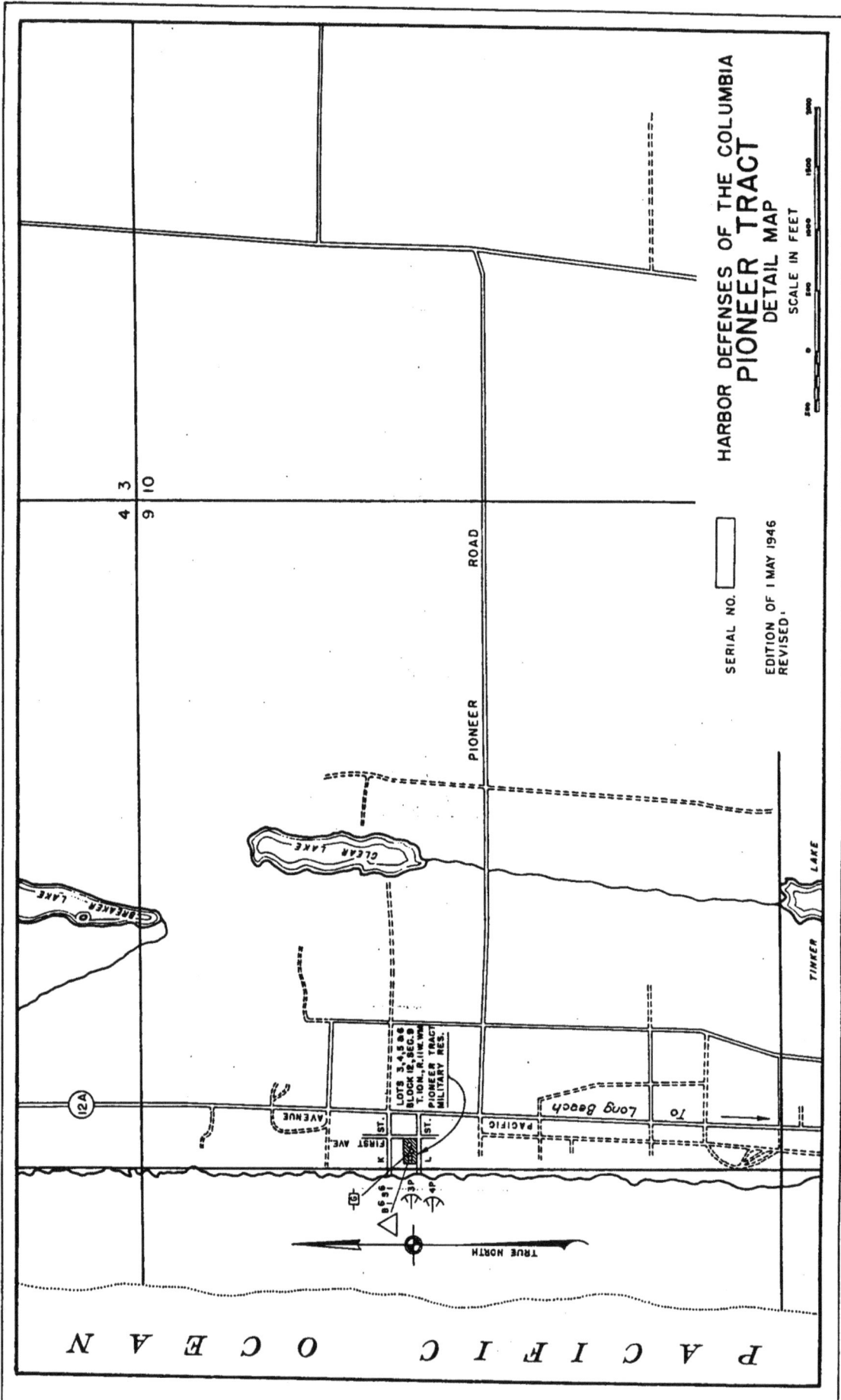

HARBOR DEFENSES OF THE COLUMBIA
PIONEER TRACT
DETAIL MAP

SCALE IN FEET

SERIAL NO.

EDITION OF 1 MAY 1946
REVISED:

PIONEER ROAD

PIONEER

CLEAR LAKE

TINKER LAKE

SNEAKER LAKE

LOTS 3,4,5,6
BLOCK 12,SEC.9
T.10N.,R.11W.WM.
PIONEER TRACT
MILITARY RES.

AVENUE

FIRST AVE.

K ST.

L ST.

PACIFIC

To Long Beach

12A

TRUE NORTH

P A C I F I C   O C E A N

BAKER BAY

ISLAND

RIVER

SAND

(OREGON)

COLUMBIA

DIKE III.2

DIKE III.8

TRUE NORTH

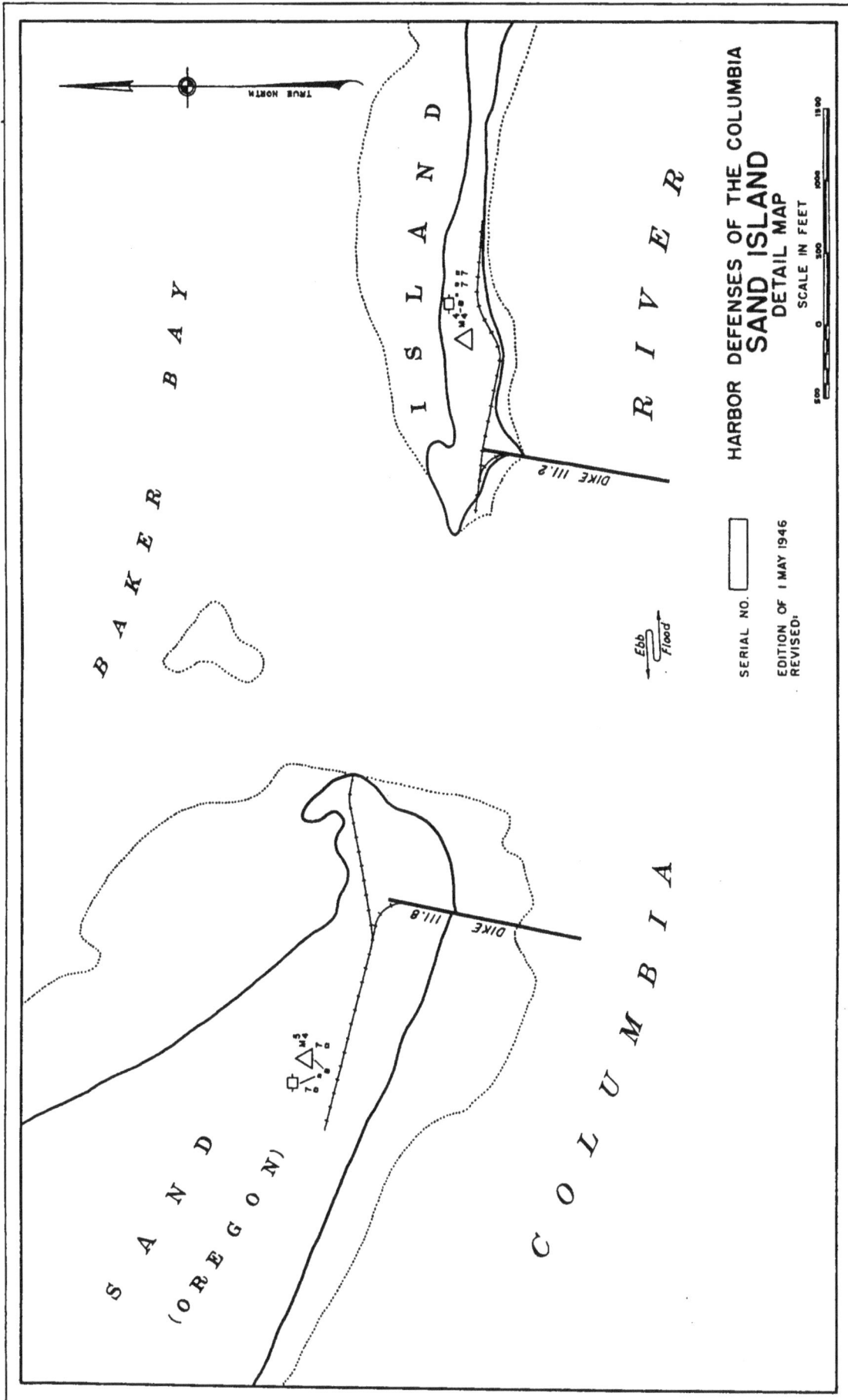

HARBOR DEFENSES OF THE COLUMBIA
SAND ISLAND
DETAIL MAP

SCALE IN FEET

SERIAL NO.

EDITION OF 1 MAY 1946
REVISED:

Ebb
Flood

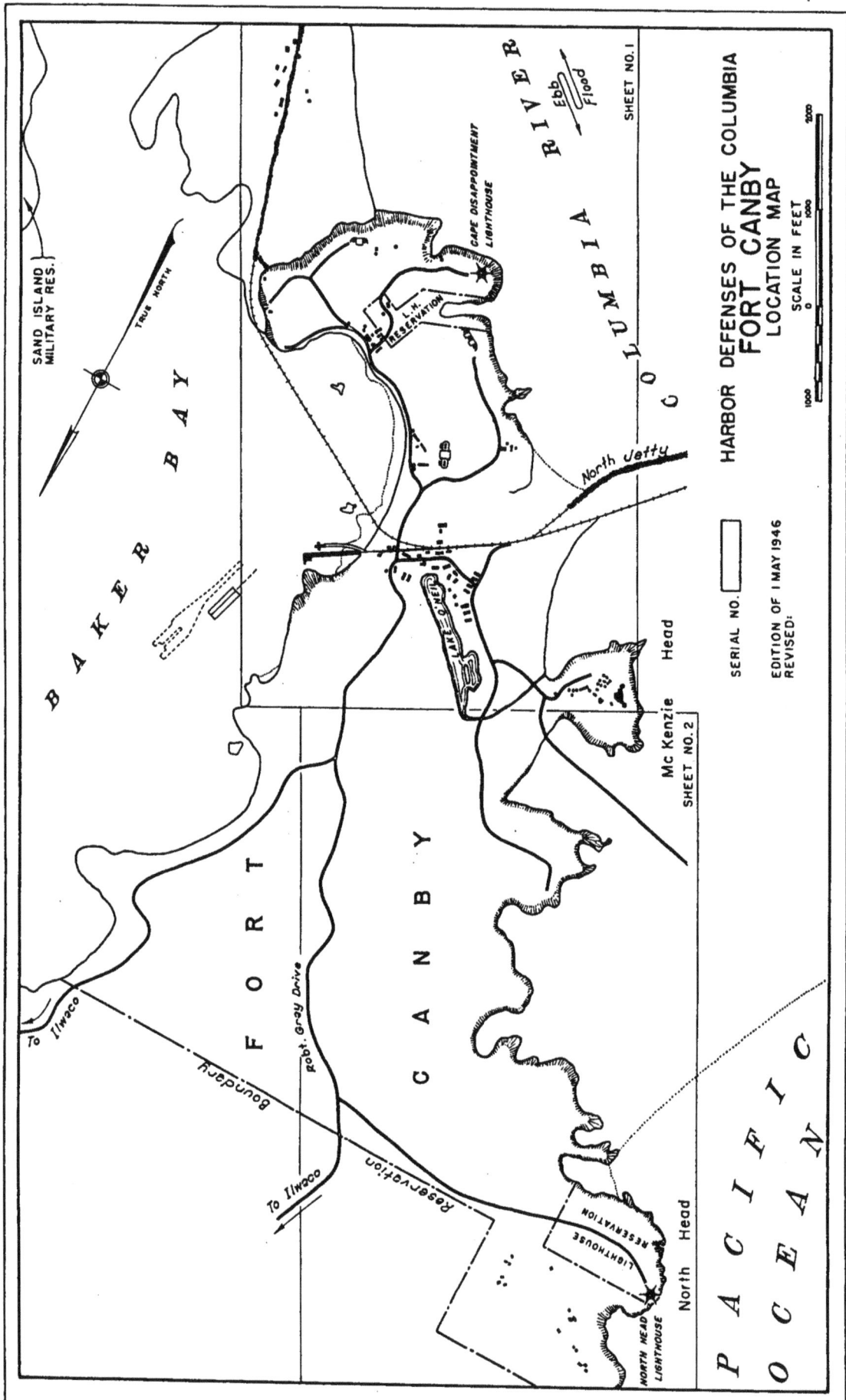

PACIFIC OCEAN

PACIFIC OCEAN

BAKER BAY

SAND ISLAND MILITARY RES.

TRUE NORTH

COLUMBIA RIVER

Ebb
Flood

SHEET NO. 1

CAPE DISAPPOINTMENT LIGHTHOUSE

L.H. RESERVATION

North Jetty

FORT CANBY

Mc Kenzie Head

SHEET NO. 2

Robt. Gray Drive

Boundary

Reservation

To Ilwaco

To Ilwaco

North Head

NORTH HEAD LIGHTHOUSE

LIGHTHOUSE RESERVATION

HARBOR DEFENSES OF THE COLUMBIA

FORT CANBY
LOCATION MAP

SCALE IN FEET

SERIAL NO.

EDITION OF 1 MAY 1946
REVISED:

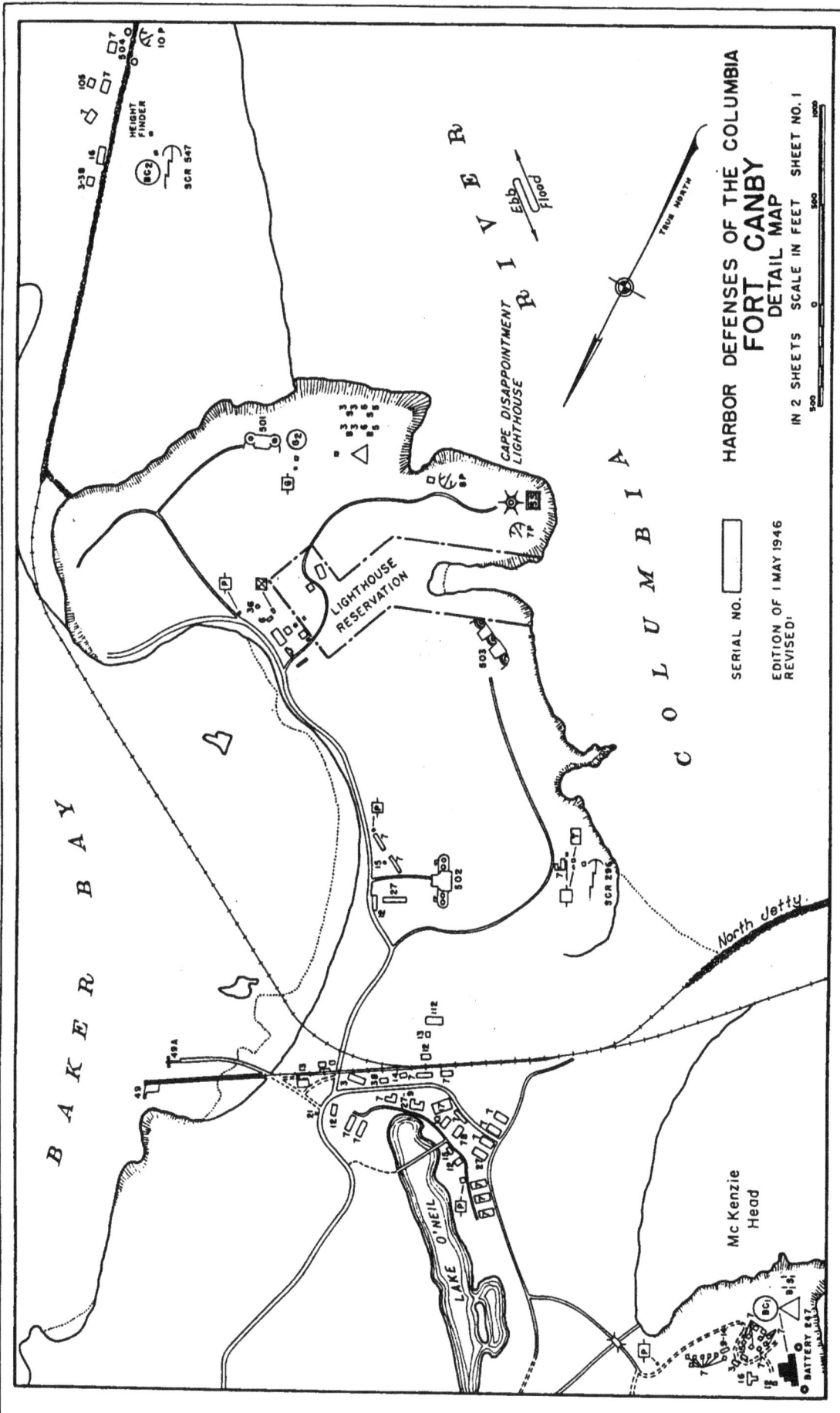

HARBOR DEFENSES OF THE COLUMBIA
FORT CANBY
DETAIL MAP
IN 2 SHEETS   SCALE IN FEET   SHEET NO. I

SERIAL NO.

EDITION OF 1 MAY 1946
REVISED:

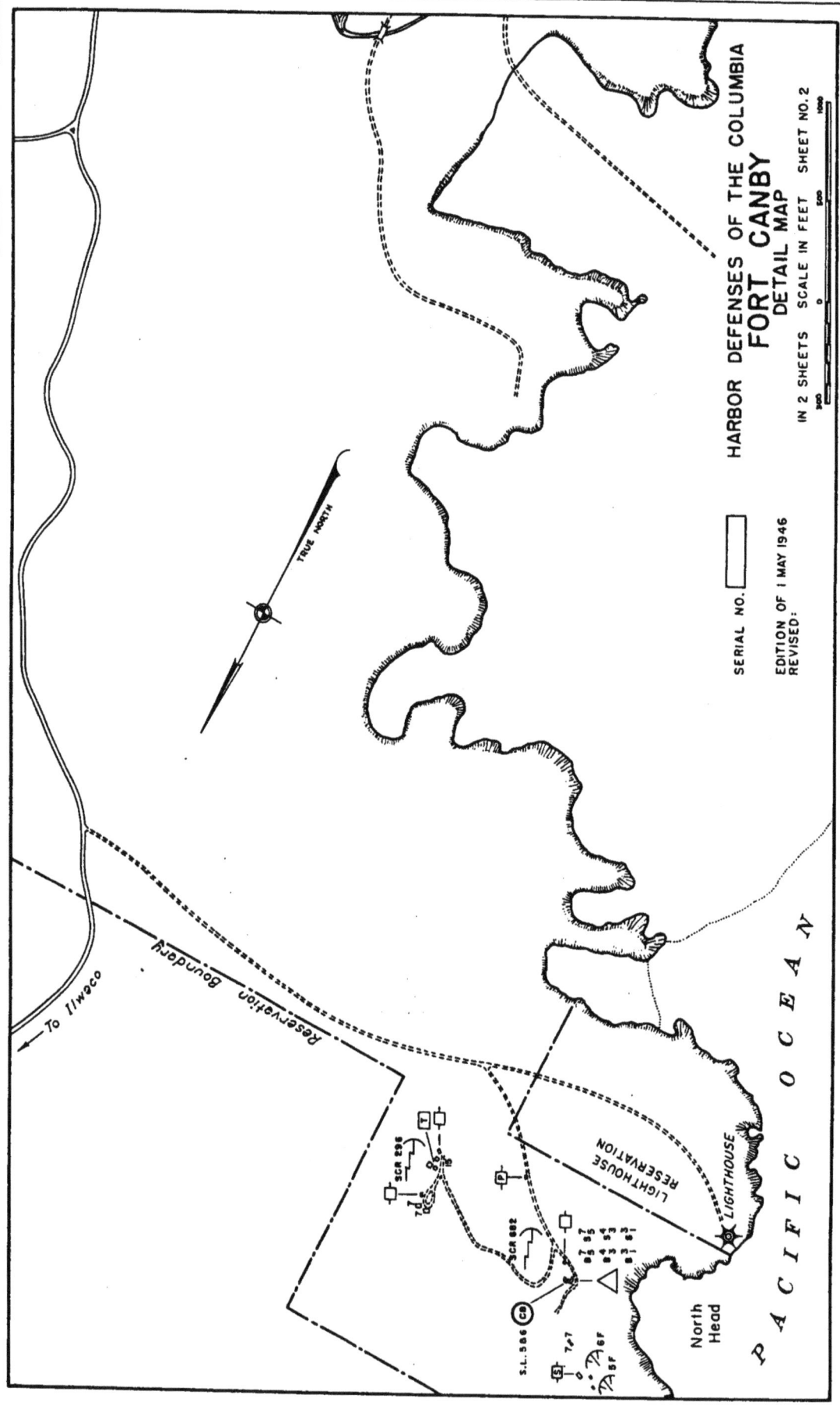

HARBOR DEFENSES OF THE COLUMBIA
FORT CANBY
DETAIL MAP
IN 2 SHEETS   SCALE IN FEET   SHEET NO. 2

SERIAL NO.

EDITION OF 1 MAY 1946
REVISED:

TRUE NORTH

Reservation Boundary

To Ilwaco

PACIFIC OCEAN

LIGHTHOUSE
RESERVATION

LIGHTHOUSE

North Head

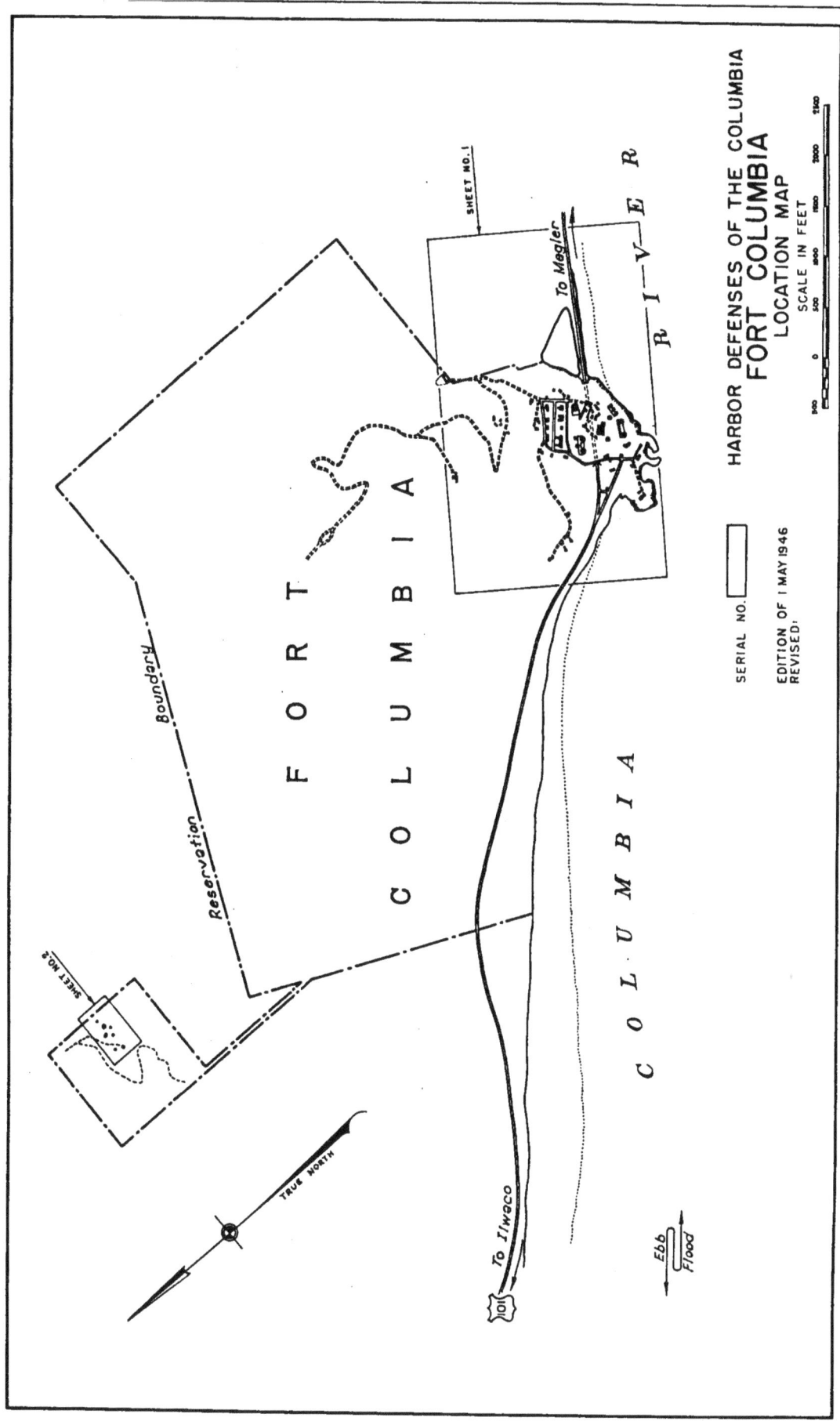

HARBOR DEFENSES OF THE COLUMBIA
FORT COLUMBIA
LOCATION MAP

SCALE IN FEET

SERIAL NO.

EDITION OF 1 MAY 1946
REVISED:

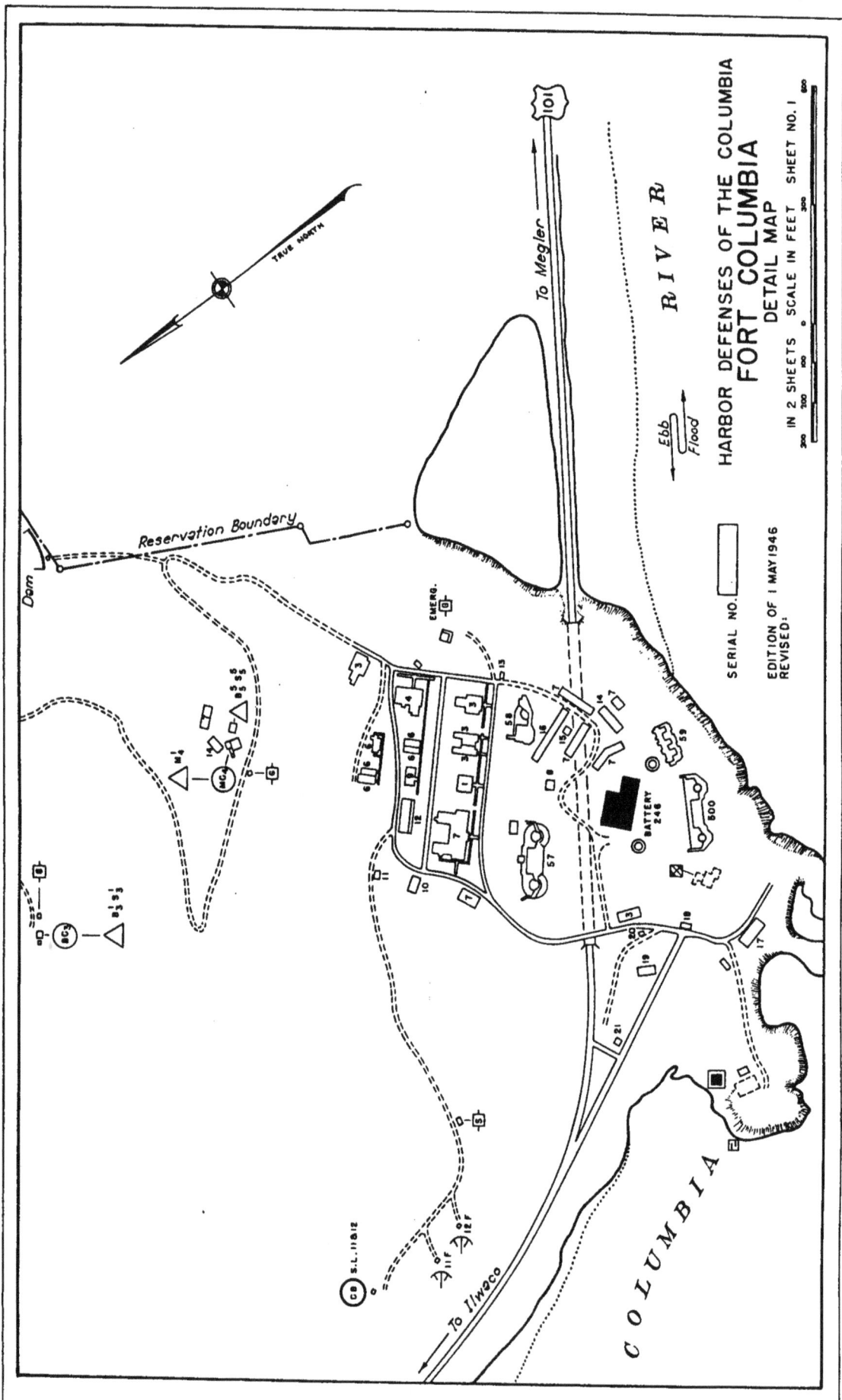

HARBOR DEFENSES OF THE COLUMBIA
FORT COLUMBIA
DETAIL MAP

IN 2 SHEETS   SCALE IN FEET   SHEET NO. I

SERIAL NO.

EDITION OF 1 MAY 1946
REVISED:

TRUE NORTH

To Megler

101

RIVER

Ebb
Flood

Reservation Boundary

Dam

EMERG.

BATTERY 246

COLUMBIA

To Ilwaco

S.L. 11812

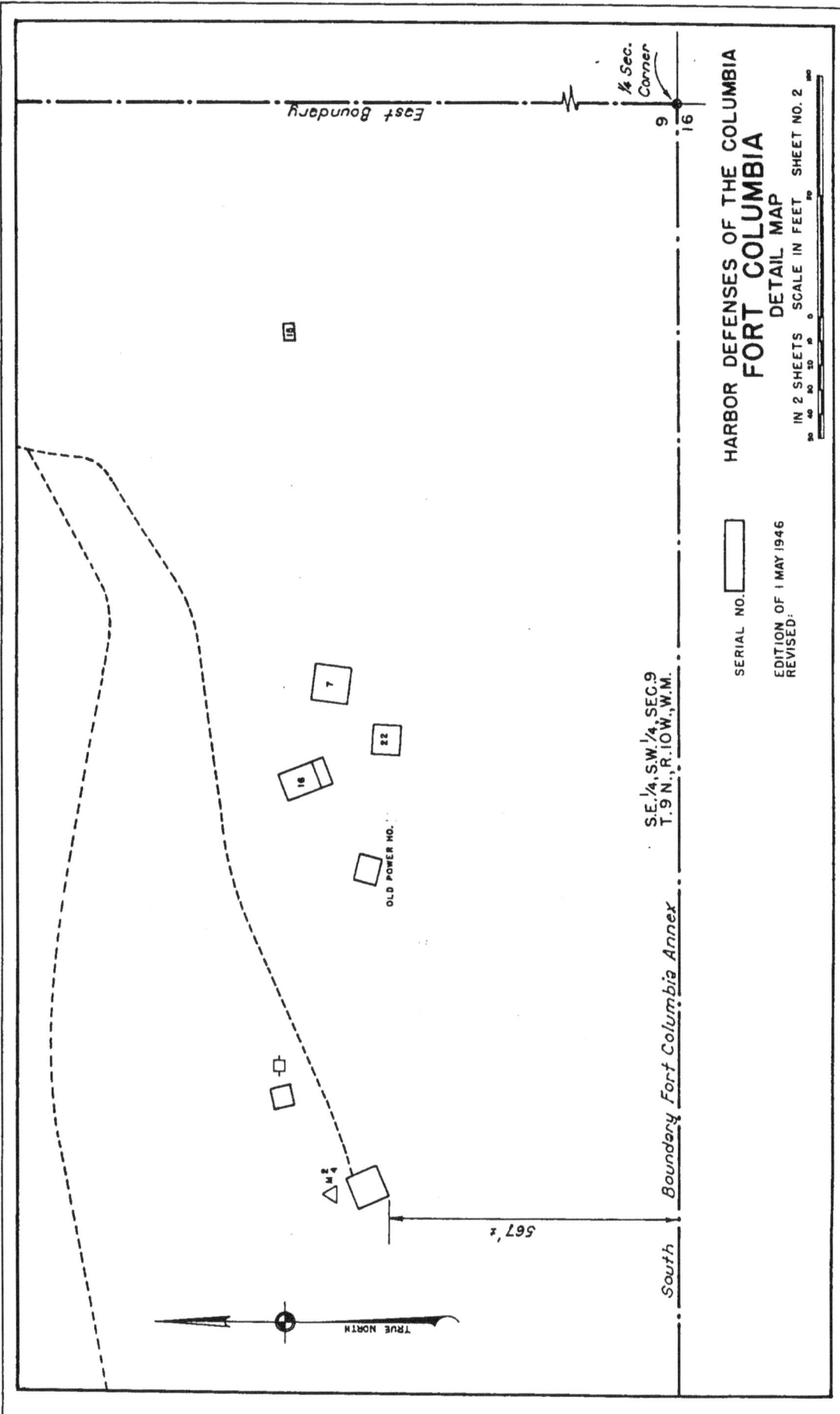

HARBOR DEFENSES OF THE COLUMBIA
FORT COLUMBIA
DETAIL MAP

IN 2 SHEETS   SCALE IN FEET   SHEET NO. 2

SERIAL NO.

EDITION OF I MAY 1946
REVISED:

S.E.¼, S.W.¼, SEC.9
T. 9 N., R.IOW., W.M.

Boundary Fort Columbia Annex

South

East Boundary

¼ Sec. Corner

9
16

OLD POWER HO.

TRUE NORTH

567'

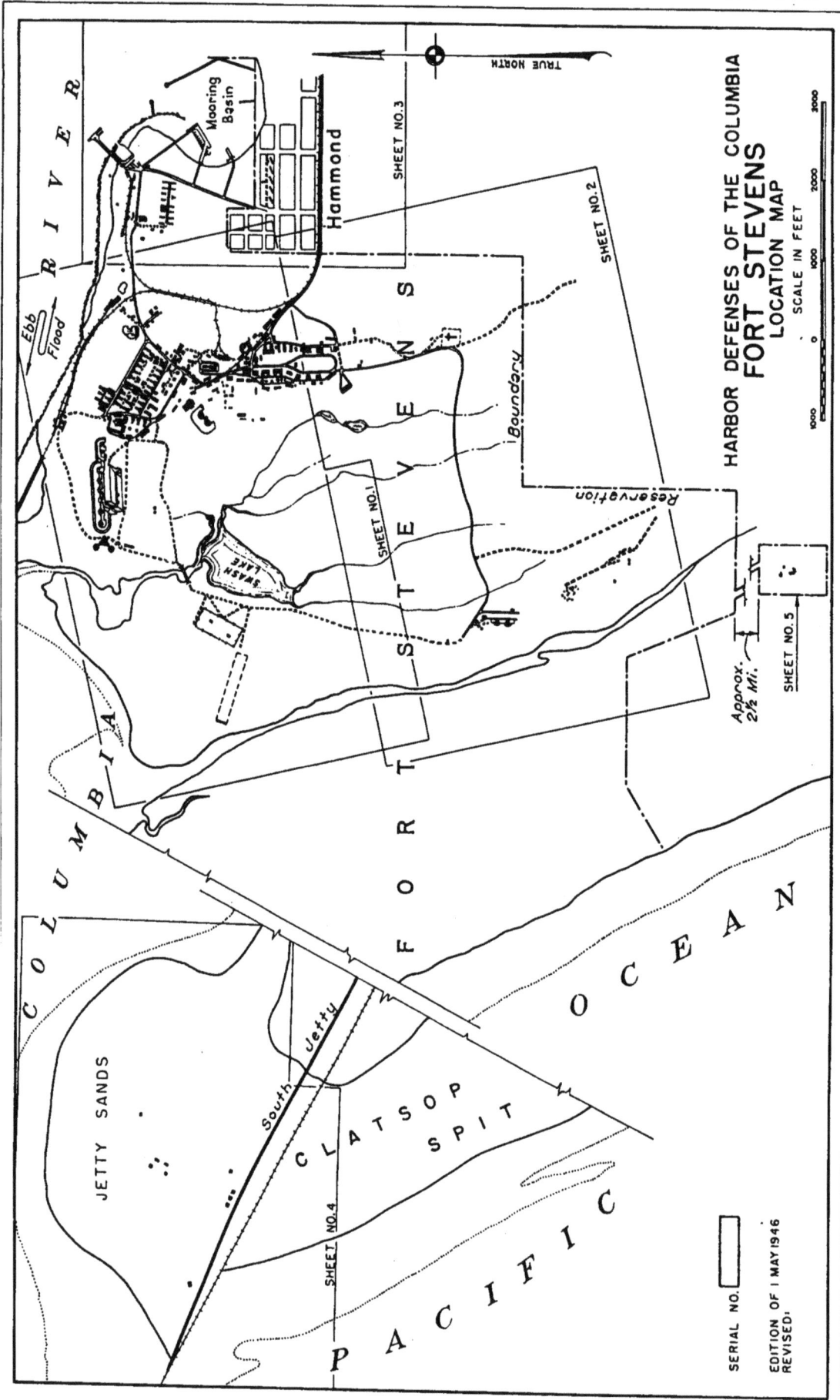

HARBOR DEFENSES OF THE COLUMBIA
FORT STEVENS
LOCATION MAP

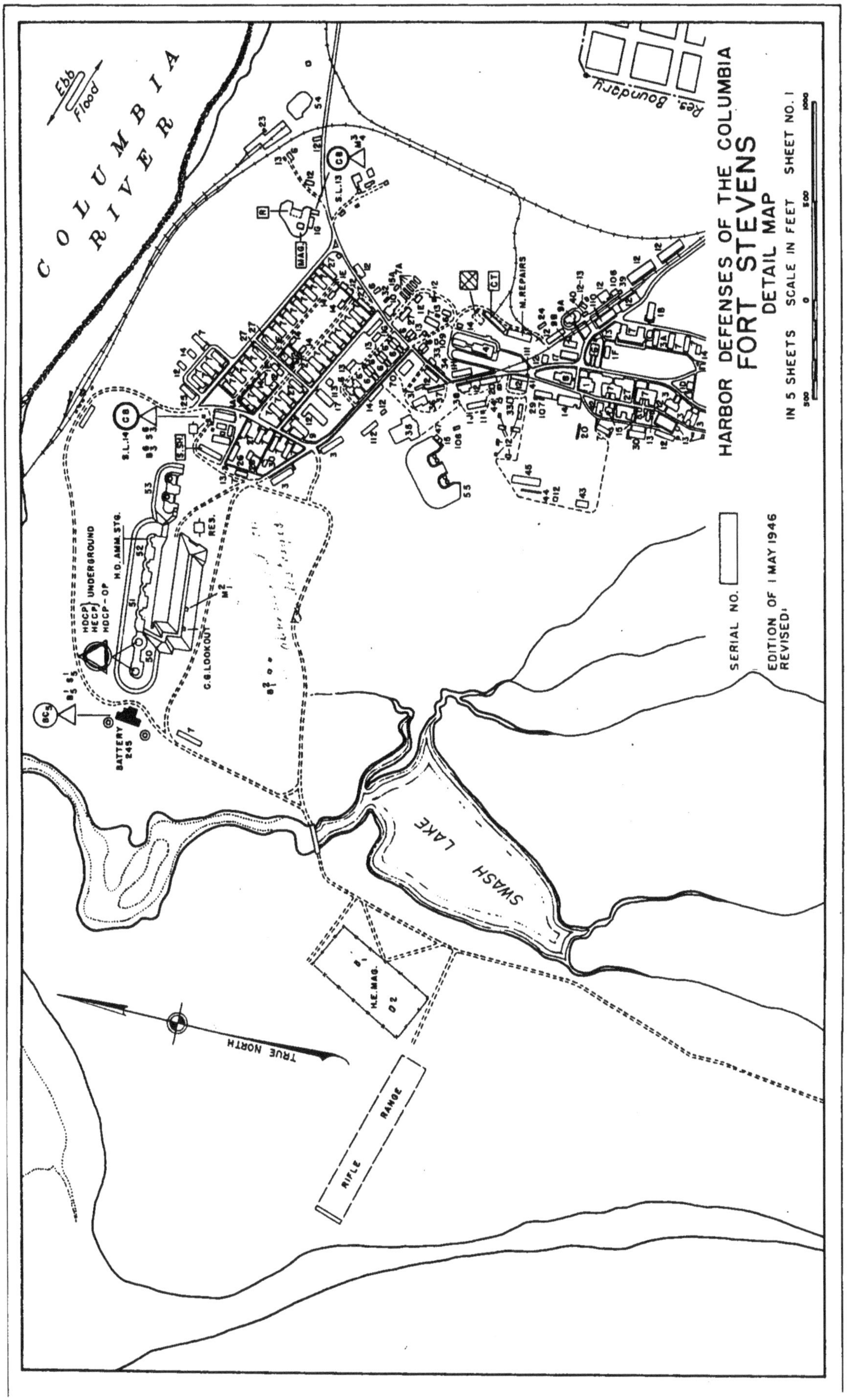

HARBOR DEFENSES OF THE COLUMBIA
FORT STEVENS
DETAIL MAP
IN 5 SHEETS SCALE IN FEET SHEET NO. 1
SERIAL NO.
EDITION OF 1 MAY 1946
REVISED:
COLUMBIA RIVER
Ebb
Flood
Res. Boundary
SWASH LAKE
TRUE NORTH
RIFLE RANGE
H.E. MAG.
BATTERY 245
HDCP
HECP
HDCP-OP
UNDERGROUND
H.D. AMM. STG.
C.G. LOOKOUT
M. REPAIRS
MAG.

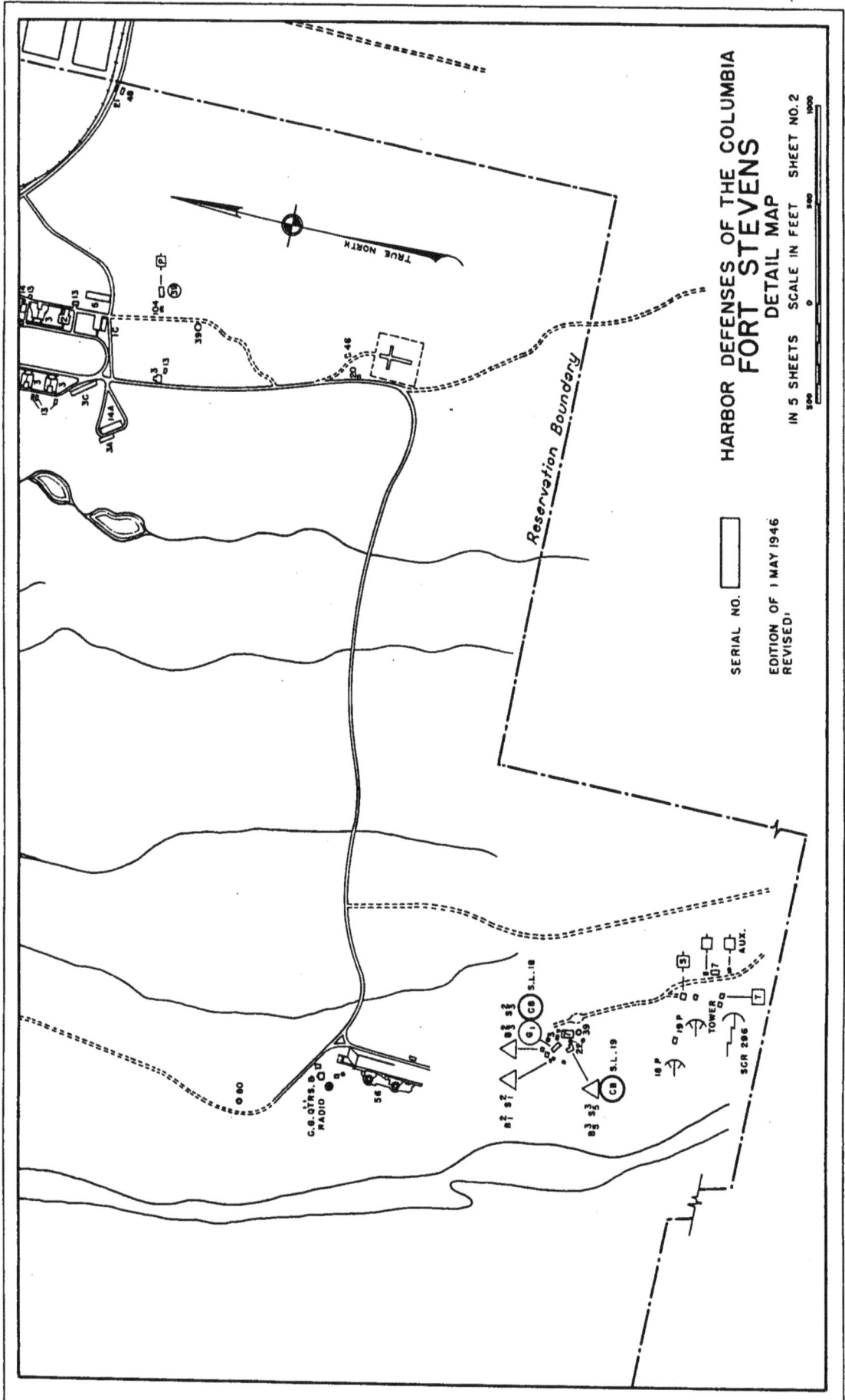

HARBOR DEFENSES OF THE COLUMBIA
FORT STEVENS
DETAIL MAP

IN 5 SHEETS   SCALE IN FEET   SHEET NO. 2

EDITION OF 1 MAY 1946
REVISED:

SERIAL NO.

TRUE NORTH

Reservation Boundary

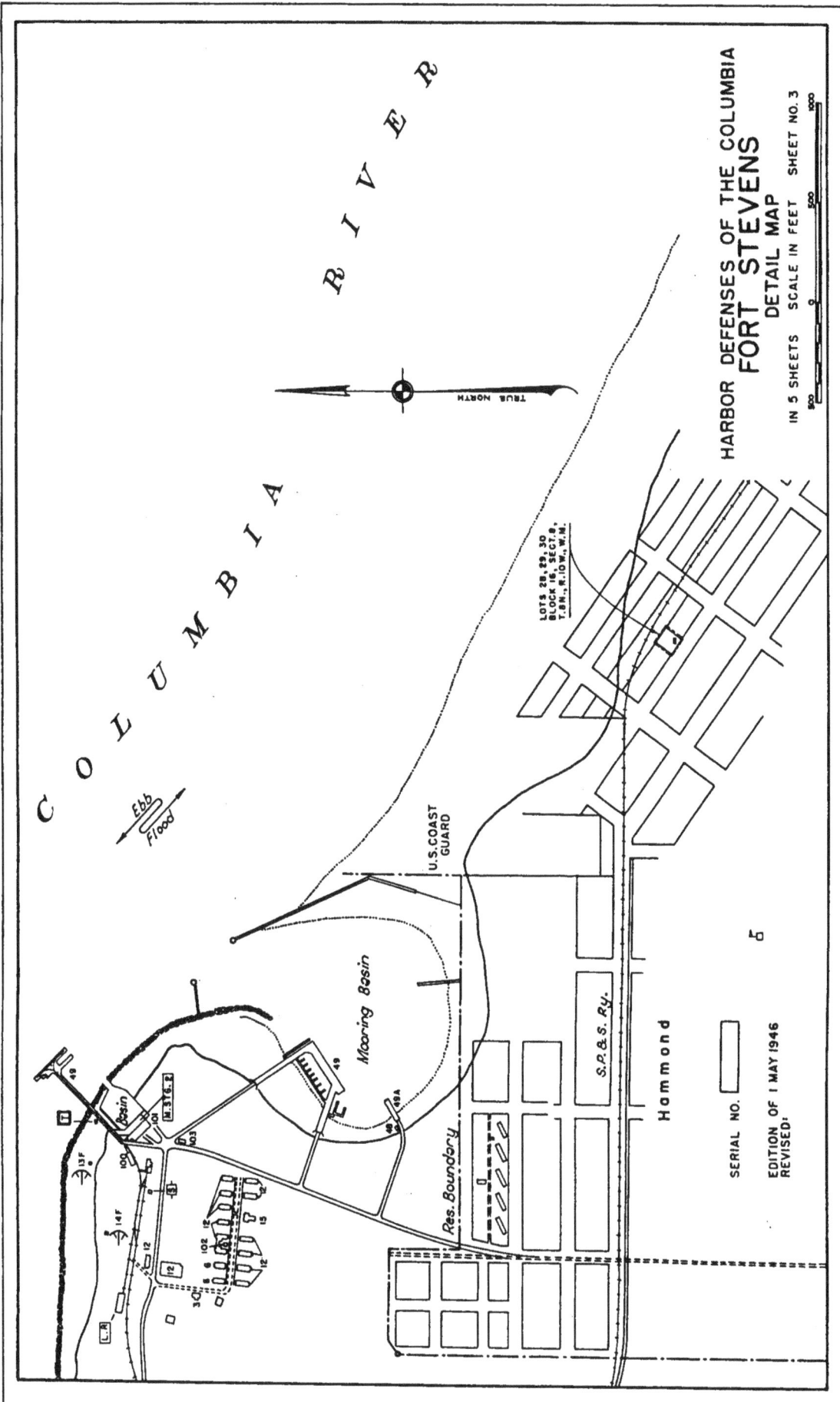

HARBOR DEFENSES OF THE COLUMBIA
FORT STEVENS
DETAIL MAP
IN 5 SHEETS   SCALE IN FEET   SHEET NO. 3

SERIAL NO.

EDITION OF I MAY 1946
REVISED:

C O L U M B I A

R I V E R

TRUE NORTH

Ebb
Flood

U.S. COAST
GUARD

LOTS 28, 29, 30
BLOCK 16, SECT. 8,
T.8N., R.10W., W.M.

Mooring Basin

Res. Boundary

Hammond

S.P.&S. Ry.

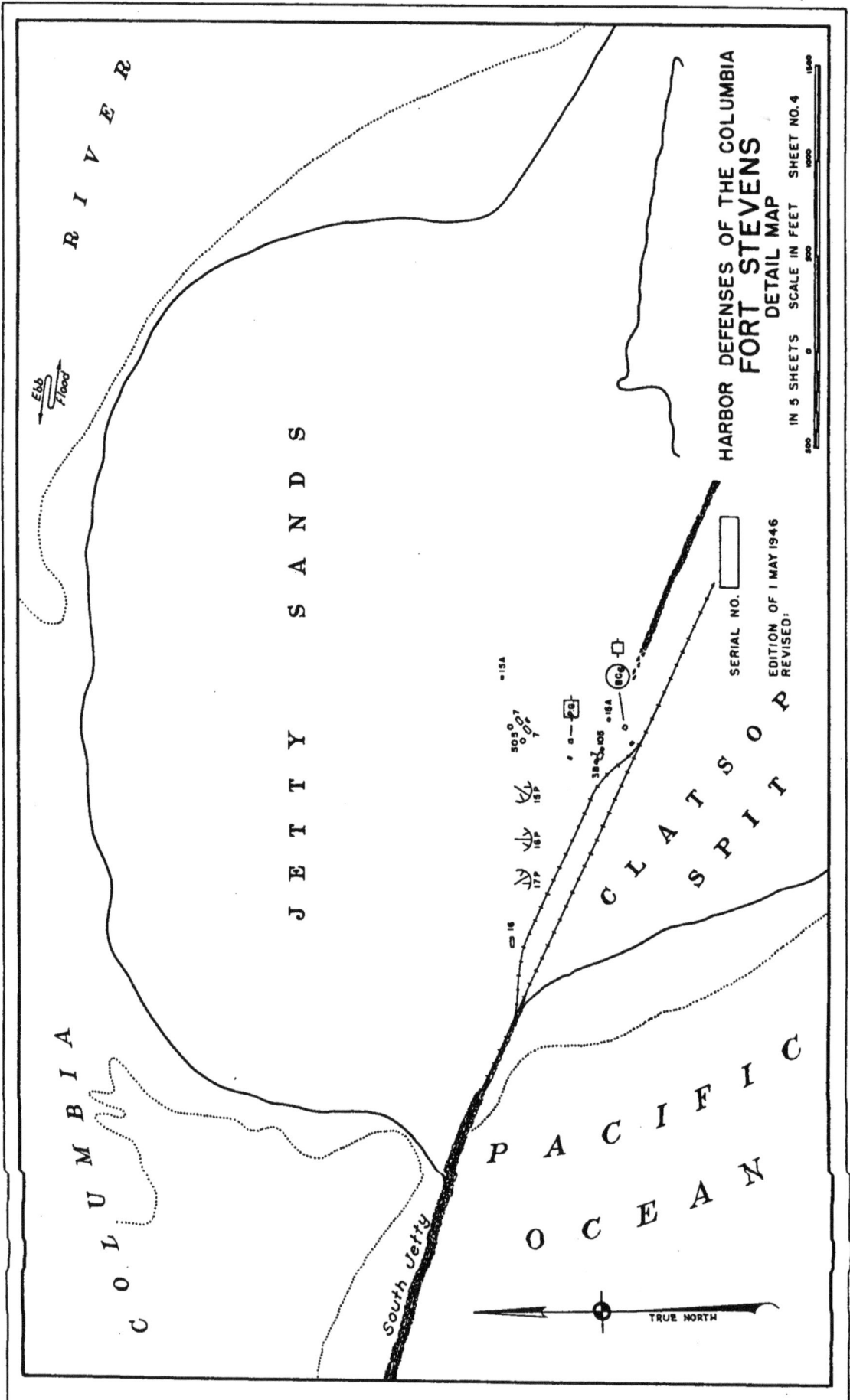

COLUMBIA   RIVER

Ebb
Flood

JETTY SANDS

COLUMBIA

CLATSOP   SPIT

PACIFIC

OCEAN

South Jetty

TRUE NORTH

HARBOR DEFENSES OF THE COLUMBIA
FORT STEVENS
DETAIL MAP

IN 5 SHEETS   SCALE IN FEET   SHEET NO. 4

SERIAL NO.

EDITION OF 1 MAY 1946
REVISED:

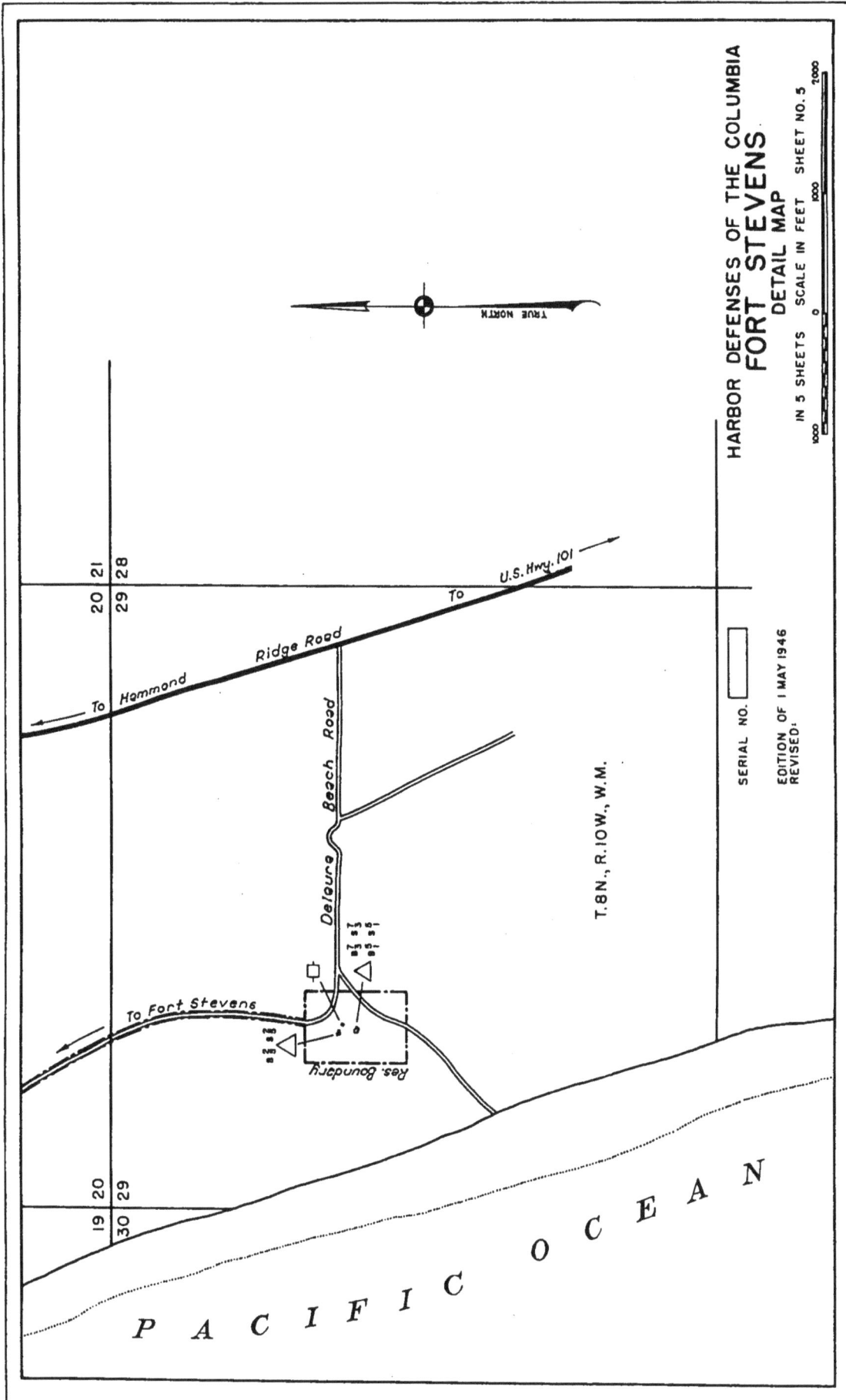

HARBOR DEFENSES OF THE COLUMBIA
FORT STEVENS
DETAIL MAP
IN 5 SHEETS   SCALE IN FEET   SHEET NO. 5

TRUE NORTH

U.S. Hwy. 101

To

Ridge Road

To Hammond

Beach Road

Deleure

To Fort Stevens

Res. Boundary

T. 8N., R. IOW., W.M.

SERIAL NO.

EDITION OF I MAY 1946
REVISED:

P A C I F I C   O C E A N

20   21
29   28

20
29

19   20
30   29

1000   0   1000   2000

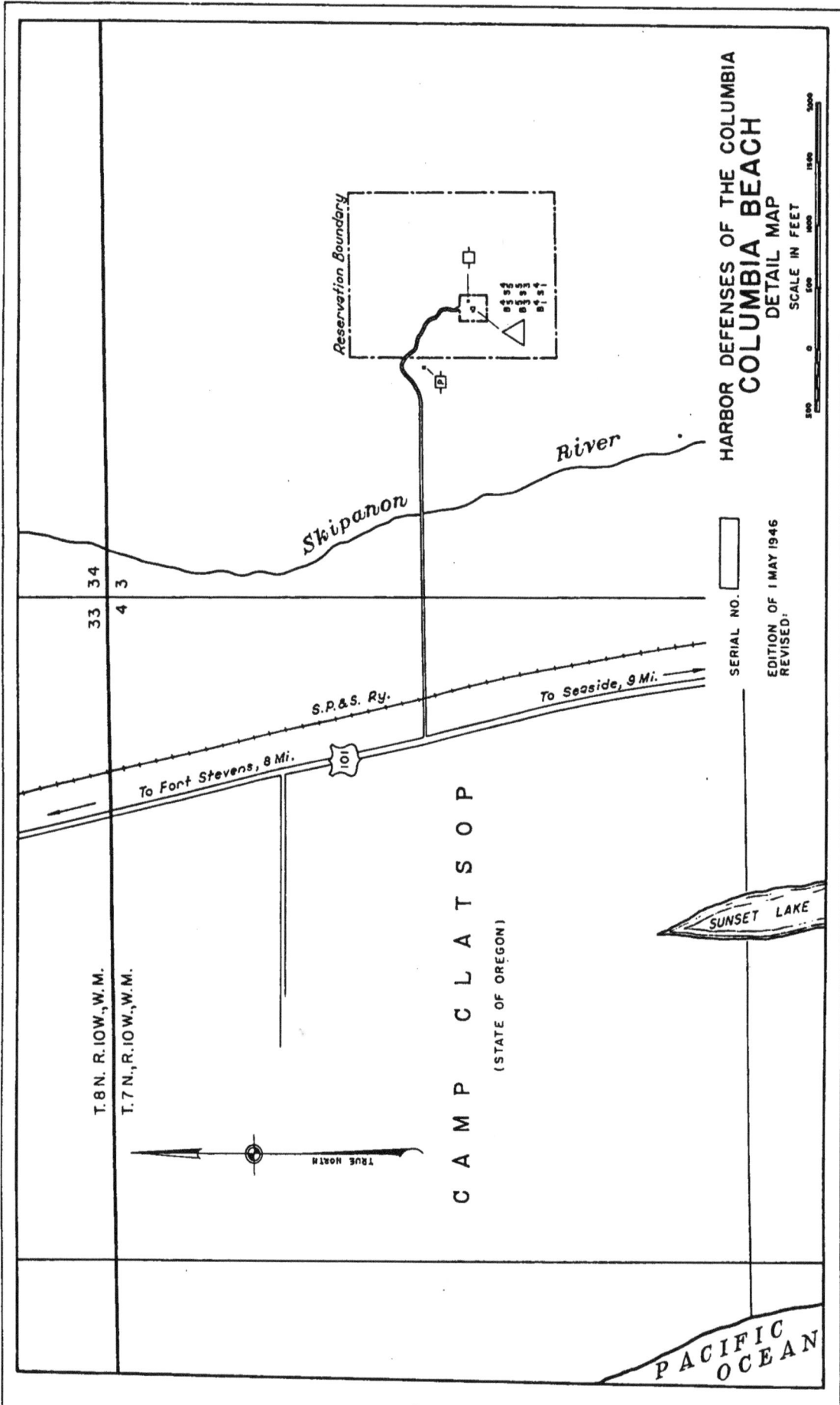

HARBOR DEFENSES OF THE COLUMBIA
COLUMBIA BEACH
DETAIL MAP
SCALE IN FEET

SERIAL NO.

EDITION OF 1 MAY 1946
REVISED:

Reservation Boundary

B.S. 91
B.S. 93
B.S. 94

Skipanon   *River*

33 34
4   3

S.P.& S. Ry.

To Fort Stevens, 8 Mi.   To Seaside, 9 Mi.

101

C A M P   C L A T S O P
(STATE OF OREGON)

SUNSET LAKE

T.8N. R.10W.,W.M.
T.7N., R.10W.,W.M.

TRUE NORTH

PACIFIC
OCEAN

WILLIPA BAY AND GRAYS HARBOR DEFENSES

Cape Shoalwater Military Reservation (1919-1932, 1942-1944) is located on the southwest Pacific coast of Washington State. The Long Beach Peninsula separates Willapa Bay from the greater expanse of the Pacific Ocean. The Cape Shoalwater M.R. was established near the community of North Cove in 1854 with a lighthouse built in 1858. Located here were two unnamed World War I batteries: Battery 1 (1919-1932) two 6-inch naval guns at the old Cape Shoalwater Lighthouse, and Battery 2 (1919) four emplacements for 12-inch mortars with post barracks, northeast of the lighthouse and north of Old North Cove. The mortars were never emplaced. The site of the old lighthouse and Battery 1 no longer exists, long eroded away in the ocean. The site of Battery 2 is located in a present-day waste dump area about halfway between North Cove and Dexter-by-the-Sea. During World War II a two-gun 155mm battery on Panama mounts (1942-1944) was built at Wash Away Beach. Also located in the area were several 75mm field guns (1942). The ruins of one Panama mount are located in the surf at the end of Tamarack Street.

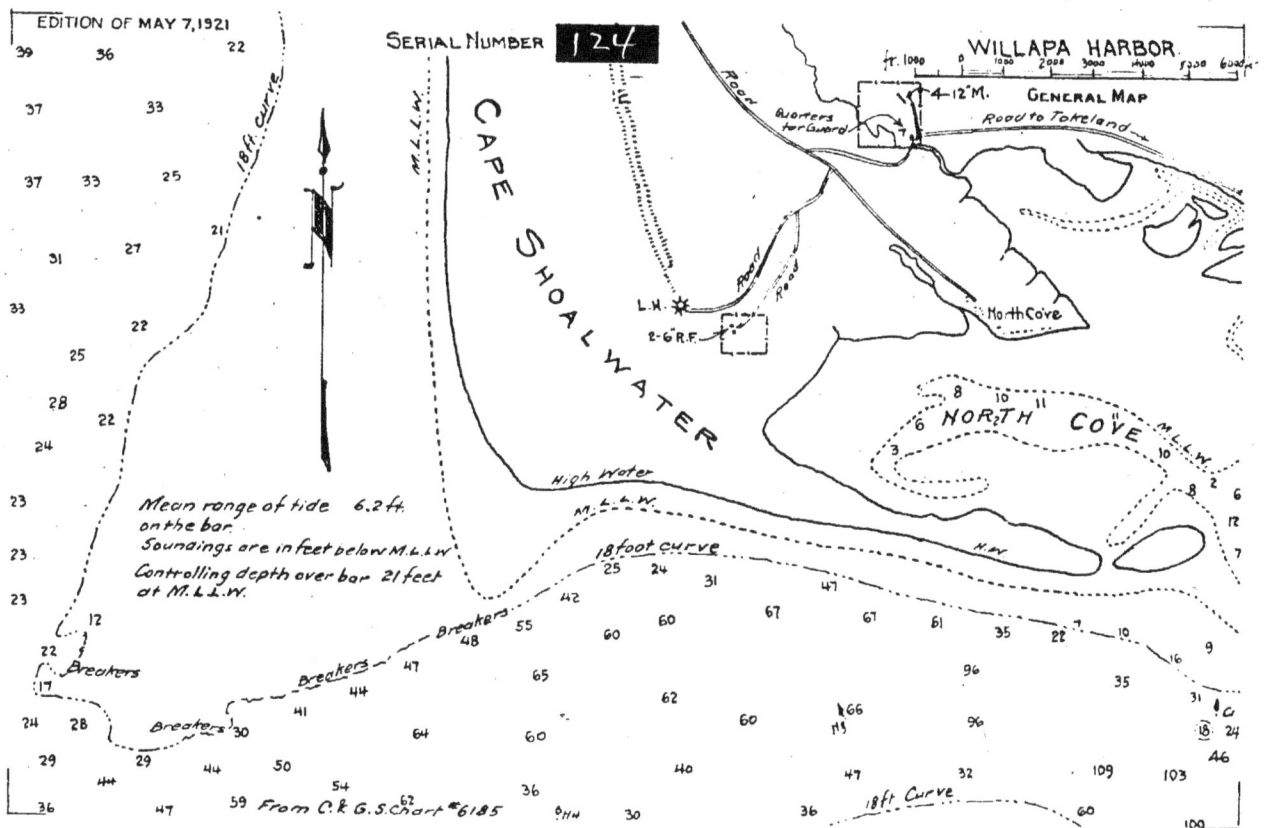

Willapa Harbor Gun Batteries

- A mortar battery approved for construction under the gun and Mortar Act of June 15, 1917. Disbursement was made on May 13, 1918 of $42,500 for twelve emplacements, including the two mortar batteries and the dual RF gun emplacements for Willapa and Gray's Harbors. Work was done on the simple gun block position from June until October 1918. Transfer to the Coast Artillery was made on October 6, 1919 at a cost of $15,235.84. The emplacements were located north of North Cape, considerably back from the beach in a ravine. The mortars were to have come from Battery McKinnon-Stotsenburg at Fort Winfield Scott, but that move was suspended on April 10, 1919 pending a study of the use of railway mortars here instead. In any event no weapons were ever actually emplaced here. The emplacement was soon abandoned, but the blocks still exist heavily overgrown near a public dump site. The battery site is open to the public.

WILLAPA HARBOR.-D1.
Scale of Feet.

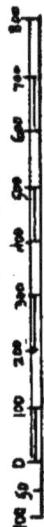

SERIAL NUMBER. 124

EDITION OF MAY 7, 1921.

LEGEND.
7. Barracks (Quarters for Guard).

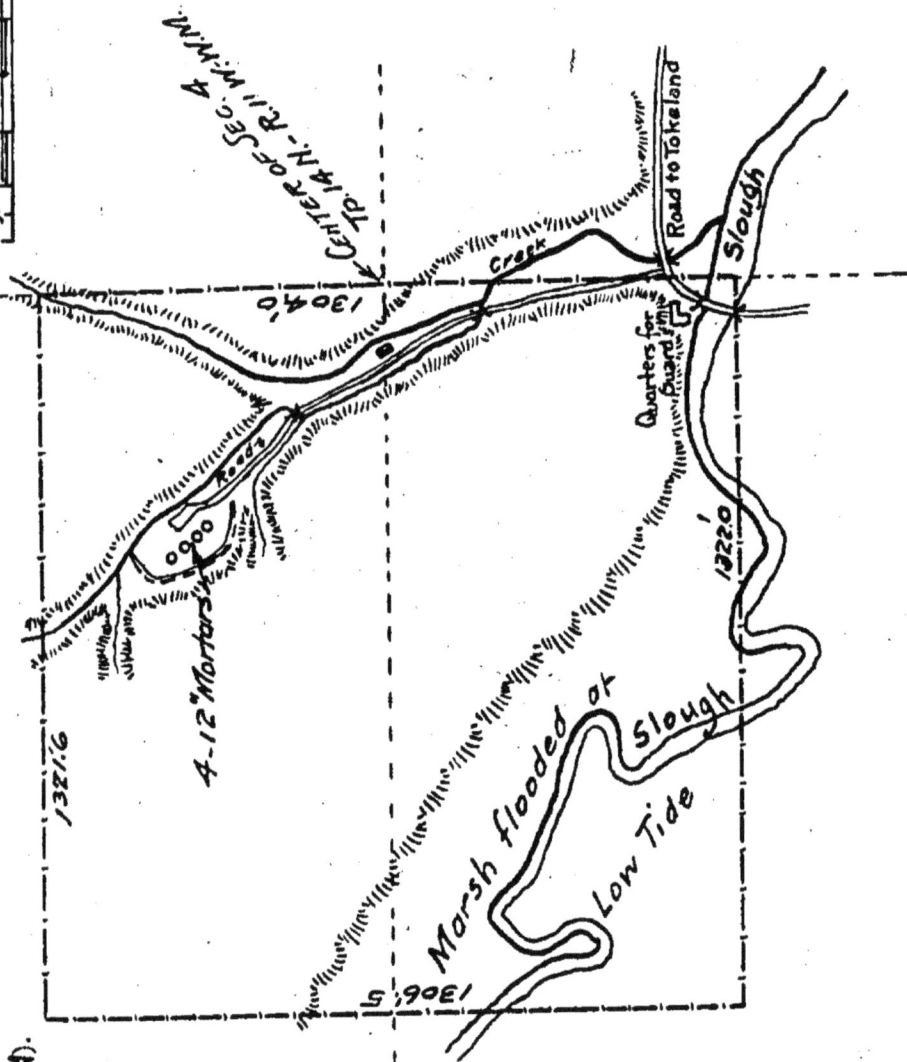

Center of Sec. 14. T.P.

1304.0 Creek

Road to Tokeland

Slough

Quarters for Guard

Roads

1322.0

1321.6

4-12"Mortars.

Marsh flooded at Slough Low Tide

1306.5

BATTERY
4-12"Mortars.

- A temporary emplacement for two 6-inch pedestal guns relocated here during World War I. They were placed just 1000-yards to the east of the lighthouse, firing to the west. The simple gun blocks were constructed from July to September 1918 for transfer on October 6, 1919 at a cost of $6,429.19. Authority for construction had come on June 15, 1917, and subsequent funds on May 13, 1918. The guns and pedestals were received and mounted on January 1, 1919. It was armed with two 6-inch Model 1900 guns and pedestals (#8/#25 and #4/#24) transferred from Battery Freeman at Fort Stevens. Unlike the other temporary batteries in Western Washington, this unit had a considerable service live. The pedestal graduate circles were overhauled and calibrated on February 16, 1925 by a detachment from Fort Stevens. An engineer order of February 7, 1923 prompted a modification to the carriage and gun block to allow the gun to fire at a 20-degree elevation (this involved removing of 125 cubic feet of concrete in April 1924). The gun was proofed and exercised again from this emplacement in June of 1930. The armament was then removed on September 28, 1932 and sent first to storage and then eventually to arm the rebuilt Battery Tolles B at Fort Worden. The blocks themselves were eventually destroyed by beach erosion, the original site is now out in the waters of the bay.

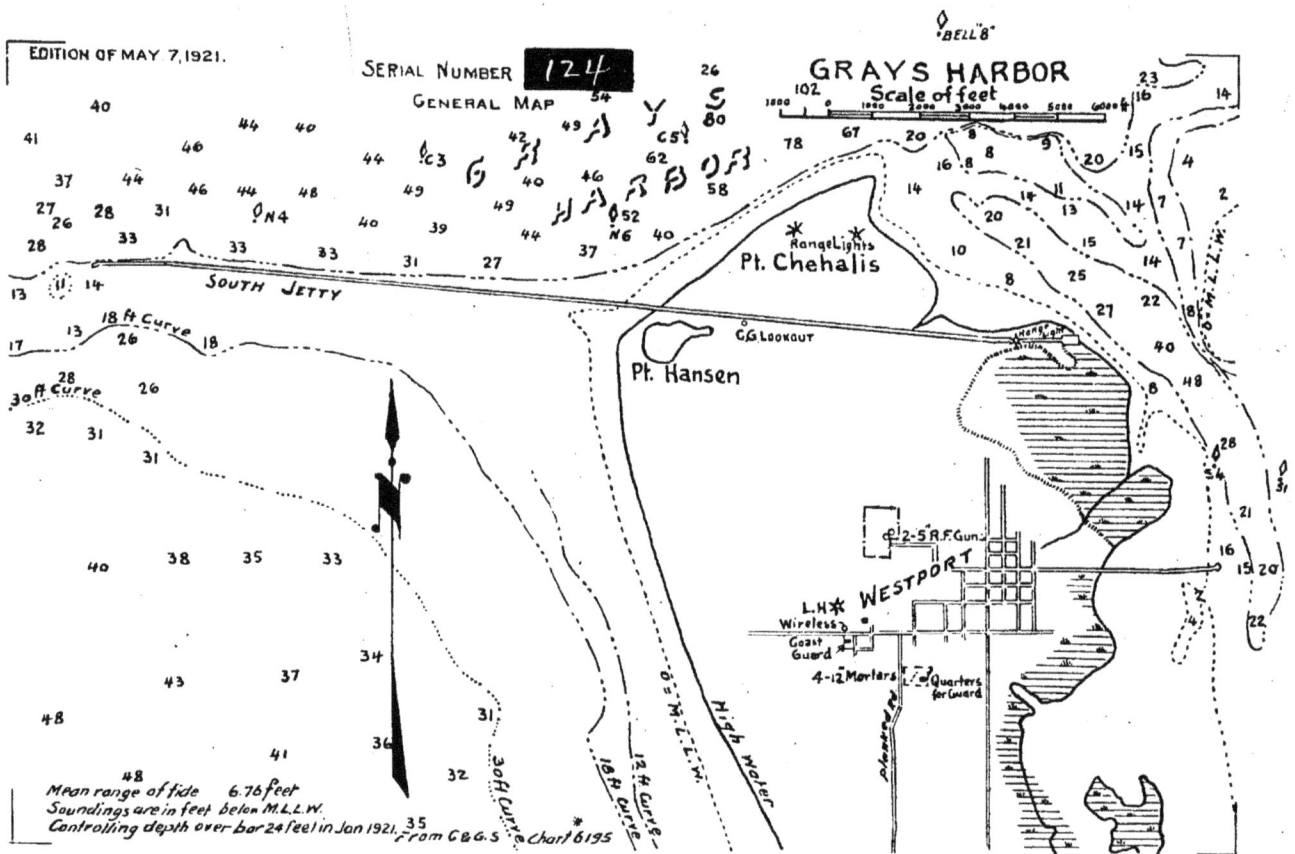

Westport Military Reservation (1918-1926, 1942-1944) is located on a peninsula on the south side of the entrance to Grays Harbor from the Pacific Ocean in Washington State. Located here were two unnamed World War I batteries. Four 12-inch mortars (1918-1919) were located on the southeast lot at West Perry Street and South Forrest Street. Two 5-inch naval guns (1918-1926) were located on the northwest lot at West Sprague Ave. and North Hoquiam Street. In January 1942 four 12" railway mortars were deployed from the Columbia River HD to Point Brown where they were emplaced and activated. The railway mortar battery at Point Brown was deactivated and replaced by the 6" BC battery. In 1942 South Battery (two 155mm guns on Panama mounts) was built at the 5-inch gun site. It was soon replaced with Battery 2B (1942-1944) two 6-inch naval guns. North Battery (two 155mm guns on Panama mounts) was built on North Baker Street near Westport City Park. It was also soon replaced by Battery 2A two 6-inch naval guns (1942-1944), located across Elizabeth Avenue from the park.

Grays Harbor Gun Batteries

- Emplacments for a mortar battery was approved for construction under the Gun and Mortar Act of June 15, 1917. Four simple blocks were constructed on a new site from July to October 1918 for transfer on October 6, 1918 at cost of $10,405.99. They were located just southwest of the town of Westport, south of the new 5-inch battery. The mortars were to have come from either a San Francisco or a Puget Sound battery, but this was never implemented. The transfer of this ordnance was suspended on April 30, 1919 pending a study for the alternative need of the guns for additional railway mounts. The block existed for quite some time, unused. They were eventually covered over and the site sold to private ownership. No trace remains today.

- A 5-inch battery was also approved by the Act of June 15, 1917 for Westport. It was located northwest of the town, firing to the northwest. Unlike most other World War I relocations, this battery used the 5-inch Model 1897 balanced pillar type. Work on the simple gun blocks was done from July to September 1918, and transfer made on October 6, 1919 at a cost of $5,050.73. The guns were actually mounted on January 1, 1919transferred from Battery Lee, Fort Flagler. They were 5-inch Model 1897 tubes on Model 1896 balanced pillar mounts (#5/#32 and #8/#31). They served only a short while, being ordered dismounted and scrapped on June 23, 1919 (though not implemented until December 1925). The site was sold into private ownership, but the blocks still exist, though endangered by local development and shore erosion.

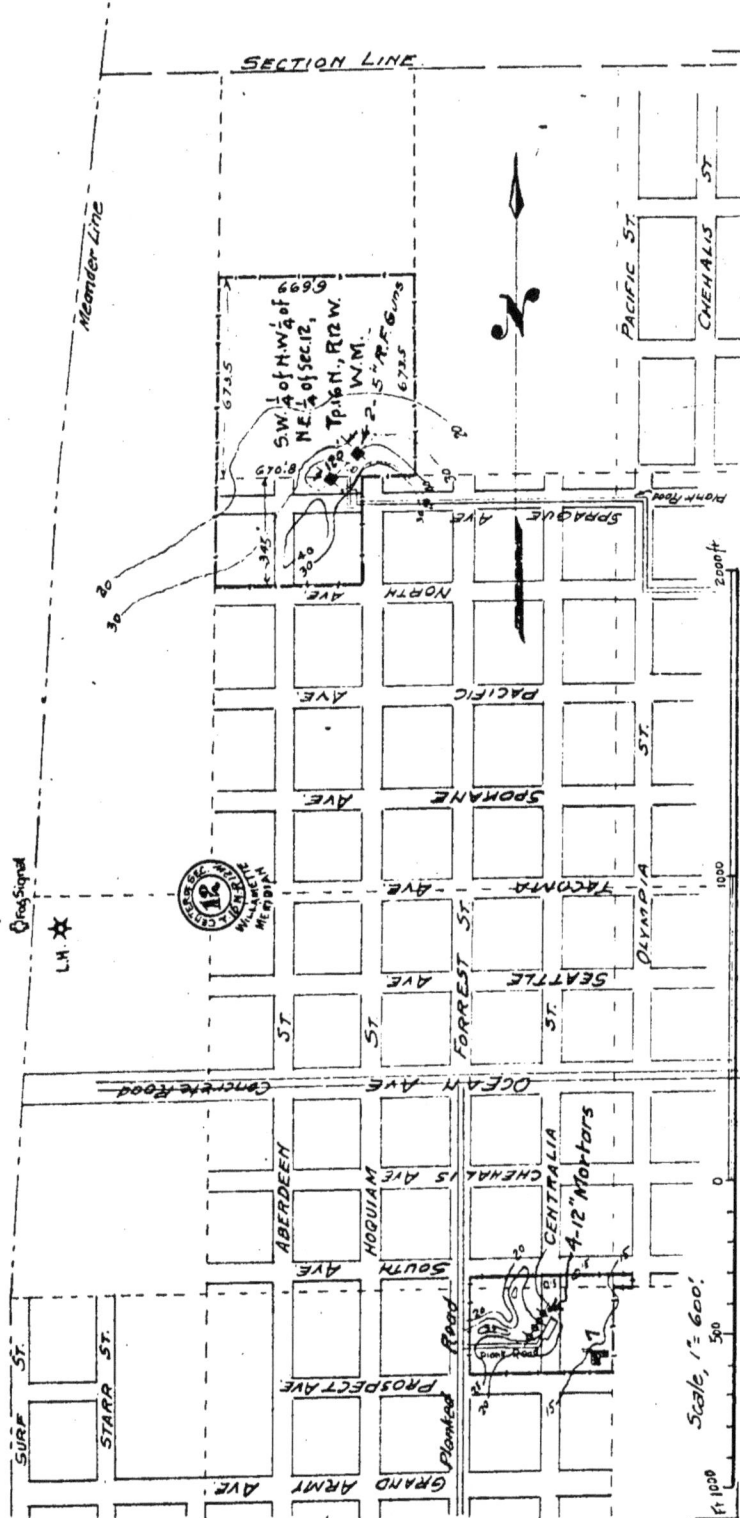

GRAYS HARBOR – DI.

WESTPORT

SERIAL NUMBER 124

EDITION OF MAY 7, 1921.

LEGEND.
7. Barracks (Quarters for Guard).

BATTERIES.
4 -12" Mortars
2 - 5" Barb'te

SECTION LINE

Meander Line

S.W. ¼ of N.W ¼ of
N.E. ¼ of sec.12,
Tp.18N., R.12W.
W.M.

4 - 2 - 5" R.F. Guns

Rock Road

N

PACIFIC ST.
CHEHALIS ST.

SPRAGUE AVE.
NORTH AVE.
PACIFIC AVE.
SPOKANE AVE.
TACOMA AVE.
SEATTLE AVE.
OCEAN AVE.

FORREST ST.
OLYMPIA ST.

Rad Signal
L.H. ☆

Concrete Road

ABERDEEN AVE.
HOQUIAM AVE.
SOUTH AVE.
CHEHALIS AVE.
CENTRALIA AVE.

4 -12" Mortars

PROSPECT AVE.
GRAND ARMY AVE.

SURF ST.
STARR ST.

Planked Road

Scale, 1" = 600'
Ft 1000   500   0   1000   2000 ft

THE HARBOR DEFENSES OF PUGET SOUND — WASHINGTON STATE

While the large network of deep-water channels was a significant defensive concern to the American government since acquisition of the Oregon Territory in 1846, the lack of connection to the rest of the United States kept it from being defended during the 3rd System period. The connection by railroad to the rest of the United States in the 1885 and the establishment of a naval shipyard in 1893 led to the construction of a significant set of defenses after 1895 of over thirty batteries at five locations. The 1940 Program brought seven planned new batteries, but only three were completed. The major posts were turned over to the state and private interests in the 1950s. The three major posts located around the Admiralty Inlet entrance to Puget Sound contain an impressive array of early modern era gun batteries, buildings, and restored coast artillery exhibits.

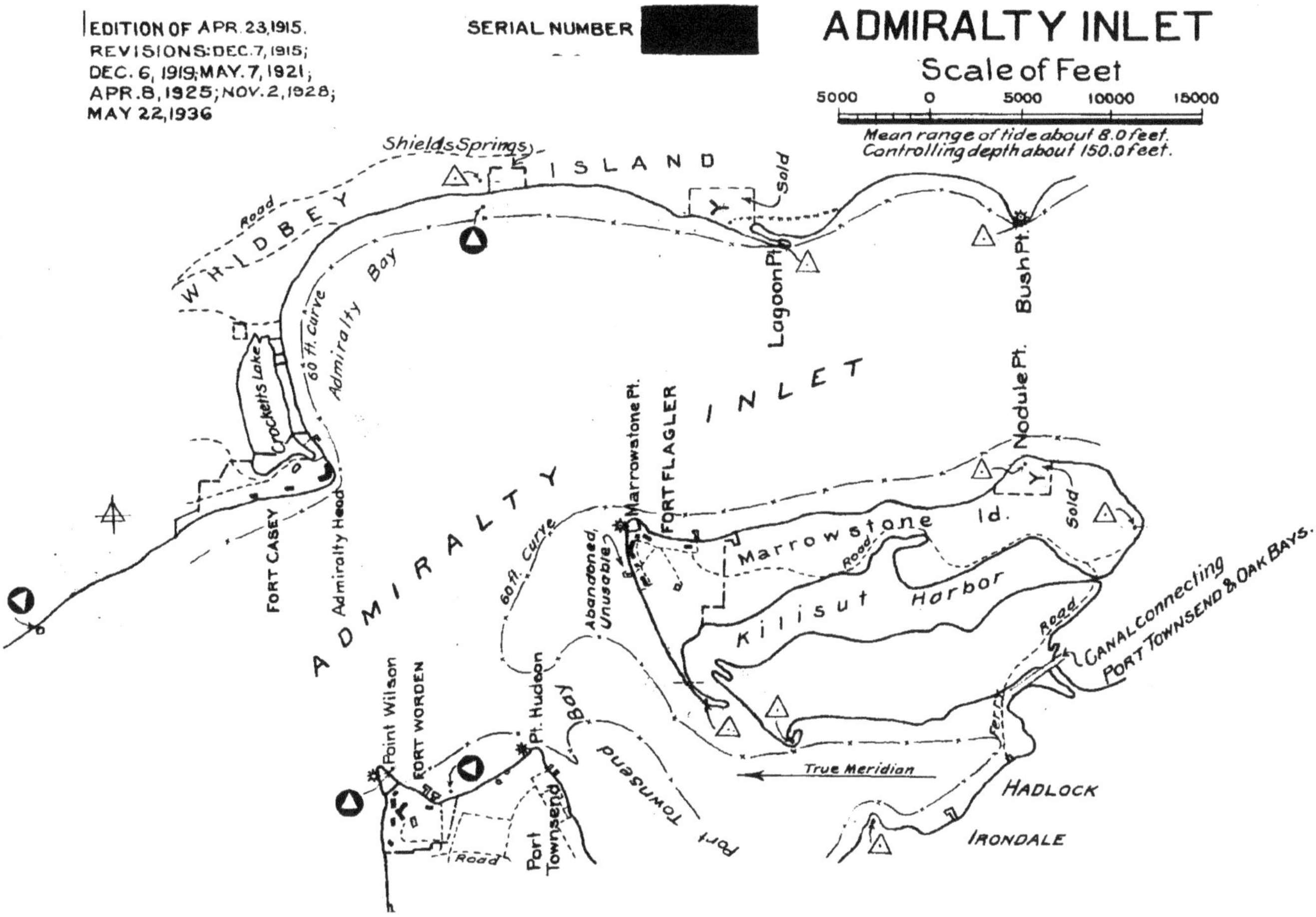

Battery Worth, Fort Casey State Park (Terry McGovern)

SERIAL NUMBER ███████

ENTRANCES TO PUGET SOUND

Scale 1/200,000

EDITION OF APR. 23, 1915.
REVISIONS: DEC. 7, 1915;
    NOV. 8, 1916; DEC. 6, 1919; MAY 7, 1921;
    APR. 8, 1925; NOV. 2, 1928; MAY 22, 1936

N.

TRUE MERIDIAN.

DECEPTION I. O   Lt.   DECEPTION PASS.   SKAGIT I.
    Hoypus Pt.   Lt.
    Lt.   Hope I.   FIDALGO IS.

SKAGIT BAY

Ilka Is.
FORT WHITMAN
GOAT ISLAND

WHIDBEY

CRESCENT HARBOR.
OAK HARBOR   Lt.
Lt.   FORBES PT.
BLOWERS BLUFF.
Pt. POLNELL.

MITCHELL BLUFF

PENN COVE.

WATSAK PT.

SARATOGA PASSAGE.

CAMANO ISLAND.

ISLAND

Light buoy

Pt. PARTRIDGE

FORT CASEY.
ADMIRALTY HEAD.

ADMIRALTY

MIDDLE POINT
    L.H.   POINT WILSON.
    FORT WORDEN.

Lt.   Pt. HUDSON.

INLET.

Shield's Springs

PORT TOWNSEND BAY.

L.H.   MARROWSTONE PT.
    FORT FLAGLER.

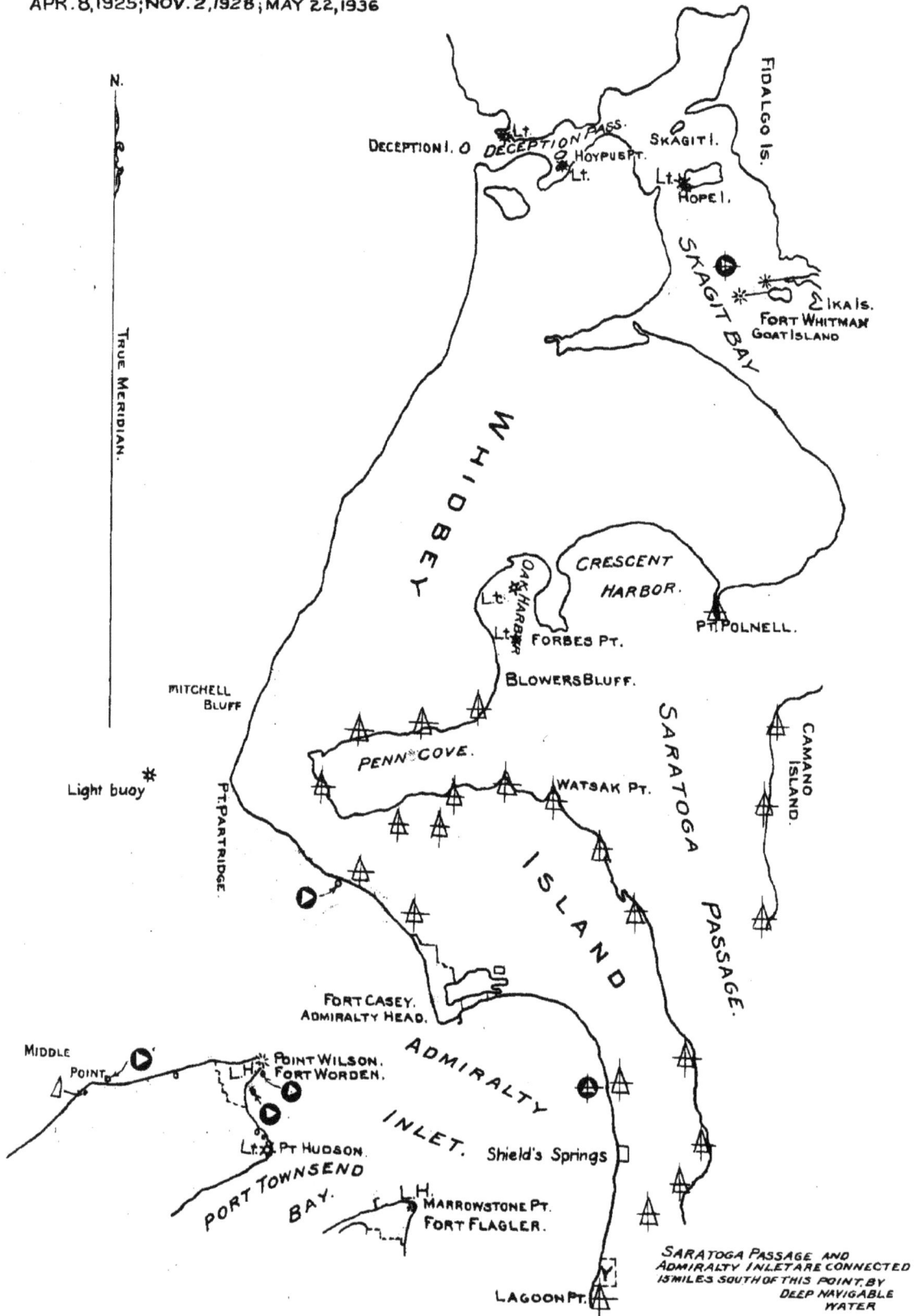

SARATOGA PASSAGE AND
ADMIRALTY INLET ARE CONNECTED
15 MILES SOUTH OF THIS POINT BY
DEEP NAVIGABLE
WATER

LAGOON PT.

Fort Worden (1898-1953) is located at Wilson Point, facing east on to Admiralty Inlet, near Port Townsend, Washington State. The 503-acre military reservation received twelve concrete batteries during the Endicott Program along with a large number of garrison buildings. It was named in General Orders 43 of 1900 for Adm. John L. Worden, USN. Fort Worden, Fort Flagler and Fort Casey were sited in a "triangle" shape to protect the entrance to Puget Sound. The post served at the headquarters for the harbor defenses of Puget Sound. The main gun line and mortar batteries were constructed from 1898 to 1901, while the secondary batteries were built from 1902 to 1906 on the high bluffs. A late Endicott Program designed 10-inch disappearing battery and a 12-inch Taft era battery were built between 1907 to 1910. In the 1920s several anti-aircraft guns were added as well as an observation balloon hangar which survives today in a modified form. The coming of World War II resulted in a major effort to modernize the defense of Puget Sound, so a 90mm AMTB battery was added in 1943. Harbor mines, anti-submarine nets, and anti-motor torpedo boat booms went across the strait to Fort Casey and Fort Flagler. The fort was declared surplus in 1953 and acquired by the Washington State Parks in 1957, while Artillery Hill remained in military service until 1970. The park is currently managed by Washington State Parks and Recreation. The State Park features camping, picnicking, and boating facilities with hiking and biking trails in and around the old coast artillery batteries. A Public Development Authority until recently managed the extensive remaining garrison buildings as a cultural arts and events center, but that is currently in transition due to financial issues. The Puget Sound Coast Artillery Museum located in one of the original 1904 barracks features several exhibits on coast artillery history and artifacts. A self-guided walking tour is available to the defenses. Today, Fort Worden State Park has one of the best remaining examples of a coast artillery post from the Endicott Program. All Washington State parks require a daily parking pass.

Fort Worden Gun Batteries

- **ASH – POWELL – QUARLES – RANDOL:** This was the main sequence of heavy guns for Fort Worden, placed high of the bluff of "artillery hill," bearing on the main channel into Puget Sound to the north and northeast. It had an unusual design history with numerous changes in armament and names. Plans were submitted on July 2, 1898 for a sequence of five 10-inch and two 12-inch guns. Due to the height of the bluff, all were to be on barbette carriages. Initial funding of $190,000 was made available on July 7th, 1898. The unusual design featured each gun on its own, raised pedestal, surrounded by a parapet wall. There was no communication or connections between the individual pedestals or between loading platforms. Three rectangular buildings were placed behind the sequence for service and relocator rooms. Ammunition was stored in magazines in traverses between guns No. 1 and 2, No. 3 and 4, No. 5 and 6, and again between No. 6 and 7. Projectiles were brought out from the magazines on carts and then raised to the loading level of the gun using the cranes attached to each carriage. Work was done on all seven emplacements from August 1898 to mid-1900. The first five emplacements were in a line, then extending with a bend to the west for the final two. The first guns were mounted on August 12, 1901. The entire battery was transferred on June 16, 1902 for a cost of $232,554.97. The emplacements were named in General Orders No. 27 of December 27, 1904. The names were Battery Randol (for Brevet Brigadier General Alanson Randol of Civil War service), Battery Quarles (for Captain Augustus Quarles killed in action at Churubusco Mexico in 1847), Battery Ash (for Brevet Colonel Joseph Ash killed in action at Todd's Tavern, VA in 1864), and Battery Powell (for Major James E. Powell killed in action at Shiloh TN in 1862 As originally configured emplacements No. 1-4 were Battery Randol, No. 5 was Battery Quarles, No. 6 was Battery Powell, and No. 7 was Battery Ash. The emplacments were armed in order: No. 1— 10-inch Model 1888 Bethlehem gun on Model 1893 barbette (#39/#7), No. 2— 10-inch Model 1888 Bethlehem gun on Model 1893 barbette (#6/#8), No. 3— 10-inch Model

PUGET SOUND

FORT WORDEN

POINT WILSON

GENERAL MAP

SERIAL NUMBER

Scale of Feet.

2000   1000   500   0   500   1000   2000

SECONDARY AND SUPPLEMENTARY
STATIONS AT SITE "C"

ADMIRALTY INLET

ACTIVE STATUS.

STRAITS OF JUAN DE FUCA.

QUARLES
ASH
BENSON
TOLLES
WALKER

VICARS
KINZIE
RANDOL
STODDARD
Q.M. WHF.
ENGR. WHF.
PUTNAM
Q.M.S.

Foot of Bluff
Bluff
ENGR. RES.

POWELL
BRANNAN

SECONDARY STATIONS
AT
SITE "B"

STRAITS OF JUAN DE FUCA

CEMETERY
RESERVATION

BOUNDARY

To reduce elevations to Mean Low Water add 9.49 ft.

EDITION OF MAY 7, 1921.
REVISIONS: APR. 8, 1925;
NOV. 2, 1928; MAY 22, 1936

N

BATTERIES

BRANNAN .... 4-12" M.
POWELL .... 4-12" M.
ASH .... 2-12" N. Dis.
KINZIE .... 2-12" Dis.
BENSON .... 2-10" "
QUARLES .... 3-10" N. Dis.
* RANDOL
TOLLES .... 2-6" Dis.
† STODDARD
† VICARS
WALKER .... 2-3" P.
PUTNAM .... 2-3" P.
A-ANTI-AIRCRAFT GUN 2-3"

* Guns removed
† Armament removed

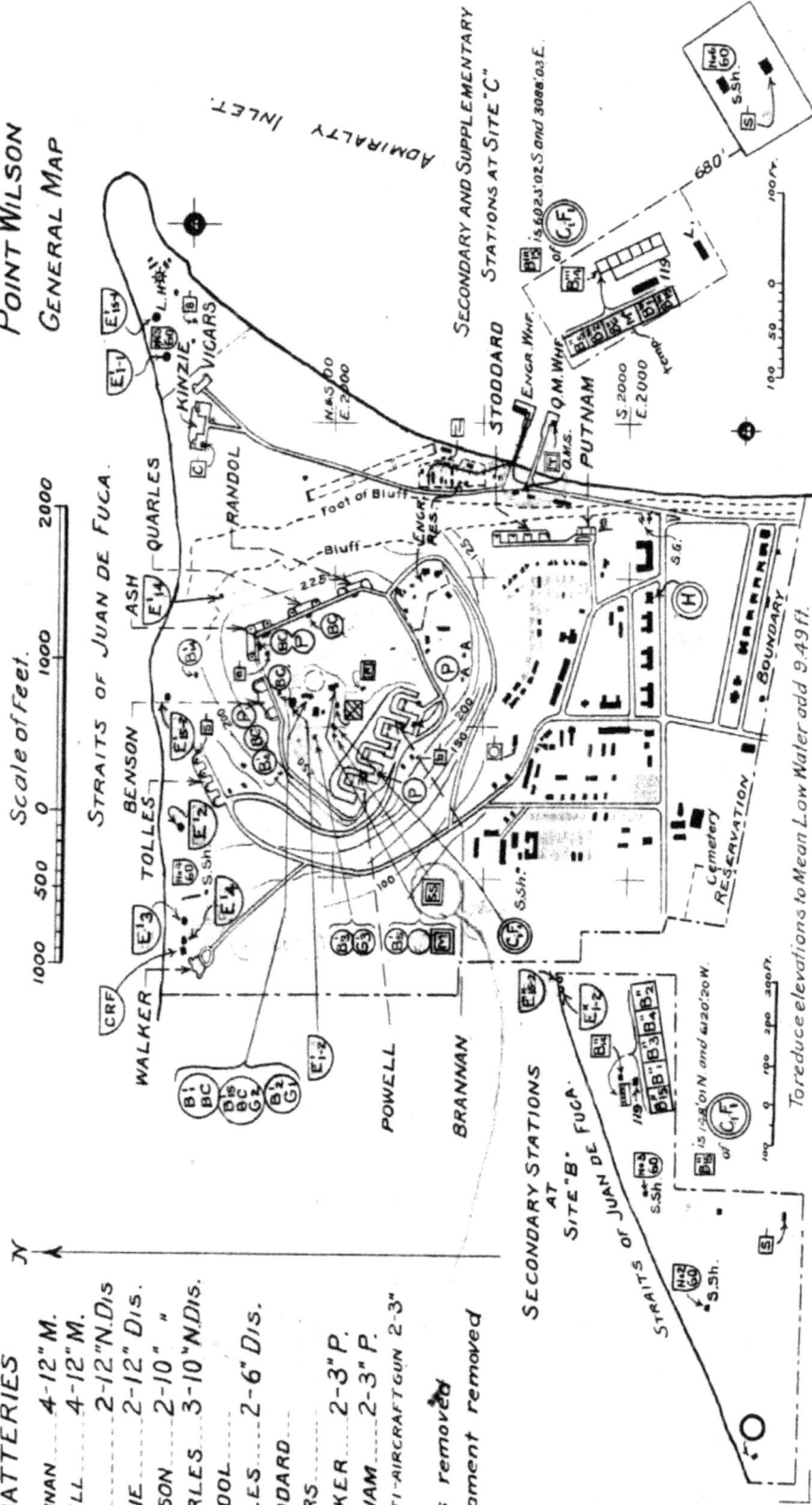

PUGET SOUND

FORT WORDEN-01

POINT WILSON

Scale of Feet

500   0   500

EDITION OF MAY 7, 1921.
REVISIONS: APR. 8, 1925,
NOV. 2, 1928, MAY 22, 1936

SERIAL NUMBER

KINZIE

N. & S. 00
E. 2000

H.W.

Foot of Bluff

ENGR. RESERVE

QUARLES

RANDOL

ASH

BENSON

BRANNAN

POWELL

TOLLES

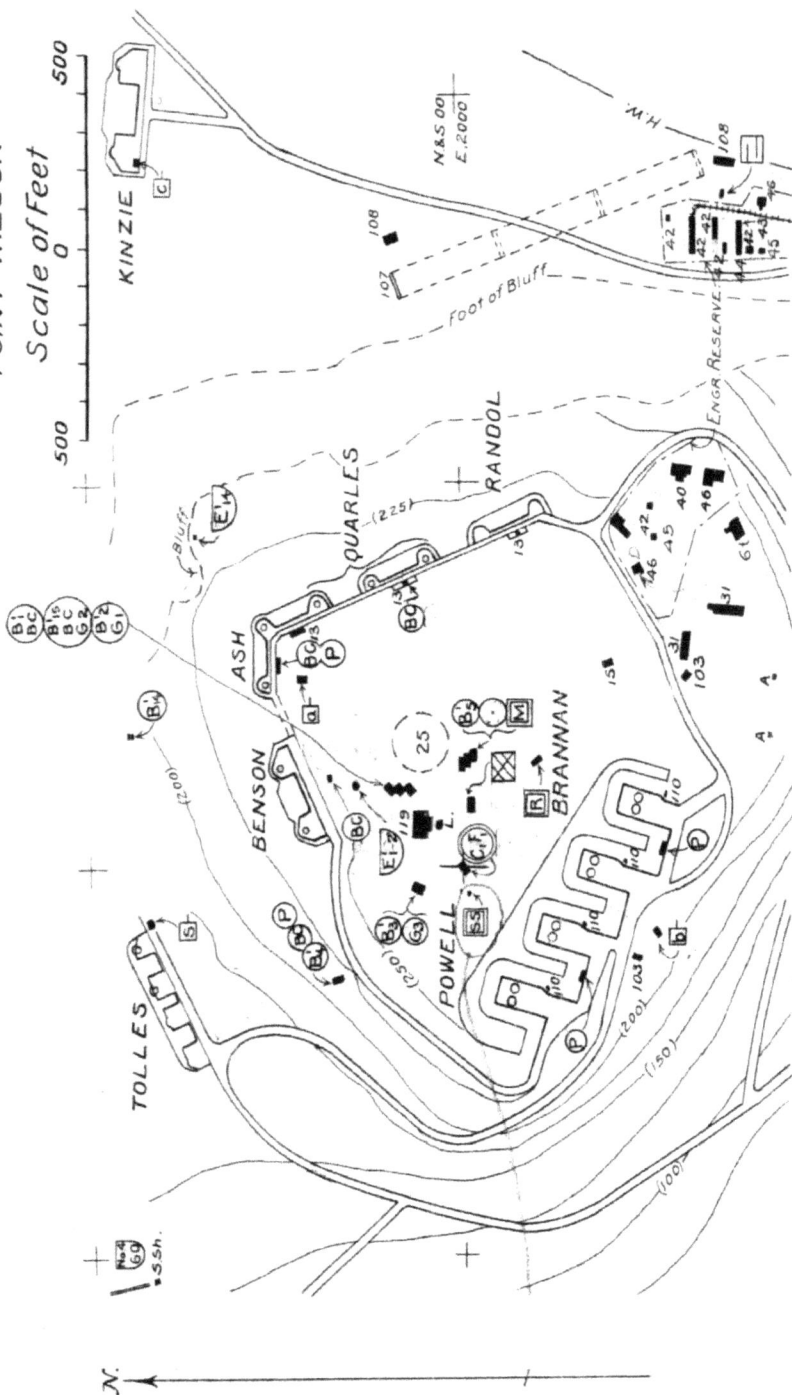

Active Status

N.

*Guns removed

LEGEND

6. N.C.OFFICERS QRS.
12.
13. CONCRETE ST. HO.
15. MILITIA STORE HOUSE.
101.
102.
107. RIFLE BUTTS.
108. TARGET SHED.
110. TELEPHONE BOOTH.
119. DORMITORY
25. Q.M. RESERVOIR.
31. ORDNANCE STOREHOUSE.
40. ENGR. OFFICE.
42. ENGR. STORE HOUSE.
43. ENGR. BLACKSMITH SHOP
44. ENGR. CARPENTER SHOP.
45. ENGR. GARAGE
46. ENGR. QUARTERS.
103. OIL HOUSE.

BATTERIES

BRANNAN .... 4 - 12"M.
POWELL ...... 4 - 12"M.
ASH .... 2 - 12"N.DIS.
KINZIE .... 2 - 12"DIS.
*RANDOL
BENSON ..... 2 - 10"DIS.
QUARLES ... 3 - 10"N.DIS.
TOLLES ...... 2 - 6"DIS.
A - Anti-aircraft gun 2 - 3"

* Guns removed

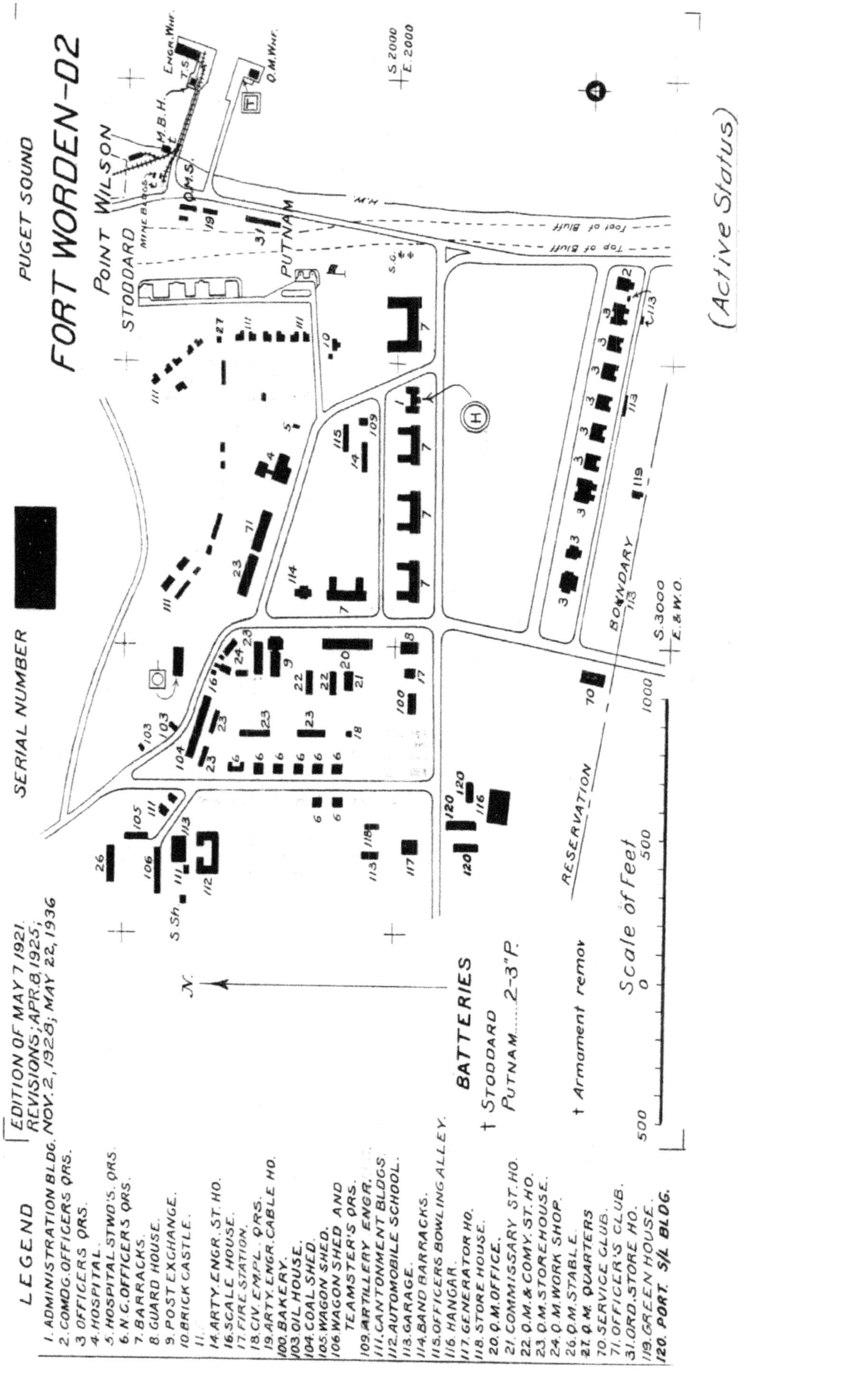

PUGET SOUND

FORT WORDEN-02

POINT WILSON

(Active Status)

SERIAL NUMBER

STODDARD

PUTNAM

Engr. Whf.
T.S.
M.B.H.
t.
Q.M.Whf.
Q.M.S.
Mine Bldg.
t.
H.W.
Foot of Bluff
Top of Bluff
Foot of Bluff
S.G.++
S. 2000 / E. 2000
S. 3000 / E. & W.O.
BOUNDARY
RESERVATION
N.

EDITION OF MAY 7 1921.
REVISIONS: APR 8 1925;
NOV. 2, 1928; MAY 22, 1936

BATTERIES
† Stoddard
Putnam ......2-3"P.
† Armament removed

Scale of Feet
500   0   500   1000

LEGEND

1. ADMINISTRATION BLDG.
2. COMDG. OFFICERS QRS.
3. OFFICERS QRS.
4. HOSPITAL.
5. HOSPITAL STWD'S. QRS.
6. N.C. OFFICERS QRS.
7. BARRACKS.
8. GUARD HOUSE.
9. POST EXCHANGE.
10. BRICK CASTLE.
14. ARTY. ENGR. ST. HO.
16. SCALE HOUSE.
17. FIRE STATION.
18. CIV. EMPL. QRS.
19. ARTY. ENGR. CABLE HO.
100. BAKERY.
103. OIL HOUSE.
104. COAL SHED.
105. WAGON SHED.
106. WAGON SHED AND
       TEAMSTER'S QRS.
109. ARTILLERY ENGR.
111. CANTONMENT BLDGS.
112. AUTOMOBILE SCHOOL.
113. GARAGE.
114. BAND BARRACKS.
115. OFFICERS BOWLING ALLEY.
116. HANGAR.
117. GENERATOR HO.
118. STORE HOUSE.
20. Q.M. OFFICE.
21. COMMISSARY ST. HO.
22. Q.M. & COMY. ST. HO.
23. Q.M. STOREHOUSE.
24. Q.M. WORK SHOP.
26. Q.M. STABLE.
27. Q.M. QUARTERS.
70. SERVICE CLUB.
71. OFFICER'S CLUB.
31. ORD. STORE HO.
119. GREEN HOUSE.
120. PORT. S/L BLDG.

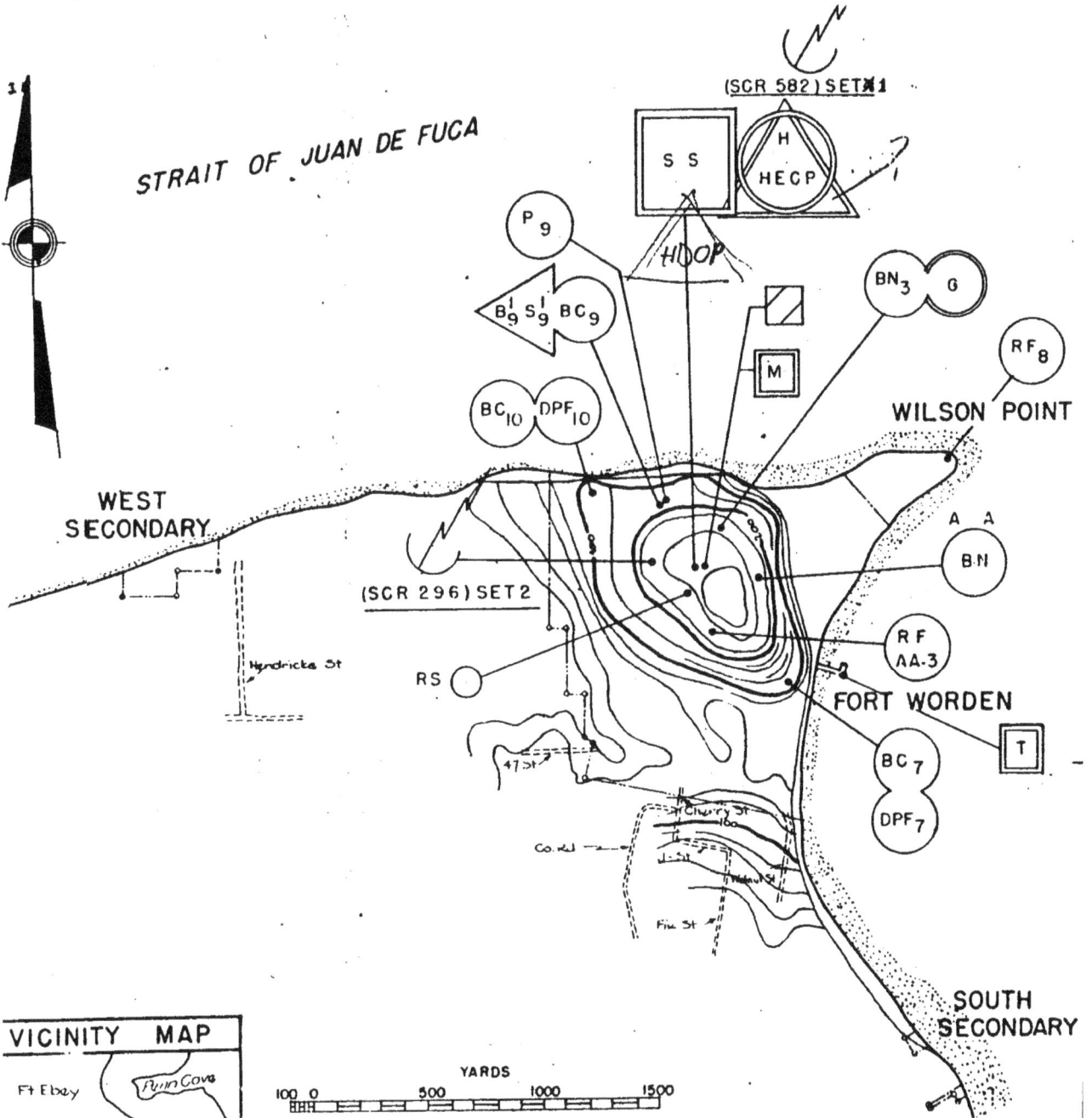

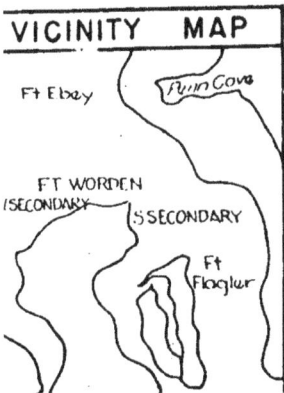

STRAIT OF JUAN DE FUCA

(SCR 582) SET 1

S S

H HECP

HDOP

P 9

BN₃ G

B⁹ S⁹ BC₉

RF₈

M

BC₁₀ DPF₁₀

WILSON POINT

WEST SECONDARY

A A

B N

(SCR 296) SET 2

R F AA-3

RS

FORT WORDEN

T

BC₇

47 St

DPF₇

Cherry St

Co. Rd

Walnut St

SOUTH SECONDARY

Fir St

VICINITY MAP

Ft Ebey

Rum Cove

FT WORDEN

(SECONDARY

S SECONDARY

Ft Flagler

YARDS

100 0   500   1000   1500

HARBOR DEFENSES OF PUGET SOUND SITE 8- SO. SECONDARY SITE 8A-WILSON PT. SITE 9- FORT WORDEN SITE 10-W. SECONDARY SHEET 1 OF 2	REVISED DATE
	1 AUG 44
PREPARED BY PLANNING OFFICE	DATE 1 JAN 44 EX NO 24B

Fort Worden 1938 (NARA)

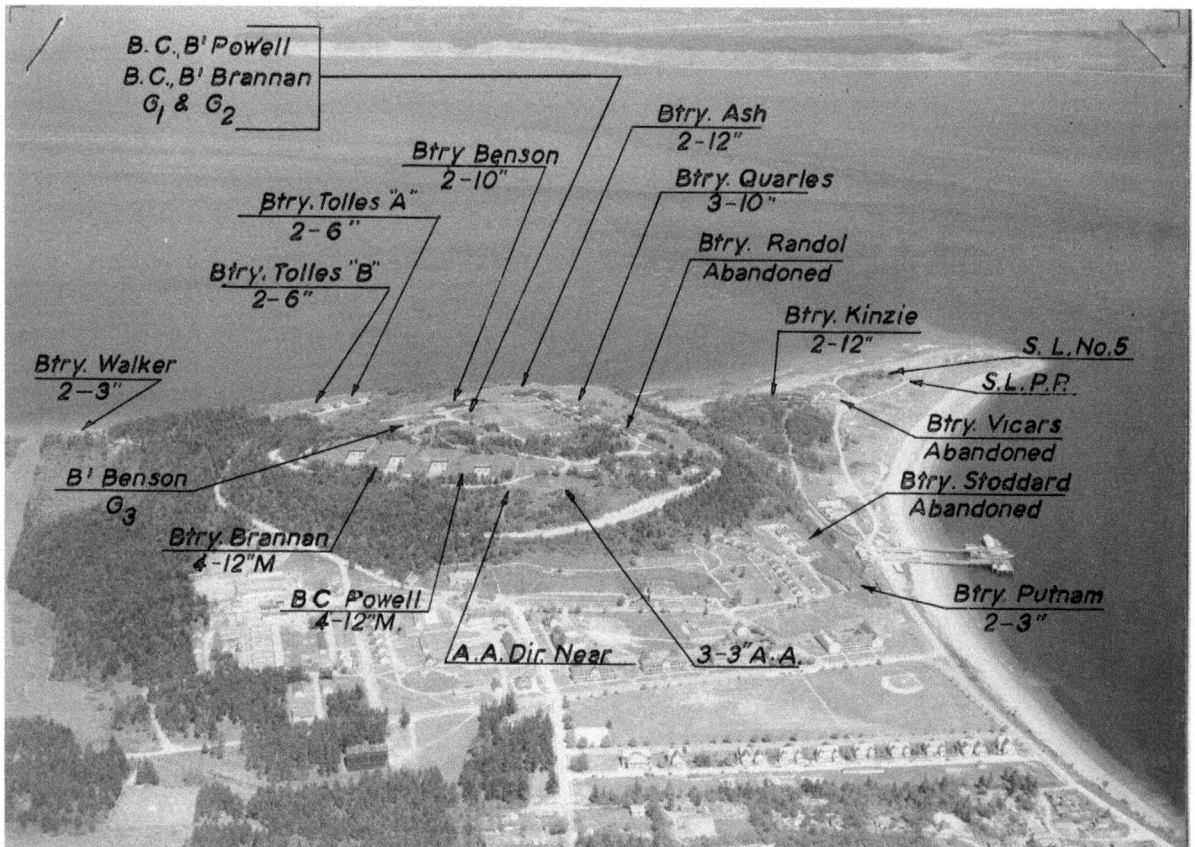

Fort Worden 1938 (NARA)

1888 Bethlehem gun on Model 1893 barbette (#36/#9), No. 4— 10-inch Model 1888 Bethlehem gun on Model 1893 barbette (#37/#1), No, 5— 12-inch Model 1888 Bethlehem gun on Model 1891 Altered Gun Lift Carriage (#9/#3), No. 6— 12-inch Model 1888 Bethlehem gun on Model 1892 barbette carriage (#10/#27) and No. 7— 10-inch Model 1888 Watervliet gun on Model 1893 barbette carriages (#12/#11). Criticism of the design came almost immediately. Lifting shells almost ten feet from the base of the pedestal to the breech was cumbersome, time consuming, and strained the cranes. Also, there was not enough room atop the pedestals to load, ram, or clean the tubes. By late 1902 (only five months after battery transfer) the entire sequence was considered unsatisfactory and suggestions were made to rebuild the battery. Plans were submitted and work done for reconstruction in 1902-03. The area around the gun pedestals was raised to form a new loading platform. Hoists were installed to lift shells and shot to the level of this platform. The relocator rooms were retained, but new guard rooms and latrines were installed, and a few years later battery commander stations were built on top. Rebuilding funds amounted to $72,500. In 1904 it was proposed to interchange the guns at emplacements No. 5 and 7, resulting in better magazine service by having five 10-inch guns adjacent, and then two 12-inch (though poorly situated emplacement No. 5 was served by the magazine shared with the 12-inch No. 6 position, subsequently it was seldom used and almost became a reserve gun). Two serious accidents also encouraged the rebuilding. On September 22, 1904 emplacement No. 6 (Battery Powell) experienced the breakage, while firing, of two base ring bolts with another five being cracked, requiring repair and remounting. Also, on November 7, 1905 the altered gun lift carriage at Battery Wilhelm, Fort Flagler was heavily damaged in a service practice. As a result, the gun lift carriage at Worden was eventually sent to Flagler as a replacement (it was actually used for parts to repair the one damaged and not mounted intact again),while a new 12-inch Model 1892 barbette carriage was obtained (Carriage No. 9, from the proof line at Sandy Hook Proving Grounds) so that the battery in this sequence had a uniform armament. By 1907, with the newest generation of chain hoists installed and the guns rearranged, the battery emerged with emplacements No. 1-2 as Battery Randol, No. 3-5 as Battery Quarles, and No. 6-7 as Battery Ash (the two 12-inch guns). The name Powell was transferred and used for one of tactically split mortar batteries on post. Battery Randol was the first disarmed, its two Model 1888 10-inch guns being dismounted and sent to Watervliet on June 7, 1918. The carriages reamained until scrapped in place, at least one as late as 1932. The entire armament of Battery Quarles was sent to Canada to be emplaced in new batteries in Newfoundland, Quebec, and Nova Scotia in 1941. Battery Ash continued to serve until World War II, authority for removal being given on October 24, 1942. The seven-gun emplacement still exists at Fort Worden State Park. These batteries are open to the public.

- **KINZIE:** After the completion of the Endicott works for the entrance to Puget Sound, there was still a need for heavy, long-range defensive fire to the north into the Strait of Juan De Fuca. The Taft Board recommended installation of a new heavy battery. While 14-inch guns were initially proposed, eventually the most modern type of 12-inch guns and emplacement was selected. On September 18, 1908 the location was selected. It was to be emplaced near the beach of Point Wilson, generally to the left flank of 5-inch Battery Vicars. Field of fire was north and just a little east. Detailed plans were submitted on December 11, 1908. It generally followed the most current type plan, but borrowing a few features from the 14-inch type plan. The distance between guns was an extended 222-feet, providing for more room in the central traverse for a plotting and battery commander station. The crest was only 38.5-feet. Each gun position was a flank emplacement allowing for maximum field of fire to the west and east. Actual construction was done from September 1908 to early 1910. Transfer was made on January 10, 1912 for a cost of $207,832.50. The battery was named in General Orders No. 245 of December 13, 1909 for Brigadier General David H. Kinzie, who had died in 1904. It was armed with two 12-inch Model 1895M1 Watervliet guns on Model 1901 carriages (#60/#21 and #73/#20). The carriages were modified in 1916-1917 for increased elevation and range. The battery was retained in service throughout the interwar period, not being removed until more modern armament had been completed in 1944. The authority to disarm and eliminate the battery came on February 15, 1944. For a while the emplacement was used for 16-inch shell storage. It still exists in good condition at the Fort Worden State Park. This battery is open to the public.

- **BENSON**: One of the last Endicott batteries built at Fort Worden. The plan was submitted on April 20, 1904. It was located to the west of the existing main gun line at the same crest height. A small tunnel was built for the road passing behind Benson to Battery Ash as part of the project. Otherwise, the battery followed the most modern type plans for 10-inch batteries and had wide loading platforms and chain hoists from its initial construction. Work was done in 1904-1906. Transfer was delayed until April 24, 1908 for a construction cost of $142,500. It was named in General Orders No. 194 of December 27, 1904 for Captain Henry Benson who was mortally wounded at Malvern Hill, VA in 1862. Along with Battery Kinzie, it was considered to be the most modern armament of the Puget Sound defenses for many years. The battery was armed with two 10-inch Model 1900 guns on Model 1901 disappearing carriages (#12/#6 and #14/#7). The battery commander's station was emplaced on the slope directly behind the emplacement. New powder hoists were added in 1911. The armament continued to serve until World War II. It was removed under authority of June 4, 1943. The removed counterweights from the disappearing carriages were sent to Fort Stevens to be used as mine anchors. The emplacement still exists at the Fort Worden State Park. This battery is open to the public.

- **BRANNAN – POWELL**: The battery for sixteen 12-inch mortars emplaced at the Point Wilson reservation of Fort Worden. Plans were submitted on July 6, 1899. It was sited some distance to the rear of the main battery, generally directed to the north. It followed the type plans but was enough later so that it incorporated the new, wider pits now recommended. Also, these pits were turned slightly from the perpendicular to better protect them from possible flank fire coming from the northwest. It had four pits arranged in a staggered line, with the magazines on the sides and under the front parapet of each pit. Ammunition service was separate for each pit, magazines were not shared. Work was done from December 1899 to March 1900. Transfer was made on June 16, 1902 for a cost of $162,102.12, Initially all four pits were named in General Orders No. 194 of December 27, 1904 for Brevet Major General John M. Brannan of Civil War service. In 1906 all four pit 16-gun mortar batteries were split, the two western pits were named for Major James Powell also of Civil War service. That name was transferred here from previous use for a 12-inch gun emplacement in the fort's main sequence. The armament was installed and proofed on July 11, 1903. It consisted of sixteen 12-inch Model 1890M1 mortars on Model 1896 mortar carriages, For Battery Brannan: Builders tubes #49/#256, #50/#258, #51/#259, Niles tubes #21/#252, #24/#254, Watervliet tubes #32/#257, #54/#256, and #56/#253. For Battery Powell: Builders tubes #27/#247, #52/#248, Niles tubes #1/#250, #15/#249, #16/#251, #20/#244, #22/#245, and Watervliet tube #105/#246. Problems were encountered with racer ring cracking during early proofing. On July 10, 1903 the racer of carriage #244 was badly cracked and that of #246 was completely wrecked when firing practice took place for zones nine and ten. Again in 1906 the top carriage of #255 was damaged. Apparently all three of these carriages were repaired and put back into service. In 1918 two mortar tubes were removed from each pit (Brannan tubes #50, #24, #32, #56 and Powell tubes #1, #15, #22, and #105). They were sent out for conversion to railway mounts, the remaining carriages were scrapped in 1920. In 1921 the unoccupied mortar pits were filled in. The remaining eight mortars served until ultimately removed in the early 1940s (Brannan under authority of February 15, 1944 and Powell on October 24, 1942). The temporary HDCP was emplaced in the battery in 1941. After removal of armament the emplacements continued to be used for anti-aircraft training and munitions storage. The batteries still exist at the Fort Worden State Park and are open to the public.

- **STODDARD:** A battery for four 6-inch disappearing guns emplaced on the east side of the reservation, on the rise of land above the beach to the east of the main cantonment which fired to the east. The plan for the first two emplacements was submitted on June 29, 1903, with the second pair on May 23, 1904. It closely followed the type plans and had four-gun pits and the larger traverses between guns No. 1 and 2 and then between No. 3 and 4 containing the shared magazines. Concrete work was done in 1903-1905. Transfer was made on May 21, 1907 for a construction cost of $91,000. It was named in General Orders No. 194 of December 27, 1904 for Major Amos Stoddard, 1st U.S. Artillery who died in action in 1813 at Fort Meigs. It was armed with four 6-inch Model 1903 Watervliet guns on Model 1903 disappearing carriages (#17/#44, #19/#43, #45/#89, and #61/#90). The gun tubes were removed on November 9, 1917 and were sent to Watervliet for conversion to field mounts, while the carriages were scrapped in place. The emplacement was later used for civil defense storage. The emplacement still exists at the Fort Worden State Park and is open to the public.

- **TOLLES:** A battery for four 6-inch disappearing guns emplaced to the west of Battery Benson. It was submitted on June 29, 1903. The battery was to protect the north and west approaches to the fort, which were not otherwise covered in the defenses. It followed standard type plans, with the platforms in line and major, shared traverse magazines between the first and second pair of guns. Concrete work was done from July 1903 to early 1904. Transfer was made on May 21, 1907 for a construction cost of $104,500. It was named in General Orders No. 194 of December 27, 1904 for Brevet Colonel Cornelius Tolles, U.S. Volunteers, who was killed in 1864 following the military action in Newton, Virginia. It was armed with four 6-inch Model 1903 Watervliet guns and Model 1903 disappearing carriages (#39/#56, #50/#58, #58/#59, and #66/#60). The guns in emplacements No. 3 and 4 were removed in November 1917 for service in the European theatre as wheeled field mounts. In June 1932 plans were made to mount two 6-inch Model 1900 guns and pedestals in the two empty emplacements. The new guns were to be transferred from a temporary battery emplaced during World War I at Willipa Harbor (which in turn had originally been mounted at Battery Freeman, Fort Stevens). The two emplacements were modified with new raised platforms and base rings during reconstruction during November 1936 to June 1937. This work was turned over on August 23, 1937 at a cost of $8,891.86 and the two guns were soon mounted (Watervliet tubes Model 1900 and pedestal carriages Model 1900 #4/#24 and #8/#25). At this time the battery was divided tactically. The two pedestal guns became Tolles B, the disappearing guns becameTolles A. During 1942 the loading platforms of Tolles A were extended. The disappearing guns of Tolles A were authorized for removal on June 11, 1943 but the pedestal guns served until after the war ended, being removed in 1946. The battery still exists at the Fort Worden State Park and is open to the public.

- **VICARS:** An emplacement for two 5-inch rapid-fire guns emplaced below the main fort reservation on a slight ridge in the sands of Point Wilson, close to the original lighthouse reservation. The plan was submitted on May 16, 1899. It followed the typical RF battery plan of the time, with two separate flank gun platforms and lower-level magazines between the guns in the traverse. Work was done essentially entirely in 1899. Transfer was made on June 16, 1902 for a cost of $11,000. It was armed with two 5-inch Model 1897 Bethlehem guns on Model 1896 balanced pillar carriages (#9/#9 and #22/#20). The battery was named in General Orders No. 194 of December 27, 1904 for 1st Lieutenant Thomas Vicars, U.S. Infantry who was killed on Bayan, the Philippines in 1902. When nearby Battery Kinzie was constructed in 1908, special effort was taken to avoid impacting the Vicars position. The armament was removed in 1918 under authority granted on November 9, 1917. For a period after World War II the emplacement was buried with earth, but that was removed, and it is open to the public at Fort Worden State Park.

- **PUTNAM:** A battery for two 3-inch pedestal mounts built adjacent to, and just to the south of Battery Stoddard on the eastern flank of the Worden reservation. It also fired to the east. Submission of the plan was made on June 6, 1903, at the same time as the other 3-inch battery for the fort, Battery Walker. It closely followed the highly standardized type plans for these later 3-inch, pedestal emplacements. Concern for possible firing from the west did prompt the addition of an earthen parados immediately behind the battery. Work was done in late 1903, but transfer was not concluded until May 21, 1907 for a cost of $12,000. It was named in General Orders No. 194 of December 27, 1904 for Colonel Haldimand S. Putnam, 7th New Hampshire Volunteers, who had died in 1863 during the American Civil War. It was armed with two 3-inch Model 1903 guns and pedestal mounts (#23/#14 and #24/#15). This armament was carried throughout the service life of the battery. By 1945 the emplacement was still actively serving as Tactical Battery No. 7. At one point there was a proposal to move these guns to Deception Pass, but it was never implemented. At the end of the war the guns were dismounted, the armament retained for a while as possible spares for Battery Walker. The emplacement still exists at the Fort Worden State Park and is open to the public.

- **WALKER:** The second dual, 3-inch battery for Fort Worden. The plan was submitted (along with the plans for sister battery Putnam) on June 6, 1903. It was located still on the Point Wilson bluff to the west of Battery Tolles, in fact near the western extreme of the reservation. It fired to the northwest to cover the possible landing and anchorage sites on that flank of the main fort. It was of standard, type design. Work was done in 1904. Transfer was made on May 21, 1907 for a cost of $12,000. It was named in General Orders No. 194 of December 27, 1904 for Lieutenant Colonel Samuel Walker, Texas Mounted Rangers who was killed in action at Huamantla Mexico in 1847 during the Mexican War. The armament was mounted in 1910 and consisted of two 3-inch Model 1903 guns and pedestal mounts (#84/#77 and #83/#76). During proof firing gun #83 developed a crack along the entire length of the jacket on June 24, 1910. The tube was returned to Watervliet Arsenal, where it was scrapped. It was replaced with Watervliet tube #10. This battery was used extensively for training at Fort Worden, the gun barrels are each recorded as having fired over 1000 rounds each by 1928. On October 30, 1943 the two guns tubes were replaced with two from Battery Wansboro at Fort Flagler (#35/#76 and #36/#77). The battery served during the war as Tactical Battery No. 10 until being disarmed shortly after the end of the war in 1946. The emplacement still exists at the Fort Worden State Park and is open to the public.

- **AMTB Point Wilson:** A 1943 Program AMTB battery of two 90mm fixed and two 90mm mobile guns. It was built from November 1, 1943 until February 24, 1944 for a cost of $26,142. It was emplaced on the rip-rap in front (to the east) of the lighthouse at Point Wilson. It consisted of two simple concrete gun blocks and BC station. The battery served as local Tactical Battery No. 8. It was disarmed in 1947. The broken and somewhat obscured gun blocks still exist. The site is open to the public, but in poor condition.

Fort Worden State Park (Terry McGovern)

Fort Flagler State Park (Terry McGovern)

Fort Flagler (1897-1953) is located at northern tip of Marrowstone Island, facing north on to Admiralty Inlet, near Port Hadlock, Washington State. The 783-acre military reservation received nine concrete batteries during the Endicott Program along with an extensive set of garrison buildings. It was named for Brig. Gen. Daniel W. Flagler, U.S. Army who was Chief of Ordnance. Fort Flagler, Fort Worden and Fort Casey were sited in a "triangle" shape to protect the entrance to Puget Sound. In the 1920s several anti-aircraft guns were added. The post was placed in caretaker status in the 1937. The post was reactivated in 1940 and upgraded with new barracks and a 90mm AMTB battery. The fort was declared surplus in 1953 and acquired by the Washington State Parks in 1955. The large and isolated state park features camping, group camps, and vacation housing in the 1,451-acre park which includes boating facilities, hiking and biking trails, picnic facilities. A seasonal concession is located at the lower campground. The park features a small museum open daily with guided historical tours of the military sites and the restored post hospital on weekends. In 1963, two 3-inch rapid-fire guns were re-installed in Battery Wansboro when they were recovered from Fort Wint in Subic Bay, Philippines. All Washington State parks require a daily parking pass.

Fort Flagler Gun Batteries

- **WILHELM:** The battery for two 12-inch barbette guns on the Marrowstone Point reservation of Fort Flagler. These were the center two guns of the six-unit main sequence located on the large bluff at the point. They fired to the north (No. 2) and northeast (No. 1). Using funds from the Act of June 6, 1896, the plan was submitted on November 30, 1896, with further modifications onJanuary 30 and April 17, 1897. The final plan was adapted to the site. The individual emplacement crests were "sunk" to ground level. The platforms and parapets around them were irregular to fit the space and desired field of fire. Magazines were on the lower level between the pairs of guns. This 12-inch battery had 188-feet between platforms, and 200-feet on either side to the adjacent left and right flank 10-inch batteries. As originally built it had balanced platform lifts for carts and individual carriage cranes. The battery was basically completed by July 1899, and the armament was mounted in early 1900. It was transferred on August 17, 1902 for a cost of $71,190. It was named in General Orders No. 16 of February 14, 1902 for Captain William Wilhelm who died in 1902 during the Philippine Insurrection. It was armed with two 12-inch Model 1888MII Watervliet guns on Model 1891 Altered Gun Lift Carriages (#35/#2 and #25/#1). These two tubes were in a fire on the beach where they were delivered in February 1899, the wood cribbing caught fire from a nearby driftwood fire. However, when the guns were examined in May of 1900, no damage was revealed. On November 7, 1905 the No. 2 emplacement carriage was badly damaged in a service firing practice accident. The AGL carriage formerly at Battery Ash, Fort Worden was transferred to Fort Flagler in late 1907 and the damaged #1 AGL was repaired using parts and pieces from the spare #3 carriage. In fact, these modified gun lift carriages were never entirely satisfactory, but the two at Battery Wilhelm did serve throughout the service life in a variety of readiness states until listed for removal by authority of October 24, 1942. Guns and carriages were sold for scrap in December 1942. The emplacement still exists at the Fort Flagler State Park and is open to the public.

- **RAWLINS:** A battery for two 10-inch barbette guns built on the right flank of the main sequence at Fort Flagler. All six emplacements of this sequence were built together in 1897-1899. These two 10-inch emplacements fired to the northeast. Using funds from the Act of June 6, 1896, the plans were submitted on November 30, 1896, with further modifications on January 30 and April 17, 1897. The plan was highly adapted to the site. The individual top crests were at ground level. The platforms and parapets around them were irregular to fit the space and desired field of fire. Magazines were on the lower level between the pairs of guns. This 10-inch battery had 156-feet between

PUGET SOUND

MARROWSTONE POINT.

FORT FLAGLER

GENERAL MAP

FORT CASEY CABLE

FORT CASEY

ADMIRALTY INLET

Ocean Cable

L.H.

RAWLINS

GRATTAN

WANSBORO

Q.M. Whf.

LEE

WILHELM

REVERE

Q.M. Whf.

DOWNES

CALWELL

BANKHEAD

Abandoned, Unusable

Flag staff

CR.F.

For details of the area
within this parallelogram
See the Map-D1.

Boundary

Country Road

Water Pipe Line

Water Pipe Line

Water Reservoir

PORT TOWNSEND BAY.

True Meridian

KILISUT HARBOR

Boundary

PLANE OF REFERENCE.
To reduce elevations to Mean Low Water add 5.37 ft.

1000   0   1000   2000   3000 Ft.

Caretaking Status.

BATTERIES.

BANKHEAD--4-12"M.
WILHELM--2-12"N.Dis
† RAWLINS ------
REVERE ----- 2-10"N.Dis
† GRATTAN ------
† CALWELL ------
† LEE ----------
DOWNES--2-3"P.
WANSBORO--2-3"P.

A-Anti aircraft gun--2-3"

† Armament removed

LEGEND.

EDITION OF APR. 23, 1915.
REVISIONS: DEC. 7, 1915; NOV. 8, 1916; DEC. 6, 1919;
MAY 7, 1921; APR. 8, 1925; NOV. 2, 1928; MAY 22, 1936

SERIAL NUMBER

1. ADMINISTRATION BLDG.
2. COMMANDING OFFICER'S QUARTERS
3. OFFICER'S QUARTERS.
4. HOSPITAL.
5. HOSPITAL STWD'S.QRS.
6. N.C.OFFICERS'QRS.
7. BARRACKS.
7a. DORMITORY.
7.
7.
7.

8. GUARD HOUSE.
9. POST EXCHANGE.
10. POST OFFICE.
11. ARTILLERY ENGR.ST.HO.
12. FIRE HOUSE.
13. BAKERY.
14. BLACKSMITH SHOP.
15.
16. LAVATORY.
17. CARPENTER SHOP.
18. OIL HOUSE.
19. FUEL SHED.
100. TELEPHONE BOOTHS.
101. RIFLE BUTTS.
102. ROOT HOUSE.
103. STABLE.
104. WAGON SHED.
105. HARNESS ROOM.
106. SHOOTING GALLERY.
21. Q.M.STOREHOUSE.
22. Q.M.STOREHOUSE.
31. ORDNANCE ST.HO.
41. ENGR. DEPT. BLDGS.
113. TEAMSTERS QRS.
70.
71. BOWLING ALLEY.
72. SERVICE CLUB.

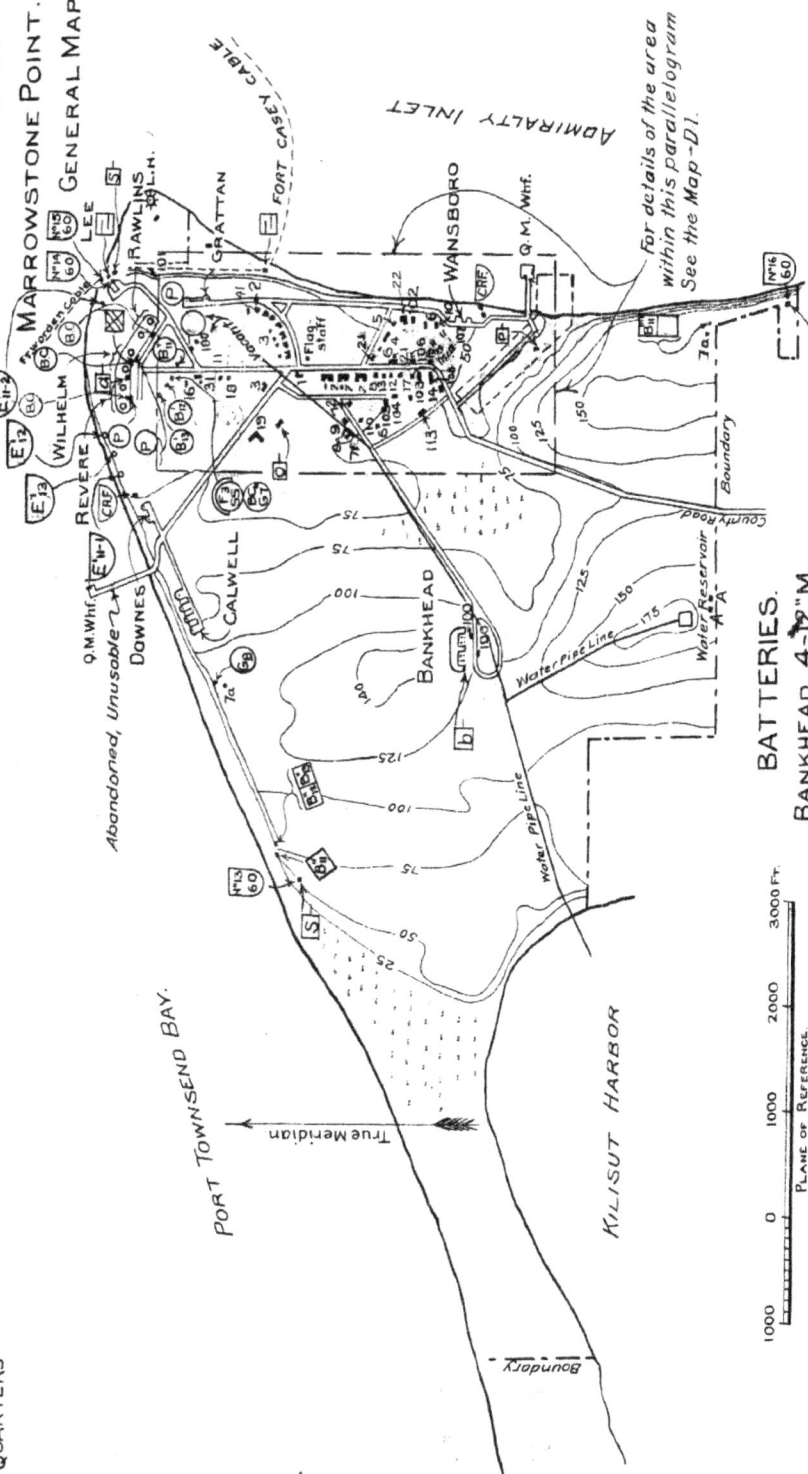

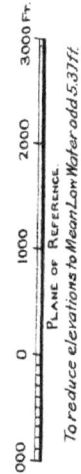

107. CANTONMENT 16 BLDGS
108. PLUMBING SHOP.
109. TOOL HOUSE.
110. PAINT SHOP.

PUGET SOUND

FORT FLAGLER-DI.

MARROWSTONE POINT.

BATTERIES

† GRATTAN
WANSBORO ...... 2-3"P.
† Armament removed

Scale of Feet.

1000

500

0

500

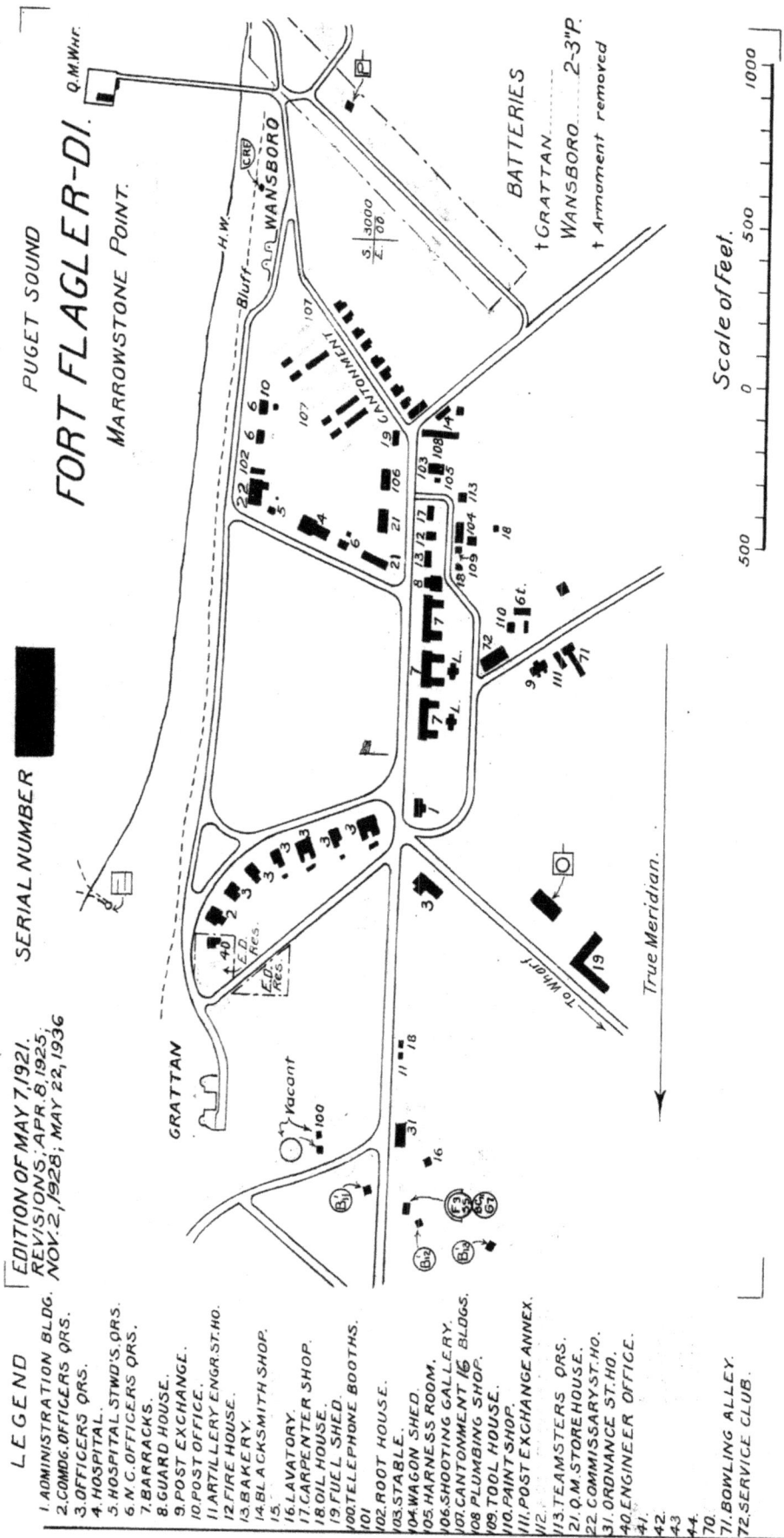

Caretaking Status

SERIAL NUMBER

EDITION OF MAY 7,1921.
REVISIONS: APR. 8,1925;
NOV. 2,1928; MAY 22,1936

GRATTAN

True Meridian.

LEGEND

1. ADMINISTRATION BLDG.
2. COMDG. OFFICERS QRS.
3. OFFICERS QRS.
4. HOSPITAL.
5. HOSPITAL STWD'S. QRS.
6. N.C. OFFICERS QRS.
7. BARRACKS.
8. GUARD HOUSE.
9. POST EXCHANGE.
10. POST OFFICE.
11. ARTILLERY ENGR. ST. HO.
12. FIRE HOUSE.
13. BAKERY.
14. BLACKSMITH SHOP.
15.
16. LAVATORY.
17. CARPENTER SHOP.
18. OIL HOUSE.
19. FUEL SHED.
100. TELEPHONE BOOTHS.
101
102. ROOT HOUSE.
103. STABLE.
104. WAGON SHED.
105. HARNESS ROOM.
106. SHOOTING GALLERY.
107. CANTONMENT 16 BLDGS.
108. PLUMBING SHOP.
109. TOOL HOUSE.
110. PAINT SHOP.
111. POST EXCHANGE ANNEX.
112.
113. TEAMSTERS QRS.
21. Q.M. STORE HOUSE.
22. COMMISSARY ST. HO.
31. ORDNANCE ST. HO.
40. ENGINEER OFFICE.
41.
42.
43.
44.
70.
71. BOWLING ALLEY.
72. SERVICE CLUB.

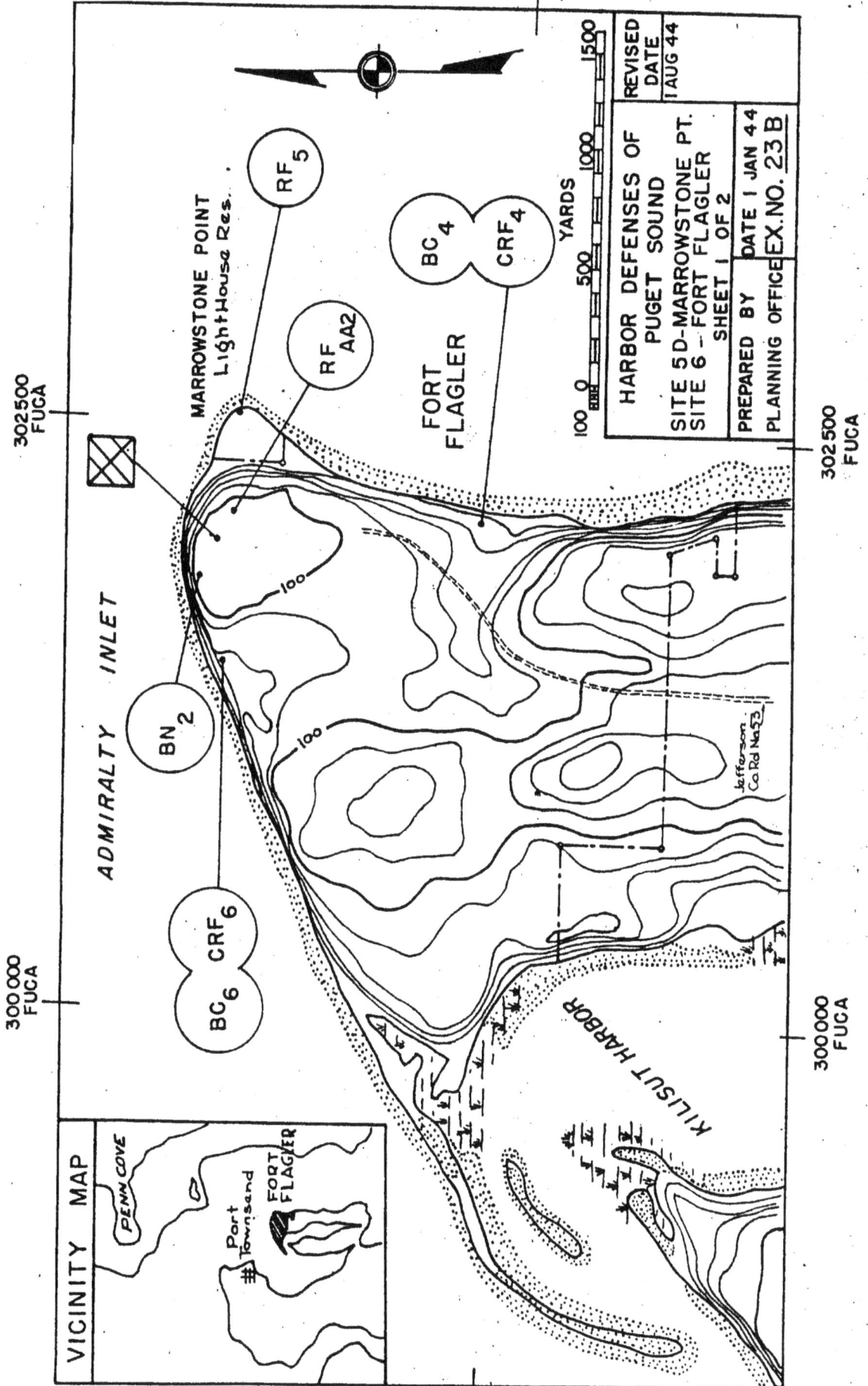

RF 5

MARROWSTONE POINT Light-House Res.

RF AA2

BC 4 CRF 4

FORT FLAGLER

302500 FUCA

ADMIRALTY INLET

100

100

BN 2

302500 FUCA

CRF 6

BC 6

300000 FUCA

300000 FUCA

KILISUT HARBOR

Jefferson Co Rd No 53

YARDS

100   0   500   1000   1500

HARBOR DEFENSES OF PUGET SOUND

SITE 5D-MARROWSTONE PT.
SITE 6 – FORT FLAGLER
SHEET 1 OF 2

| PREPARED BY | DATE 1 JAN 44 | REVISED DATE |
| PLANNING OFFICE | EX. NO. 23 B | 1 AUG 44 |

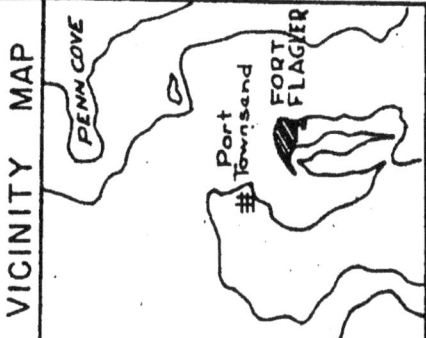

VICINITY MAP

PENN COVE

Port Townsend

FORT FLAGLER

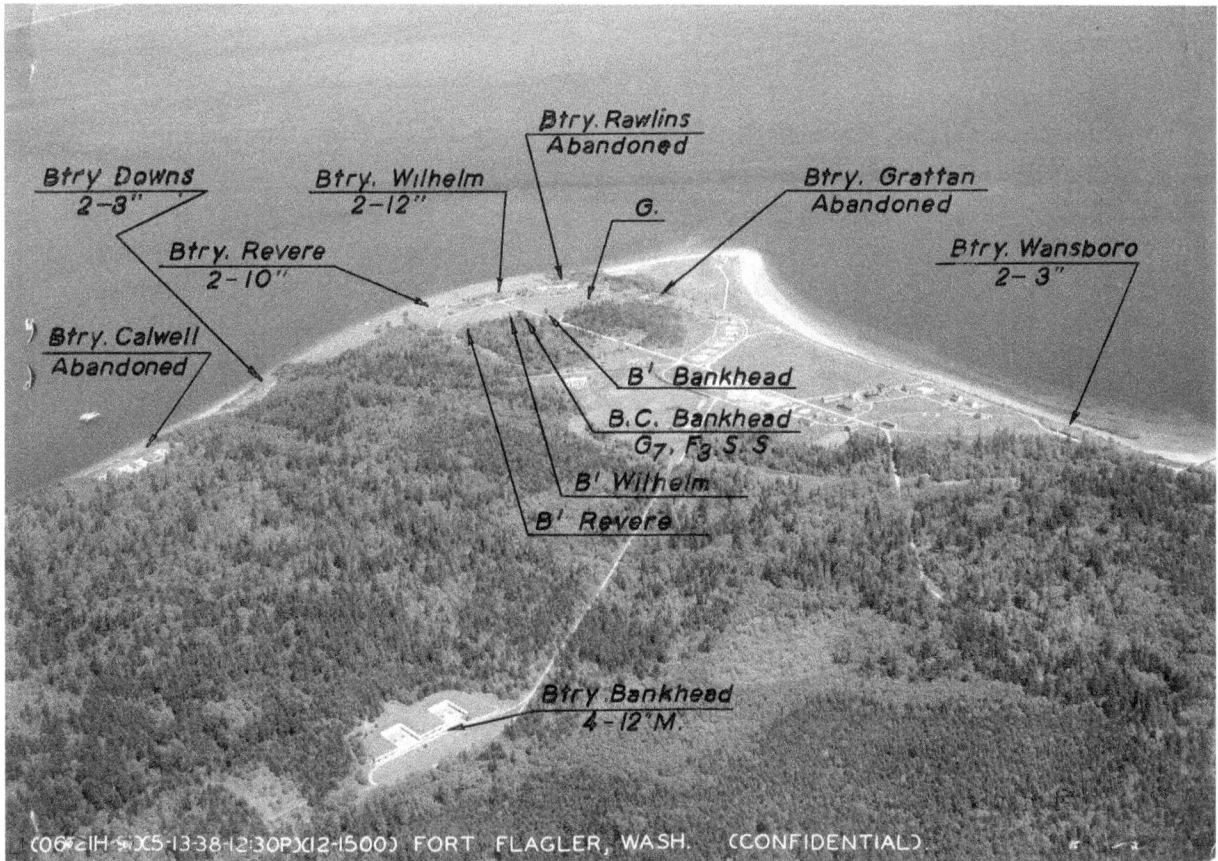

Fort Flagler 1938 (NARA)

Fort Flagler 1932 (NARA)

platforms, and 200-feet from the adjacent 12-inch battery. As originally built it had balanced platform lifts for carts and individual carriage cranes. Transfer was made on August 17, 1902 for a cost of $60,000. The battery was named in General Orders No. 20 of January 25, 1906 for Brigadier General John Aaron Rawlins, former Chief of Staff and Secretary of War. It was armed with two 10-inch Model 1888MII Bethlehem guns on barbette carriages Model 1893 (#35/#10 and #28/#3). This armament was listed for removal on May 29, 1918 for fitting into railway mounts, and were removed by June 7, 1918. The carriages were scrapped by December 10, 1920. In later years the battery emplacements were modified to hold a 3-inch anti-aircraft guns (shifted from a location at the south of the fort reservation). The pits were filled-in, and new base rings laid to mount 3-inch Model 1917 AA guns on December 9, 1942 (a third gun was placed a short distance in front of the two in the old emplacements). These were dismounted after the war, but the emplacement as modified still exists at the Fort Flagler State Park and is open to the public.

- **REVERE:** A battery for two 10-inch barbette guns built on the left flank of the main sequence at Fort Flagler. All six emplacements of this sequence were built together in 1897-1899. Using funds from the Act of June 6, 1896, the plans were submitted on November 30, 1896, with further modifications on January 30 and April 17, 1897. The plan was highly adapted to the site. The aprons were at ground level, loading platforms sunk below them. The platforms and parapets around them were irregular to fit the space and desired field of fire. Magazines were on the lower level between the pair of guns. This 10-inch battery had 156-feet between platforms, and 200-feet to the adjacent 12-inch battery. They were oriented to fire to the north. As originally built it had balanced platform lifts for carts and individual carriage cranes. The battery was transferred on August 17, 1902 for a cost of $60,000. It was named in General Orders No. 20 of January 25, 1906 for Colonel Paul J. Revere who was mortally wounded at Gettysburg, PA in 1863. The battery was armed with two 10-inch Model 1888MII Bethlehem guns on Model 1893 barbette carriages (#29/#4 and #30/#5). While it was anticipated at one point during World War I to remove this armament, that action was rescinded on July 18, 1918 before anything could be accomplished. Later this armament was removed and sent to Canada and re-emplaced on April 11, 1941 in their defenses. The emplacement was not subsequently used for armament. It still exists at the Fort Flagler State Park and is open to the public.

- **BANKHEAD:** A two-pit mortar battery emplaced at the Fort Flagler reservation of Marrowstone Point. Plans were submitted on June 25, 1900. It faced due north, and the engineers calculated it could be built at an estimated cost of a little under $80,000. The pit centers were 175-feet apart and followed the general type plans. Work was done in 1900-1901. It was transferred on August 17, 1902 for a final cost of $89,585. It was named in General Orders No. 194 of December 27, 1904 for Brevet Brigadier General Henry C. Bankhead, U.S. Army officer serving during the Civil War. It was armed with eight 12-inch Model 1890M1 Watervliet mortars on Model 1896 carriages (#114/#269, #90/#272, #77/#270, #133/#271, #134/#273, #137/#276, #67/#274, and #128/#275). Telephone booths were added in 1905, and capability for electrical firing was installed soon after. Four mortars were authorized for removal on May 3, 1918. This was accomplished and the four selected units shipped away on August 24, 1918 to Morgan Engineering in Alliance, Ohio for conversion to railway mounts. The empty carriages were scrapped on July 26, 1920. The final four mortars served until removed under authority of October 24, 1942 and sold for scrap in December 1942. The emplacement still exists at the Fort Flagler State Park and is open to the public.

- **CALWELL**: A set of emplacements for four 6-inch disappearing guns located to the west on the main reservation, on a separate access road. Their plans were submitted on June 8, 1903. At first just two guns were authorized, with two more soon added. It was of conventional mimeograph design. The emplacement was strictly in line, with flank emplacements on either end or interior emplacements in the center. There were large, shared magazines in the traverses between each of the two pairs, with a smaller traverse of support room between the pair themselves. The battery fired to north, northwest. Work was done in 1903-1904. Transfer was made on May 15, 1907 for a construction cost of $99,500. The battery was named in General Orders No. 194 of December 27, 1904 for Captain James H. Calwell, who was killed in action at Paso Ovejas Mexico in 1847. It was armed with four 6-inch guns Model 1903 on Model 1903 disappearing carriages (#48/#53, #60/#54, #65/#55, and #70/#72). All four guns were listed for removal on August 24, 1917, which was implemented on November 28, 1917. The carriages were sold for scrap on December 13, 1920. During World War II the emplacement was used for 90mm ammunition storage. The emplacement structure still exists at the Fort Flagler State Park and is open to the public.

- **GRATTAN**: A second battery for a pair of 6-inch disappearing guns emplaced to the east of the main battery on the bluff at Marrowstone Point. Its plans were submitted at the same time as Battery Calwell on June 8, 1903. It was of conventional mimeograph design and had two flank emplacements and a single central traverse with shared magazine. The access road ended at the western end of the battery, which was recessed below the bluff crest. Work was done in 1903-1904. Transfer was made on April 23, 1907 for a cost of $48,000. It was named on General Orders No. 194 of December 27, 1904 for Brevet 2nd Lieutenant John Grattan, 6th U.S. Infantry who was killed during the Indian Wars in Nebraska in 1854. The guns were mounted in March of 1907, armed with two 6-inch guns Model 1903 on Model 1903 disappearing carriages (#68/#73 and #69/#45). These guns were slated for removal in August 1917 for use on wheeled field carriages. That was implemented on November 9, 1917. The carriages were scrapped in place in 1920. The emplacement served in later years as a U.S. Navy underwater sound-ranging facility. It still exists at the Fort Flagler State Park and is open to the public.

- **LEE**: A battery for two 5-inch rapid-fire guns on balanced pillar mounts emplaced at Fort Flagler. Plans were submitted on November 7, 1898. It was emplaced on the bluff slope, in front and some-what to the east of the main battery sequence, and at a lower level so the larger guns could fire over them. Platforms were separated by 53-feet. It was very similar to the dual 5-inch batteries also at Fort Worden and Casey. The emplacement was of conventional design and had two adjacent platforms and a covered central traverse with two magazine rooms. Work was done in 1899, and transfer made on August 17, 1902 for a cost of $15,000. It was named in General Orders No. 16 of February 24, 1902 for Lieutenant Walker Lee, engineer, killed in action in 1901 during the Philippine Insurrection. The guns were mounted in 1902 and consisted of two 5-inch Model 1896 Bethlehem guns on Model 1897 balanced pillar mounts (#5/#31 and #8/#32). In turn these were listed for removal on July 15, 1918 and sent to Fort Stevens for disposal on October 8, 1918. Subsequently the emplacement was modified to allow access to a new searchlight emplacement mounted directly in front of the battery on the bluff slope. As modified the emplacement still exists, though at times badly overgrown, at the Fort Flagler State Park and is open to the public.

- **DOWNES**: One of two dual 3-inch pedestal emplacements assigned to Fort Flagler. The plan was submitted on June 9, 1903. It was consistent with type plans, and highly regular in detail to the current mimeographs. It was located beyond the left flank of Battery Wilhelm, and thus between that battery and Battery Calwell. It fired to the northwest. Work was done in 1903-1904. It was transferred on April 25, 1907 for a construction cost of $12,000. It was named in General Orders No. 194 of December 27, 1904 for 1st Lieutenant Edward E. Downes, 1st U.S. Infantry who was killed on Samar in the Philippine Insurrection in 1901. It was armed on October 25, 1909 and July 7, 1910 with two 3-inch Model 1903 guns on pedestal carriages (#27/#78 and #16/#79). In the early 1920s a new, but small uncovered CRF station was built on the east flank. It served for a long period, not ultimately being removed after World War II in 1946 or 1947. The emplacement still exists at the Fort Flagler State Park and is open to the public.

- **WANSBORO**: The second dual 3-inch pedestal emplacement assigned to Fort Flagler. The plan was submitted on June 9, 1903. It was consistent to type plans and conformed in detail to the current mimeographs. It was located on the southeast of the reservation, quite a distance from the main batteries. It was placed near the south dock, firing into the channel to the east. Work was done in 1903-1904. Transfer was made on April 25, 1907 at a cost of $15,000. The battery was named in General Orders No. 194 of December 27, 1904 for 2nd Lieutenant Thomas Wansboro, 7th U.S. Infantry who was killed in Cuba in 1898. Its original armament was two 3-inch Model 1903 guns and pedestal mounts mounted in October 1904 (#35/#16 and #36/#17). These served until World War II. A CRF station was built nearby in 1921. The original armament was exchanged for that of Battery Walker at Fort Worden on October 30, 1943. Wansboro then was armed with guns/carriages #84/#16 and #10/#17. The battery served during the war as local Tactical Battery No. 4 until removed in 1947. In the 1960s the battery received two display guns, a 3-inch Model 1903 gun and pedestal mount and a navy 3-inch pedestal gun, relocated from armament abandoned at Fort Wint in the Philippines. The emplacement at Fort Flagler State Park is open to the public.

- **AMTB Marrowstone Point**: A 1943 Program battery of two 90mm fixed and two 90mm mobile guns emplaced at the Marrowstone Point, near the lighthouse at the Coast Guard reservation. It was built from October 8 to December 20, 1943 for transfer on March 25, 1944 at a construction cost of $8,842. It consisted of just its two concrete gun blocks, magazine and BC being of temporary wood construction. The mobile guns were received on February 25, 1943 but shipped back to Benicia Arsenal on March 29, 1943. Known within Puget Sound Harbor Defense as Tactical Battery #5. The fixed guns served apparently until early postwar when they were removed. Broken remains of the blocks and aprons still exist on the lighthouse reservation. The battery is open to the public.

Fort Casey (1897-1953) is located at Admiralty Head, facing west on to Admiralty Inlet, near Coupeville, Washington State on Whidbey Island. The military reservation received ten concrete batteries during the Endicott Program along with a large set of garrison buildings. It was named in General Orders 134 of 1899 for Brig. Gen. Thomas L. Casey, U.S. Army who was a Chief of Engineers. Fort Casey, Fort Flagler and Fort Worden were sited in a "triangle" shape to protect the entrance to Puget Sound. The main gun line and mortar batteries were constructed from 1898 to 1904, while the secondary batteries were built from 1901 to 1907 on the low bluffs. In the 1920's several anti-aircraft guns were added. The post was placed in caretaker status in 1937 with many buildings being removed. During World War II, the fort returned to active status and new cantonment buildings were constructed. A submarine net and other defense measure were installed across the Admiralty Inlet to Fort Worden. The fort was declared surplus in 1953 and about half was acquired by the Washington State Parks in 1957. Seattle Pacific University owns the old garrison area of the military reservation, which is used for youth camps and adult retreats. The campus is generally open for exploration and walking. Fort Casey Inn on Whidbey Island, Washington, features 10 units in the old officers and NCO quarters, just north of Fort Casey State Park. Fort Casey State Park features camping, picnicking, and boating facilities as well as a historic lighthouse. It features one of the premier coast artillery battery restoration and interpretive programs centered round Battery Worth and its two 10-inch disappearing carriage guns, recovered from Battery Warwick, Fort Wint, Subic Bay, Philippines in 1963. Also, two 3-inch rapid fire guns were recovered from Fort Wint and placed in Battery Trevor. Battery Worth has been extensively renovated with restored shell hoists, a magazine, a plotting room, a barracks room, and a battery commander's station. Fort Casey State Park is the only place in United States that one can view larger caliber disappearing guns from the Endicott Program. Guided tours of the gun batteries are offered during the summer months. The park is within a short walk from the ferry landing connecting Whidbey Island to Port Townsend. All Washington State parks require a daily parking pass.

Fort Casey Gun Batteries

- **SEYMOUR – SCHENCK:** An emplacement of four pits and sixteen 12-inch mortars emplaced at Admiralty Head. It was sited just to the north and behind the hill on the northeast side of the fort Casey reservation. Plans were submitted on November 16, 1897. The design featured the four pits arranged linearly, but with the pits angled in so that they were in echelon. As the ground for the parapet level was sloping while the pits were all at the same level, the depth of the parapet differed widely. The pits were still of narrow design and separated in centers by 175-feet. Magazines wrapped around the sides and under the front parapet for each pit, the pits not being connected to each other. The construction contract was let on February 28, 1898. Work was done in 1898-1899. Transfer was made on June 16, 1902 for a cost of $93,207. It was named in General Orders No. 84 of June 12, 1903 for Major Truman Seymour, 5th U.S. Artillery who served in the Mexican and Civil Wars. In General Orders No. 20 of January 25, 1906 the southern two pits were named Battery Schenck for Lieutenant Colonel Alexander Schenck of Civil War service. It was armed with sixteen 12-inch Model 1890 and Model 1890M1 mortars on Model 1896 carriages. For Battery Seymour: Watervliet tubes M1890 #5/#117, M1890 #8/#138, #59/#134, #60/#136, Bethlehem tube #53/#135, Niles tubes #17/#114, #18/#113, #19/#115. For Battery Scheck: Watervliet tube #55/#112, Builders tubes #45/#293, #47/#137, #48/#140, Bethlehem tubes #48/#141, #54/#126, #55/#139, and Niles tube #14/#127. Telautograph booths were added later, as were separate standing plotting rooms in 1914. On May 3, 1918 four guns were removed from Battery Seymour, but none from Battery Schenck. All twelve guns then served until the 1940s. They were authorized for removal under authority of October 24, 1942, which was implemented in 1943. The emplacement still exists at the Fort Casey State Park. This battery is open to the public.

- **WORTH:** A battery for two 10-inch disappearing guns built as the first approved emplacement for the Fort Casey reservation at Admiralty Head. Plans were submitted on March 26, 1897 for these guns in emplacements No. 1 and 2 on the right flank of the sequence. These were of conventional type plan design, with the No. 1 emplacement a flank emplacement with its field of fire mostly to the west, and No. 2 an interior emplacement bearing more to the southwest. There was 124-feet between gun centers, and trunnion elevation was about 84-feet. Magazines were on the lower level to the left of each platform, projectile service performed by hoists. Work was done in 1897-1898. Transfer was made on June 16, 1902 for a cost of $103,898.40. It was named on General Orders No. 20 of January 25, 1906 for Brigadier General William J. Worth of Mexican War service. They were armed with two 10-inch Model 1895 Watervliet guns on Model LF 1896 disappearing carriages (#12/#59 and #15/#60). In 1915 a new BC station on a concrete tower to the rear of the central traverse was added, along with a commander's walk connecting it to the main work. While at one point proposed for disarmament in 1918, this was rescinded on July 18, 1918. The battery was placed in deletion status in 1932. The battery retained its guns until ultimate authority to disarm on October 24, 1942. At that time the tubes were stored on post and the carriages scrapped. After

PUGET SOUND

FORT CASEY

ADMIRALTY HEAD

GENERAL MAP.

STRAITS OF JUAN DE FUCA

Caretaking Status.

SERIAL NUMBER

True Meridian.

PLANE OF REFERENCE
To reduce elevations to Mean Low Water subtract 5.72 ft.

3000 FT
1000 FT   0   1000   2000   3000 FT

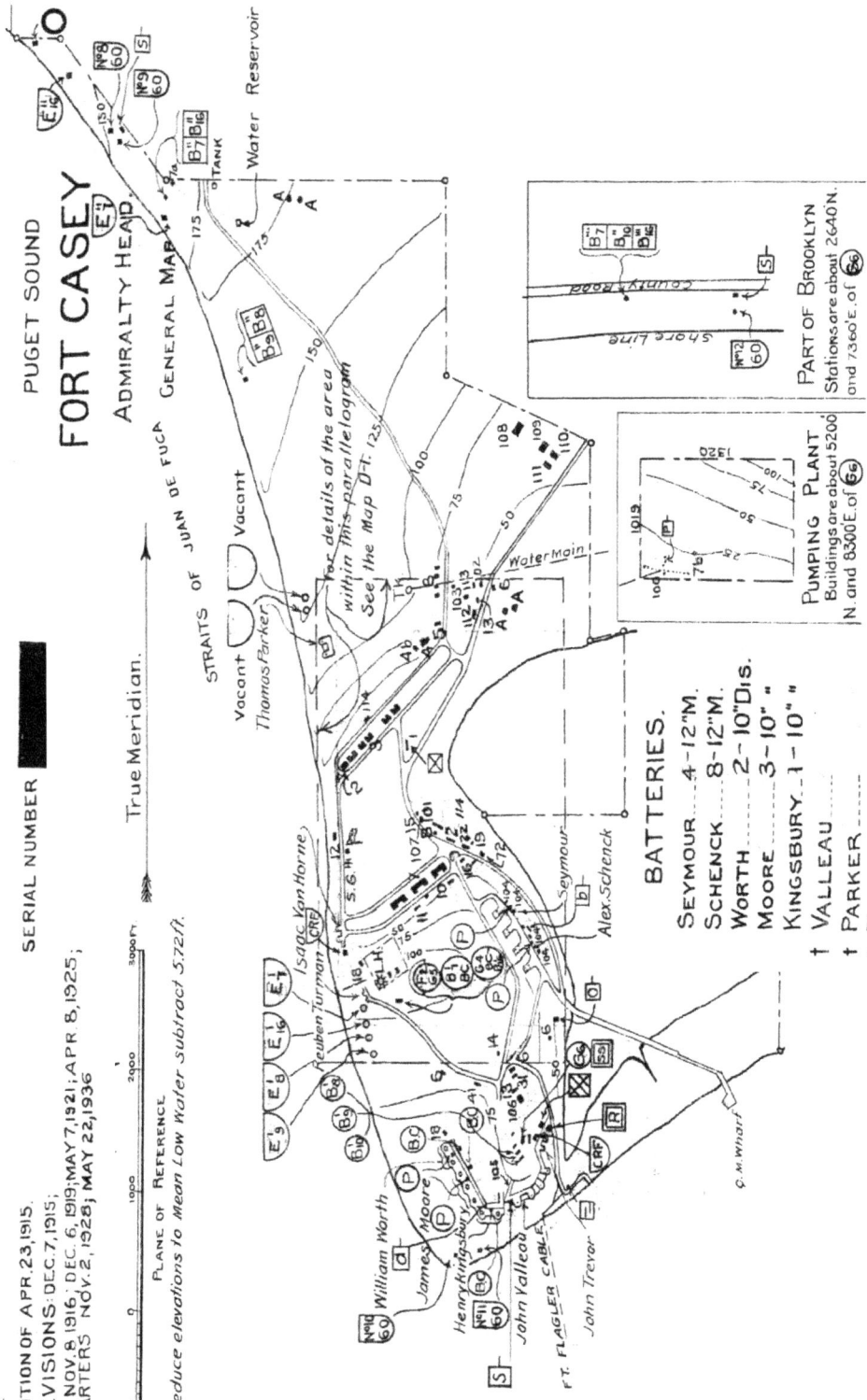

EDITION OF APR. 23, 1915.
REVISIONS: DEC. 7, 1915;
NOV. 8, 1916; DEC. 6, 1919; MAY 7, 1921; APR. 8, 1925;
QUARTERS NOV. 2, 1928; MAY 22, 1936.

Water Reservoir

TANK

Part of Brooklyn
Stations are about 2640 N.
and 7360 E. of 66

Pumping Plant
Buildings are about 5200
N. and 8300 E. of 66

For details of the area
within this parallelogram
See the Map Dr't.

Water Main

Vacant
Vacant
Thomas Parker
Vacant

Isaac Van Horne
Reuben Turman
William Worth
James Moore
Henry Kingsbury
John Valleau
John Trevor
Alex. Schenck
Seymour

FT. FLAGLER CABLE.
Q.M. WHARF

BATTERIES.

SEYMOUR .... 4-12" M.
SCHENCK .... 8-12" M.
WORTH .... 2-10" Dis.
MOORE .... 3-10" "
KINGSBURY .. 1-10" "
† VALLEAU
† PARKER
† TURMAN
VAN HORNE .. 2-3" P.
* TREVOR .... 2-3" P.
[A-Anti-aircraft gun 4-3"

† Armament removed
* Tubes removed

LEGEND.

1. ADMINISTRATION BLDG.
2. COMMANDING OFFICER'S QUARTERS.
3. OFFICER'S QUARTERS.
4. HOSPITAL.
5. HOSPITAL STWD'S QRS.
6. N.C. OFFICERS' QRS.
7. BARRACKS.
7a. DORMITORY.
7b. PUMPING ENGINEER'S QRS.
8. GUARD HOUSE.
9. POST EXCHANGE & P.O.
10. GYMNASIUM.
11. SHOOTING GALLERY.
12. RIFLE BUTTS.
13. STOREHOUSE.
14. QUARTERS (t.)
15. FIRE HOUSE.
16. BAKERY.

18. LAVATORY.

100. WELLS.
101. COAL SHED.
102. STABLE.
103. CIVILIAN TEAMSTER'S QRS.
104. TELEPHONE BOOTH.
105. PARADOS.
21. Q.M. & SUB. STOREHOUSE.
22. Q.M. WORKSHOP.
31. ORDNANCE ST. HO.
41. ENGINEER BUILDINGS.
51.
71.
72. SERVICE CLUB.
106. ART. ENG. OFF. & STO. HO.
107. GASOLENE FILLING STA.
108. HANGAR BALLOON SERVICE.
109. GENERATOR HO." "
110. GARAGE BALLOON SERVICE.
111. STOREHOUSE " "
112. MACHINERY SHED.
113. FARM MACH. BUILDING.
4b. MORGUE.
114. GARAGE.

PUGET SOUND

FORT CASEY-D I.

ADMIRALTY HEAD

PARKER

SERIAL NUMBER 124

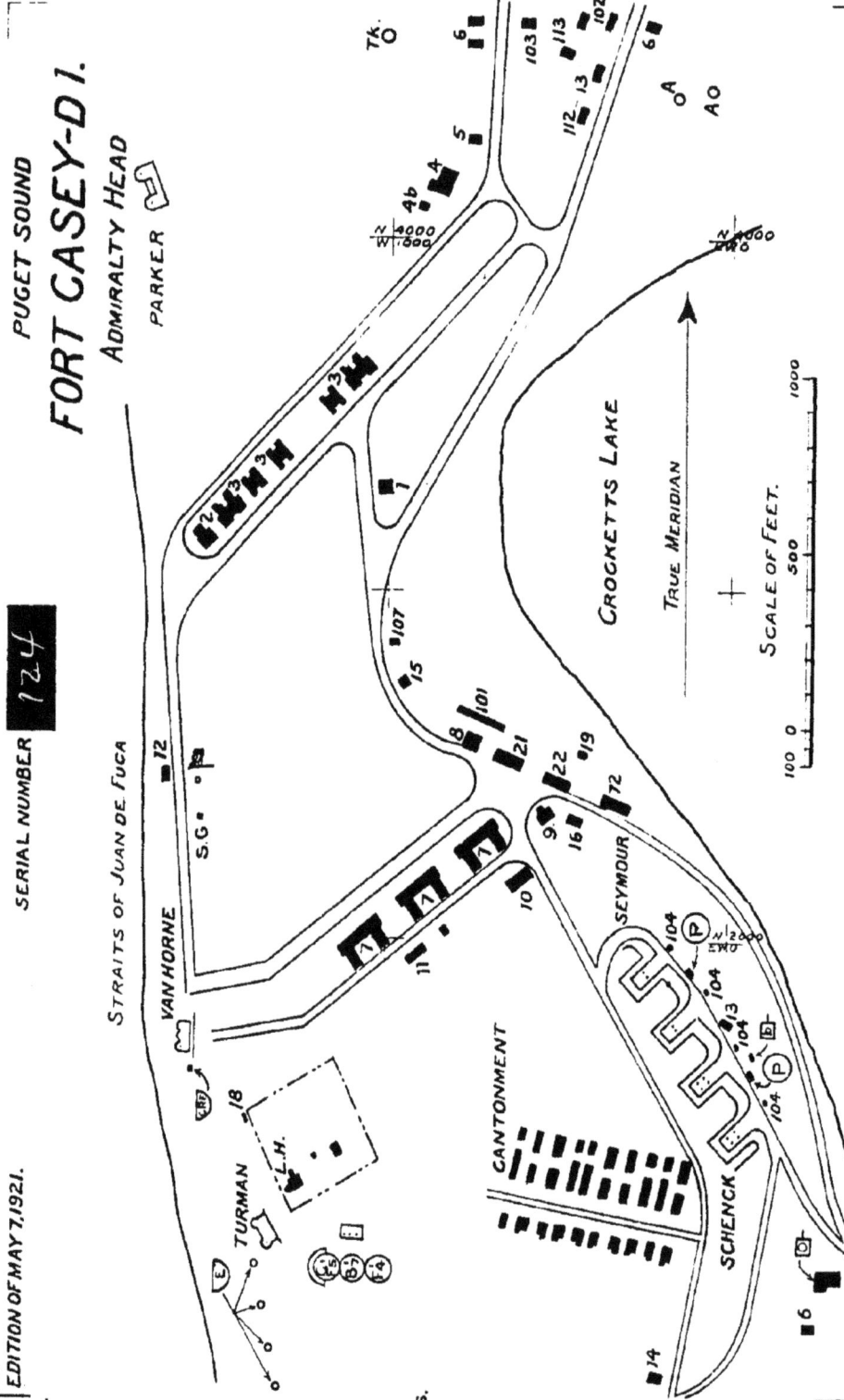

EDITION OF MAY 7, 1921.

STRAITS OF JUAN DE FUCA

VAN HORNE

TURMAN

CANTONMENT

SCHENCK

SEYMOUR

CROCKETTS LAKE

TRUE MERIDIAN

SCALE OF FEET.

100   0   500   1000

LEGEND

1. ADMINISTRATION BLDG.
2. COMM. OFFICER'S QRS.
3. OFFICER'S QRS.
4. HOSPITAL.
4b MORGUE.
5. HOSPITAL STWD'S. QRS.
6. N.C.OFFICER'S QRS.
7. BARRACKS.
8. GUARD HOUSE.
9. POST EXCHANGE & P.O.
10. GYMNASIUM.
11. SHOOTING GALLERY.
12. RIFLE BUTTS.
13. STORE HOUSE.
14. QUARTERS (t).
15. FIRE HOUSE.
16. BAKERY.
18. LAVATORY.
19. OIL HOUSE.
101. COAL SHED.
102. STABLE.
103. CIVILIAN TEAMSTER'S QRS.
104. TELEPHONE BOOTH.
107. BAND STAND.
112. MACHINERY SHED.
113. FARM MACH. BUILDING.
21. Q.M. & SUB. STOREHOUSE.
22. Q.M. WORK SHOP.
72. SERVICE CLUB.

BATTERIES

SEYMOUR ...... 4-12"M.
SCHENCK 8-12
PARKER ....
TURMAN ....
VAN HORNE ... 2-3"P.

A-Anti-aircraft gun 2-3"

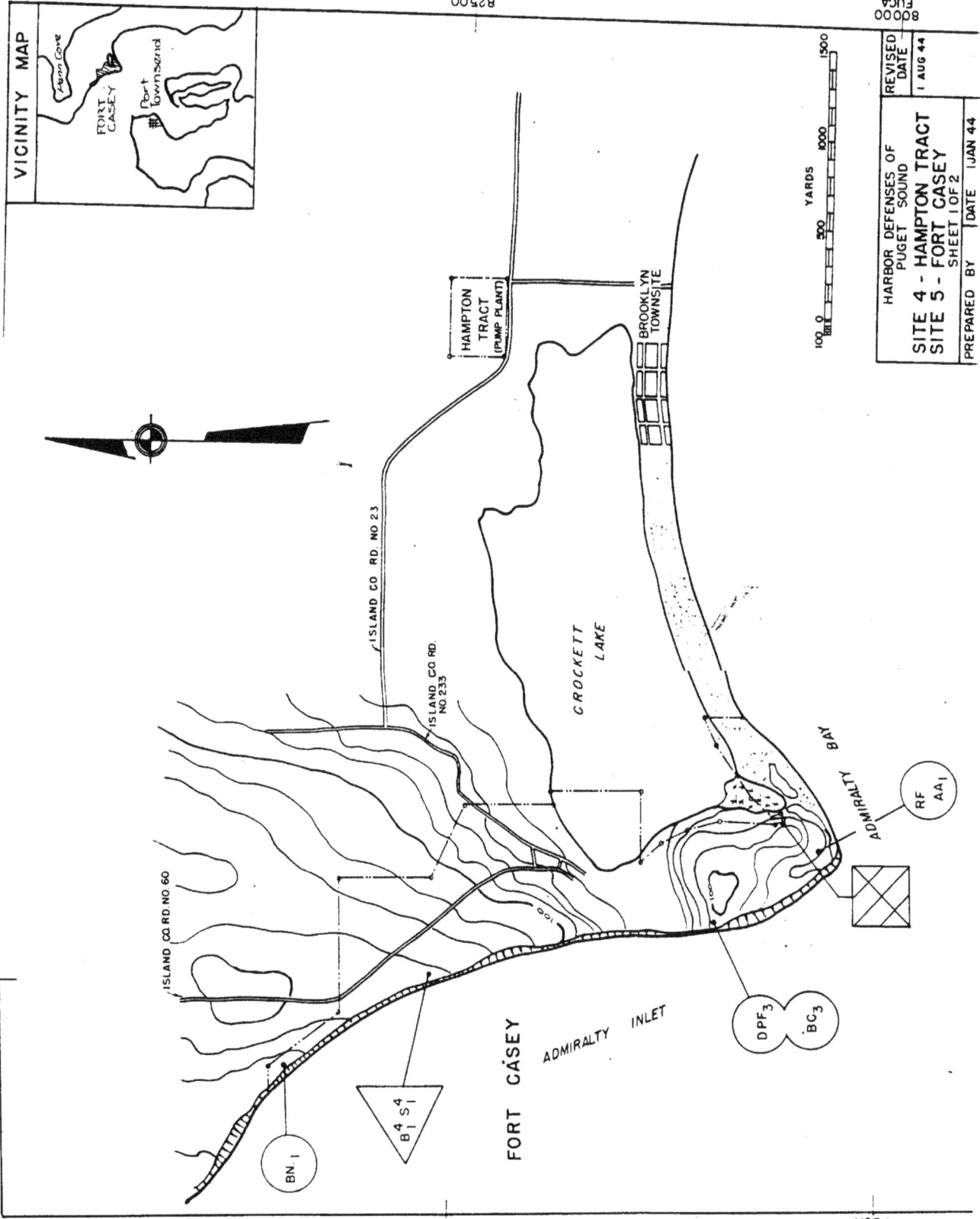

Fort Flagler 1932 (NARA)

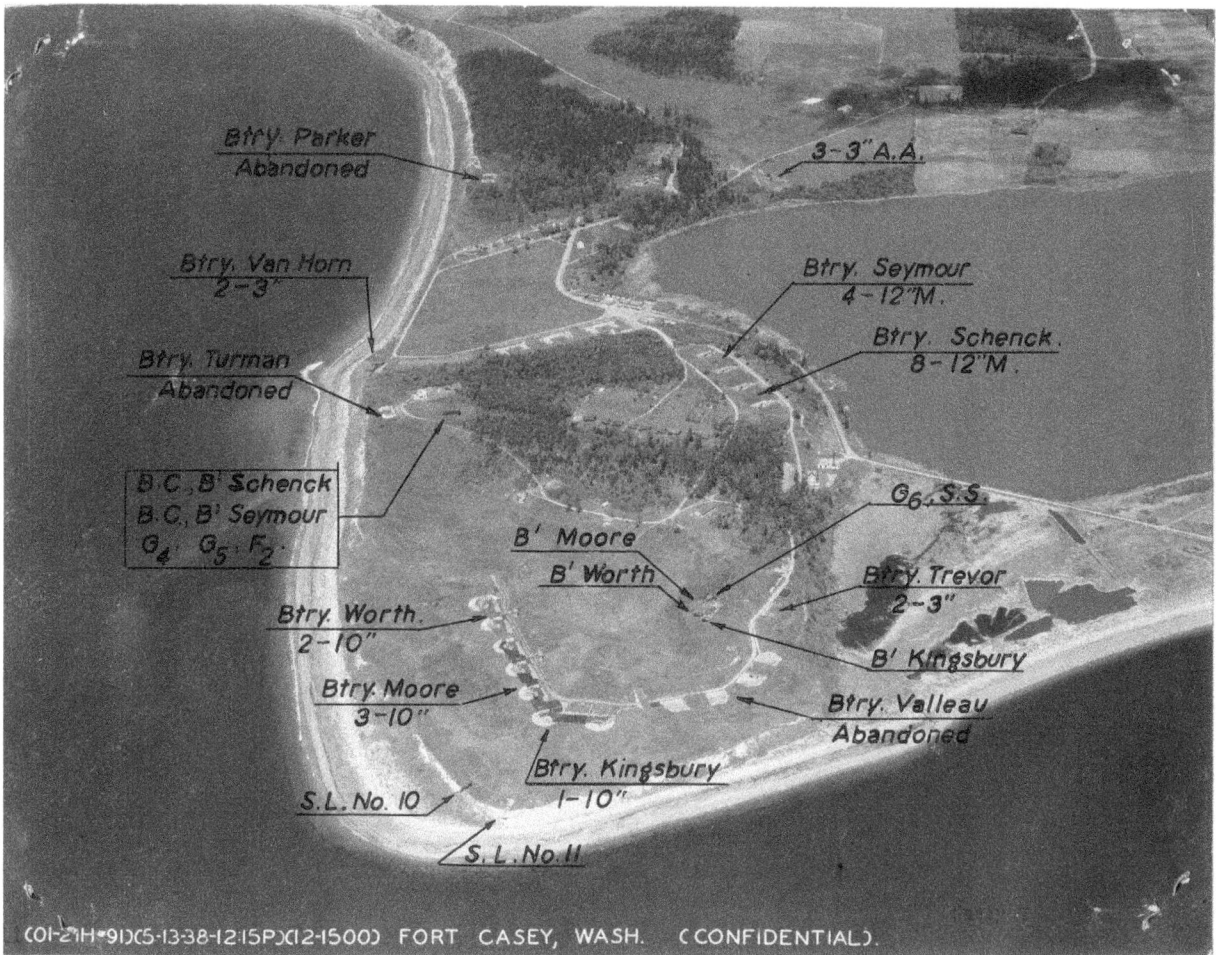

Fort Casey 1938 (NARA)

Main gun battery line at Fort Casey State Park (Terry McGovern)

Battery Worth Fort Casey State Park (Terry McGovern)

the site became a Washington State Park, the battery received two 10-inch guns Model 1895 on Model 1901 disappearing carriages (consequently they don't quite fit the foundation) in the mid-1960s from Battery Warwick at Fort Wint, Subic Bay the Philippines for display purposes. The battery's shell hoists have also been restored as the plotting room. Fort Casey State Park has one of the best restored Endicott-era batteries in rthe United States. The battery is open to the public with the rooms open for seasonal tours and events.

- **MOORE**: The second half of the five-gun disappearing battery built on the central bluff overlooking Admiralty Head. Submission on March 26, 1897 had been for three guns, a fourth was added on June 2, 1897, and then on July 13, 1900 two more guns (all of these progressively added to left) were added. Eventually Battery Moore was organized with three 10-inch disappearing guns, using parts of three separately submitted, authorized emplacements. These were all for interior emplacements, all aligned in the same direction. Crest was kept at 85-feet, emplacement intervals at 124-feet, and magazines were all on the left, lower flank (though the later, last emplacement had a different room arrangement). All guns fired to the southwest. Work was done in 1897-1898 for the first two emplacements (No. 1 and 2) and transferred on June 16, 1902 with Battery Worth for a construction cost of $103,894.49. Emplacments No. 1 and No. 2 were armed with 10-inch Model 1895Watervliet guns on Model LF 1896 disappearing carriages (#16/#61 and #11/#64). The third emplacement was started in June 1901 and finished by July 1902. It was transferred with the two guns of Battery Kingsbury on November 27, 1905 for a cost of $138,392.41. It was armed originally with an older Model 1888 10-inch Bethlehem gun on a Model 1896 disappearing carriage (#40/#72). This gun was removed in February 1909 and re-emplaced in Battery Kingsbury. Battery Moore's No. 3 gun was replaced with a Model 1895 Watervliet gun (#29) received in 1908. The battery was named in General Orders No. 194 of December 27, 1904 for Brigadier General James Moore, a Revolutionary War officer. In 1915 a new concrete BC station on a tower and connecting commander's walk was added behind the traverse between emplacements No. 1 and 2 of the battery. Like Worth it was proposed to remove this armament in World War I for use on railway carriages, but this was rescinded on July 18, 1918. The battery was placed in deletion status in 1932. The battery was disarmed with authority given on October 24, 1942. In 1943 a 3-inch AA gun emplacement was built into the pit of Emplacement No. 3, to complement the two at nearby Battery Kingsbury. As modified the emplacement still exists at the Fort Casey State Park and is open to the public.

- **KINGSBURY**: The final 10-inch battery of two guns of the Fort Casey main sequence. Plans for three new emplacements were submitted on June 25, 1900—the emplacement on the right flank to complete Battery Moore, and the two emplacements to become Battery Kingsbury. These guns changed the angle of the sequence, No. 1 emplacement at the bend, and No. 2 starting a new line running to the east. The guns fired to the south. They were at the same crest height as the rest of the line but spaced 143-feet apart for gun centers. Work was done in 1901-1902, and transfer was made on November 27, 1905 for a cost (with Moore No. 3) of $138,392.41. It was named in General Orders No. 194 of December 27, 1904 for Colonel Henry W. Kingsbury, 5th U.S. Artillery who died in 1862 at the Battle of Antietam. The No. 1 emplacement was armed with a Model 1888 10-inch gun on Model 1896 disappearing carriage (#43/#71). This carriage had been involved in a damaging accident during transit of May 21, 1902, but was subsequently repaired. The tube was mistakenly designated a Model 1895 gun in some early documents. The No. 2 emplacement received a new Model 1900 Watervliet gun on Model 1901 carriage (#9/#8). This was not mounted until September 1907 and served only a short time before being removed and retained at the fort.

In February 1909 it was replaced with a Model 1888 gun (#40) previously at Battery Moore. In 1915 a new battery commander's station was added to the emplacement. On May 29, 1918 Battery Kingsbury was authorized for disarming, and that was done on June 7, 1918 to release guns for railway carriages. On July 22, 1919 authority was granted to re-arm and keep the No. 2 emplacement with the Model 1900 tube #9 on January 5, 1921, making Battery Kingsbury a single-gun battery.. The battery was placed in deletion status in 1932. In 1943 the No. 2 emplacement was again disarmed under authority of October 24, 1942. The empty emplacements were modified for 3-inch anti-aircraft guns which relocated here. The battery still exists at the Fort Casey State Park and is open to the public.

- **PARKER**: An emplacement for two 6-inch disappearing guns emplaced to the north of the main sequence, almost at the limit of the fort reservation in that direction. Plans were submitted on June 10, 1903. The battery was sited to protect the northern approaches to the strait and was pointed to fire due west. It was of conventional design, closely following the proscribed mimeographs, but with a parados in the rear to protect from possible fire from Admiralty Bay. Work was done from August 1903 to 1904, for transfer on May 22, 1907 at a construction cost of $50,380. It was named in General Orders No. 194 of December 27, 1904 for Brevet 1st Lieutenant Thomas D. Parker who was killed in action at Gaines Mill, VA in 1862. Carriages and guns were received in 1907 and mounted by mid-1908. Itwas armed with two 6-inch Model 1905 Watervliet guns on Model 1903 disappearing carriages (#3/#74 and #12/#75). The guns were removed for overseas service under authority of November 9, 1917 and the carriages scrapped soon after. The emplacement was not thereafter used for armament but still exists on the property of Seattle Pacific University's Camp Casey. This battery is accessible to the public.

- **VALLEAU**: An emplacement for four 6-inch disappearing guns emplaced to the flank of Battery Kingsbury as a general continuation of the sequence of guns at the bluff of Admiralty Head. Submission of the plans was made on April 30, 1903, and modified on June 12, 1903. They called for a battery (authorized separately in units of two guns each) of two guns each bent in the center. Each pair would consist of almost separate two-gun emplacements with all flank emplacements and a shared magazine between guns No. 1 and 2 and then again between No. 3 and 4. It fired to the southeast and east. Work was done from October 1903 to late 1904. Transfer was made on May 22, 1907 for a cost of $92,125. It was named in General Orders No. 194 of December 27, 1904 for 1st Lieutenant John Valleau, 13th U.S. Infantry, who was killed in Canada during the War of 1812. It was armed with four 6-inch Model 1905 Watervliet guns on disappearing carriages Model 1903, all mounted during 1908 (#18/#76, #20/#77, #21/#78, and #22/#83). The guns were listed for removal for use overseas under authority of August 24, 1917. This was done by November 9, 1917 with the corraiges being scrapped soon after. The emplacement still exists at the Fort Casey State Park. The battery is open to the public.

- **TURMAN**: A battery for two 5-inch rapid-fire guns emplaced near the lighthouse, north of the main battery line at Fort Casey. Plans were submitted on November 7, 1898 for all three similar 5-inch dual batteries (one each for Forts Worden, Flagler, and Casey) and were built at about the same time. The plan was typical for early RF batteries, featuring two deep platforms for the balanced pillar guns, and two magazines under the center traverse between the guns. Service for ammunition was by hand. Work was begun on August 19, 1899, and almost completed by the end of the following June, though then suspended due to lack of funding. Difficulties of the site and excavation of rock added to the cost overruns. It was finally transferred on June 30, 1902 for $18,850. The battery was named in General Orders No. 194 of December 27, 1904 for 2nd Lieutenant Reuben S. Turman,

6th U.S. Infantry who was killed in 1898 in Cuba. It was armed with two 5-inch Model 1897 guns tubes on Model M1896 balanced pillar carriages (Bethlehem tubes #12/#18 and #25/#19). The armament was listed for removal in 1918 to be re-located at Goose Rock in Deception Pass, but that was never implemented. Removal did come on July 18, 1918 for use on wheeled mounts, and the carriages were scrapped in place within a couple of years. The emplacement still exists at the Fort Casey State Park and is open to the public.

- **TREVOR**: One of the two dual, 3-inch pedestal batteries for Fort Casey. Location sites for both 3-inch batteries were submitted and fixed on September 2, 1902. This unit was to go on the left flank of the new 6-inch battery on the very far left of the main gun line. Plans were submitted on May 14, 1903. It was of standard, mimeograph plan with its two-gun platforms, short parapets, and one service magazine in the traverse between the guns. Work was done from August 1903 to mid-1904, completing electrical installations in June of 1905. It was transferred on May 22, 1907 for a construction cost of $15,800. The battery was named in General Orders No. 194 of December 27, 1904 for 1st Lieutenant John Trevor, 5th U.S. Cavalry who was killed in action at Winchester VA in 1864. It received its armament in early 1908—two 3-inch Model 1903 guns and pedestal carriages (#39/#20 and #50/#21). These tubes served until removed on November 16, 1933 and shipped as replacements to Fort Mills in the Philippines. The emplacement still exists at Fort Casey State Park. In the 1960s the state obtained two 3-inch Model 1903 guns and pedestal mounts from Fort Wint, Subic Bay, the Philippines for re-emplacement in Battery Trever as a publuc display.

- **VAN HORNE**: One of the two dual, 3-inch pedestal batteries for Fort Casey. The location was selected on September 2, 1902 for a site near the lighthouse, on the northern half of the reservation, north of Battery Turman and on the downward slope of the Admiralty bluff. Plans were submitted on May 21, 1903. It was designed to assist in covering the bay south of Partridge Point and to cover the various beaches to the north, firing to the west. It was of conventional type plans, having two platforms spaced by 62-feet for the guns, and a traverse space for two magazines and a storeroom. Construction was done from August 1903 to June of 1905. Transfer was made on May 22, 1907 for a cost of $14,695. The battery was named in General Orders No. 194 of December 27, 1904 for Captain Isaac Van Horne, 19th U.S. Infantry who was killed at Fort Mackinac in 1814. It was armed with two 3-inch Model 1903 guns and pedestal carriages (#37/#18 and #38/#19) mounted in 1908. The armament was carried throughout the battery's service life. It served as Tactical Battery No. 3 during World War II and was not removed until 1945. The emplacement still exists at the Fort Casey State Park. The battery is open to the public.

Fort Whitman (1909-1947) is located on Goat Island, two miles southwest of La Conner, Washington State, on Skagit Bay. The military reservation received one concrete battery and a controlled mine casemate during the Taft Program. It was named in General Orders 245 of 1909 for Dr. Marcus Whitman, a pioneer missionary killed by Cayuse Indians in 1847. Fort Whitman was constructed to protect the shallow water entrance to Puget Sound through Deception Pass. The post was placed in caretaker status in 1930s. During of World War II the fort was again manned and provided with a 37mm AMTB battery. The fort was transferred to Washington State in 1947 as a wildlife refuge. The state game reserve is only accessible by boat.

Fort Whitman Gun Battery

- **HARRISON**: A Taft Program battery for four 6-inch disappearing guns built on the Goat Island military reservation of Fort Whitman. While plans for a battery here had been planned throughout the Endicott Period, funding and definitive planning did not take place until the early Taft period.

PUGET SOUND
SKAGIT BAY.

SERIAL NUMBER 124

EDITION OF APR 23,1915.
REVISIONS: DEC 7, 1915:
DEC. 6, 1919, MAY 7, 1921.

FORT WHITMAN

PUGET SOUND

GOAT ISLAND.

Scale

2000' 1000' 500' 0 1000'

BATTERIES

HARRISON ___ 4 - 6"Dis.

Caretaking Status.

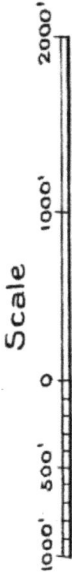

SERIAL NUMBER

EDITION OF APR. 23, 1915.
REVISIONS: DEC. 7, 1915;
NOV. 8, 1916; DEC. 6, 1919; MAY 7, 1921.
APR. 8, 1925; NOV. 2, 1926; MAY 22, 1936

LEGEND.

1
2
3
4
5
6
7
8
9
10 WAREHOUSE. (Q.M.)
40 OFFICE 4 QRS. (Q.M.)
41
42
43 ENGINE HOUSE. (Q.M.)
44
45

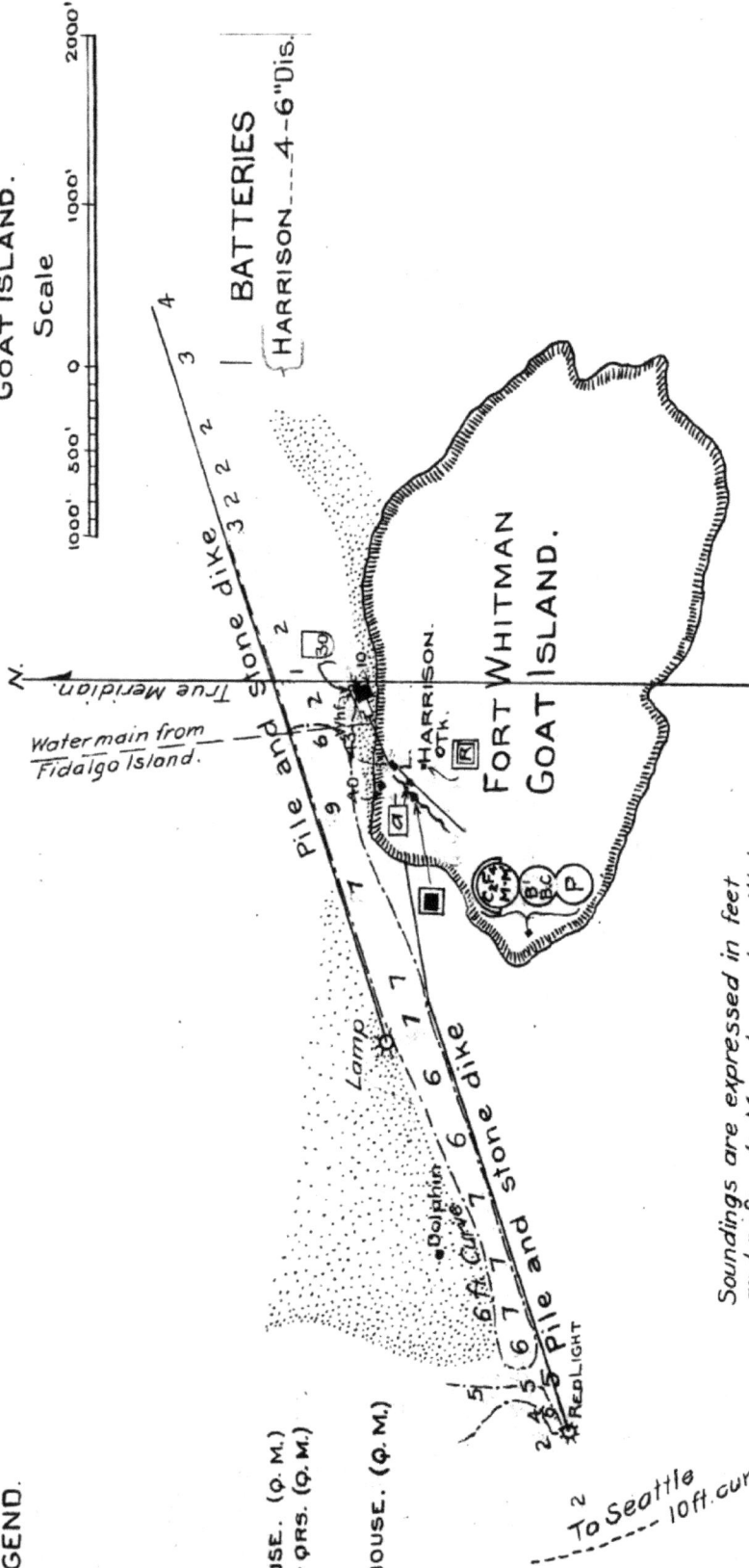

N True Meridian

Water main from Fidalgo Island.

Pile and stone dike

FORT WHITMAN GOAT ISLAND.

HARRISON QTR.

Lamp

Dolphin

6 ft. Curve

Pile and stone dike

Red Light

To Seattle

10 ft. curve

Soundings are expressed in feet
and refer to Mean Lower Low Water
Mean range of tide = 8.0 ft.
Controlling depth in Saratoga Passage = 36.0

S. L. Shelter

Btry. Harrison
4 – 6"

(VI-2IG-9I)(5-18-38-10:30A)(12-4000) DECEPTION PASS, WASH. (CONFIDENTIAL)

Fort Whitman 1938 (NARA)

Btry. Harrison
4 – 6"

(04-2IG-9I)(5-18-38-10:35A)(12-1000) DECEPTION PASS, WASH. (CONFIDENTIAL)

Fort Whitman 1938 (NARA)

Site Plans were submitted on October 10, 1908. It fired to the northwest to cover mines and the passage from Deception Pass down the east side of Whidbey Island. Revised plans were submitted on March 1, 1909. It closely followed recommended mimeograph type plans. The four platforms were in a straight line, two central interior emplacements and on the ends flank emplacements. Shared magazines were in the traverses between emplacements No. 1 and 2 and then between No. 3 and 4. The smaller center traverse held a storeroom and plotting room. The mining casemate for the post was built into the traverse near emplacement No. 2. The battery was on the high ground of the island, at a crest elevation of 98-feet. Work was done from early 1909 to 1910, for transfer on May 9, 1911 at a cost of $92,000. It was armed with four 6-inch Model 1908 Watervliet guns on Model 1905M1 disappearing carriages (#1/#14, #2/#15, #3/#16, and #4/#17). It was named in General Orders No. 245 of December 13, 1909 for Colonel George F. E. Harrison, U.S. Army, who died on March 26, 1909. The battery had a long service life, though usually in out of service status as Fort Whitman was usually just garrisoned with a small caretaker detachment. The armament was finally authorized for removal on July 17, 1943, actual dismounting and scrapping of the carriages occurring in 1944. The emplacement still exists on Goat Island, part of the public Skagit Wildlife Area. The battery is open to the public, the island is accessible only by small boats with no docking facilities.

Fort Ebey (1942-1946) is located on the shoreline of Whidbey Island, facing west toward the Strait of Juan de Fuca, about two miles north of Coupeville, Washington State. The 226-acre military reservation was established as part of the major effort to modernize the defense of Puget Sound for the coming of World War II. It was named in General Orders 6 of 1943 for Isaac N. Ebey a local pioneer militia volunteer killed by Indians in 1857. The 1940 Program saw the addition of a #200 Series battery with two 6-inch barbette guns (Battery Construction #248) and a 90mm AMTB battery at Ebey's Landing. Several fire control stations were also constructed. After World War II, Fort Ebey transferred to the U.S. Navy. Fort Ebey is now a 651-acre Washington state park, acquired from the federal government in three parcels between 1965 and 1991 and opened to the public in 1981. The park features three picnic areas, at the gun battery with a large open grass area, the beach access area, and at Point Partridge. Battery 248 is in excellent condition, open and accessible, as is the battery commander's station in front of the battery. All Washington State parks require a daily parking pass.

Fort Ebey Gun Batteries

- **Battery #248**: A standard 1940 Program dual 6-inch barbette battery emplaced on Whidbey Island at a new reservation named Fort Ebey. Work was approved in early 1942, and foundation concrete construction done from May 18, 1942 to February 1, 1944. Transfer was made on March 22, 1944 for a total cost of $349,000. It was of standard 200-series design. The battery was sited on the bluff to the east of the shore, firing to the southwest. It was armed with two 6-inch guns Model 1905A2 on Model M1 barbette carriages (#13/#54 and #17/#55). The battery served from 1944 until deactivated and dismounted in 1948. During the war it was referred to as Tactical Battery No. 1. The battery was never named, being known during construction just as Battery Construction No. 248. The emplacement still exists at the Fort Ebey State Park. The battery is open to the public.

- **AMTB Deception Pass**: A variety of proposals were made over time to arm the narrow Deception Pass entry at the north end of Whidbey Island. In 1918 it was proposed to move two 5-inch guns here to Goose Rock, but that was never implemented. In 1940 the relocation of two 3-inch guns from Battery Putnam had been suggested, but in fact was not implemented either (there is some

REPRODUCED AT THE NATIONAL ARCHIVES

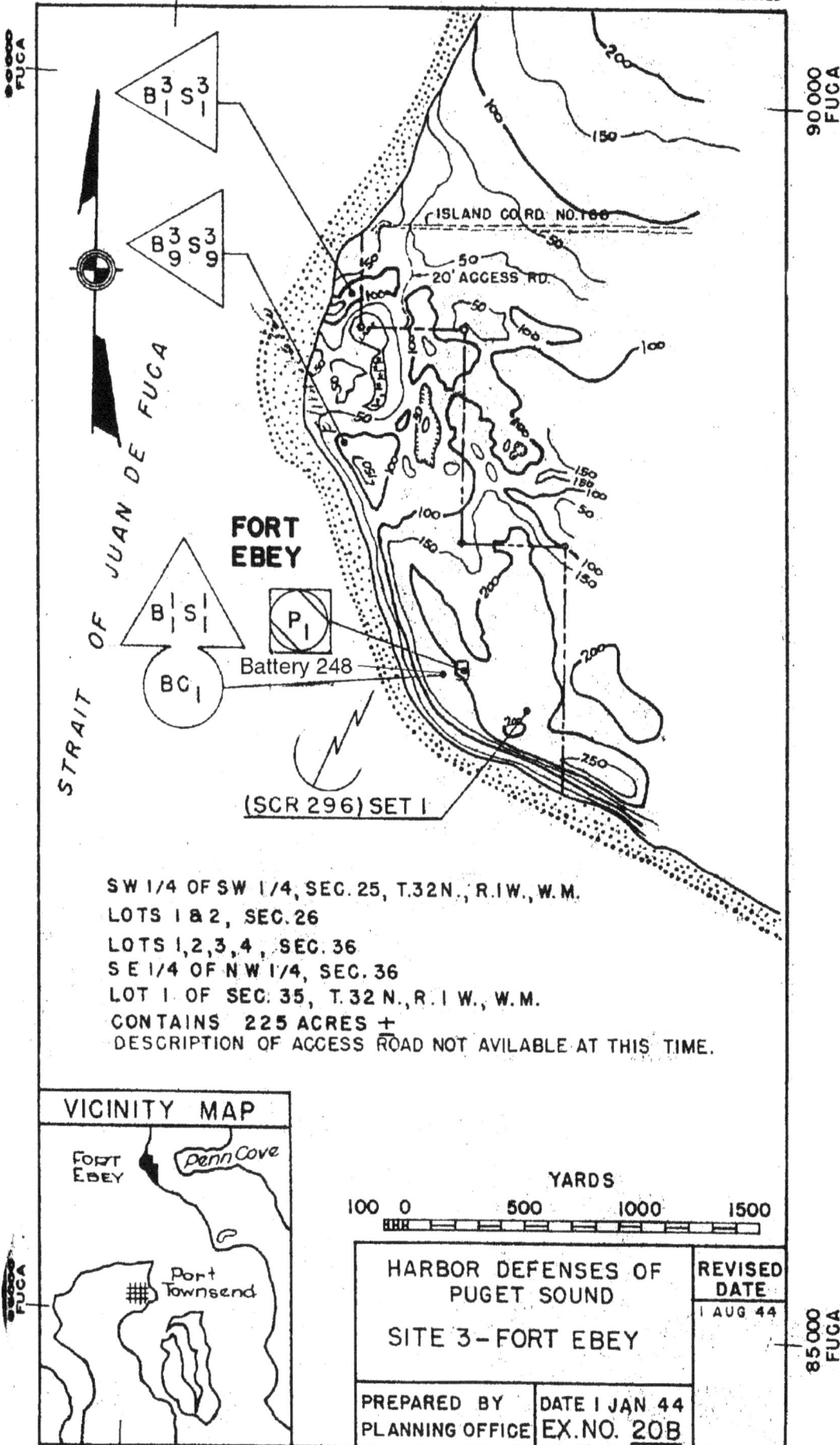

90 000 FUGA

90 000 FUGA

$B_1^3 S_1^3$

$B_9^3 S_9^3$

200

150

100

ISLAND CO. RD. NO. 100

50

50

20' ACCESS RD.

100

100

100

150
150
100
50

100

FORT EBEY

STRAIT OF JUAN DE FUCA

B_1 S_1

P_1

150

200

100
150

BC$_1$

Battery 248

200

(SCR 296) SET 1

250

250

SW 1/4 OF SW 1/4, SEC. 25, T.32N., R.IW., W.M.

LOTS 1 & 2, SEC. 26

LOTS 1,2,3,4, SEC. 36

S E 1/4 OF N W 1/4, SEC. 36

LOT I OF SEC. 35, T. 32 N., R. I W., W. M.

CONTAINS 225 ACRES ±

DESCRIPTION OF ACCESS ROAD NOT AVILABLE AT THIS TIME.

VICINITY MAP

FORT EBEY

Penn Cove

Port Townsend

85 000 FUGA

85 000 FUGA

YARDS			
100 0	500	1000	1500

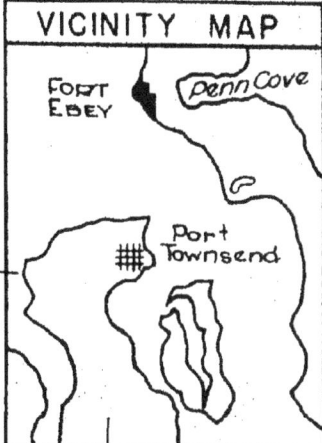

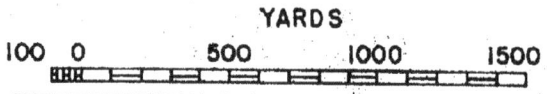

HARBOR DEFENSES OF PUGET SOUND	REVISED DATE
SITE 3 - FORT EBEY	I AUG 44

| PREPARED BY PLANNING OFFICE | DATE I JAN 44 EX. NO. 20B |

suggestion that the blocks may have in fact been poured but never received armament). Finally, during World War II, two 90mm mobile guns were emplaced on the south shore, at West Point as local Tactical Battery No. 1A. Battery construction started in 1942 and was completed in 1942. Concrete pads were provided for the mobile guns. Two earth-covered magazines provided protection for the ammunition. Electric power was provided by M-7 portable generators. The equipment and the guns for the battery were normally stored at Fort Casey and emplaced for alerts and practice. Deactivated in 1945 at the end of the war. No remains of any sort have been located at the pass.

- **AMTB Ebey's Landing**: A 1943 AMTB battery erected north of Fort Ebey on a low site along the coast of Whidbey Island. Work was done from January 26 to February 29, 1944 on the gun blocks, ammunition shelter and battery commander's station. It was transferred on March 25, 1944 for a cost of only $11,312. It consisted of two fixed 90mm guns on M3 barbette carriages, the blocks were spaced at 120-feet. It is likely the two mobile guns of this battery were the ones placed at Deception Pass to the north. It fired to the southwest. The battery served until right after the end of the war. Today the blocks are either buried or heavily overgrown, no trace of them has been located at the site.

Battery Construction Number 248, Fort Ebey State Park (Terry McGovern)

Fort Ward (1901-1928) is located at Bean Point, on the east side of the Rich's Passage, on Bainbridge Island, Washington State. Four concrete batteries and a controlled mine complex with a mine casemate were built during the Endicott Program on the 375-acre military reservation. It was named in General Orders 84 of 1903 for Bvt. Brig. Gen. George H. Ward who died of wounds received during the Battle of Gettysburg in 1863. Part of Fort Ward was the Middle Point Military Reservation (1901-1928) located six miles from Port Orchard, Washington State, on the west side of the Rich's Passage, on Bremerton Peninsula. The 325-acre military reservation saw the construction of one concrete battery and a controlled mine complex with a mine casemate during the Endicott Program. Fort Ward and Middle Point Military Reservation across the passage were constructed to protect the approaches to the Bremerton Naval Yard. Both posts were decommissioned as coast artillery posts in the late 1920s but were taken over for U.S. Navy use with the coming of World War II. The U.S. Navy used Fort Ward as a radio station and communication school until 1958. The U.S. Navy used Middle Point as a fuel supply depot and fire-fighting training school. The fort at Beans Point was sold to private interests except for a 137-acre section purchased by the Washington State Parks in 1960. The park was transferred to the City of Bainbridge Island in the mid-2000s. Today, two of the Endicott batteries are private property and two are part of the city park. Most of the Fort Ward garrison area was subdivided into private lots; most of the remaining post buildings becoming private homes which are on the National Register of Historic Places. A section of the old parade ground was designated as a city park in 2005. In 2017 work began on a privately funded project to restore the post bakery building and turn it into a community hall and visitor center which was completed in 2022. The U.S. Navy transferred a 111-acre section of the Orchard Point reservation to the Washington State Parks in 1960 and developed into Manchester State Park, which retains the original mine storehouse, mine casemate and the emplacement for two 3-inch guns. Washington State Parks require a parking permit.

Fort Ward Gun Batteries

- **NASH**: A battery for three 8-inch disappearing guns emplaced on the bluff running to the east of the shore at Fort Ward on Bainbridge Island. Plans were submitted on September 2, 1899. The emplacements were on the bluff firing to the south. It was of mimeograph type plan, with gun separation distance of 116-feet, magazines on the lower left flank and service of ammunition by hoists. The No. 1 emplacement was of flank type, the other two of interior arrangement. Work was done between April 5, 1900 and November 1, 1901. Transfer came on January 18, 1904 for a cost of $120,433. It was named in General Orders No. 194 of December 27, 1904 for Brigadier General Francis Nash mortally wounded at Germantown, PA in 1777. It was armed with three 8-inch Model 1888 guns on LF 1896 disappearing carriages (West Point Foundry tube #6/carriage #36, Watervliet tube #42/carriage #37, and West Point Foundry tube #2/carriage #38). During the mounting of the gun in emplacement No. 2 there was an accident which killed a laborer on March 26, 1903 and slightly delayed completion of the armament. Everything was in service on April 20, 1903. Modifications were made to the emplacement's loading platforms about 1910. The gun tubes were listed for removal on August 24, 1917, which was carried out by October 9, 1917. The carriages were scrapped on October 12, 1918. The battery was subsequently abandoned. It still exists on private property. The battery is closed to the public.

- **WARNER**: A battery for two 5-inch guns on pedestal mounts located directly on the shore of Rich Passage near Bean Point, firing to the southwest. Submission of plans was made on November 22, 1899 to use two new wire-wound 5-inch guns. The plan followed general type recommendations with two internal platforms and magazines covered by traverses on the lower left of each. Ammunition service was by hand. Gun centers were 45.5-feet apart. The location was some 130-feet below

PUGET SOUND
FORT WARD
BEAN POINT

BATTERIES

NASH..........
WARNER....2-5" P.
THORNBURG....
VINTON......

SERIAL NUMBER

EDITION OF: DEC. 7, 1915;
REVISIONS: NOV. 8, 1916; DEC. 6, 1919
MAY 7, 1921.

LEGEND

1. ADMINISTRATION BLDG.
2. COMMANDING OFFICER'S QUARTERS.
3. OFFICER'S QUARTERS.
4. HOSPITAL
5.
6. N.C. OFFICERS' QRS.
7. BARRACKS.
7a TEAMSTERS' QRS.
8. GUARD HOUSE.
9. POST EXCHANGE AND GYMNASIUM.
10. FIRE STATION, AND POWER HOUSE
11. BAKERY.
12. LAUNDRY.
13. OIL HOUSE.
14. COAL SHED.
15. STABLE.
16. WAGON SHED.
21. Q.M. & COMMISSARY ST. HO.
22. Q.M. SHOP.
40. E.D. OFFICE.
41. E.D. QUARTERS.
71. CLUB HOUSE
72. HALL.

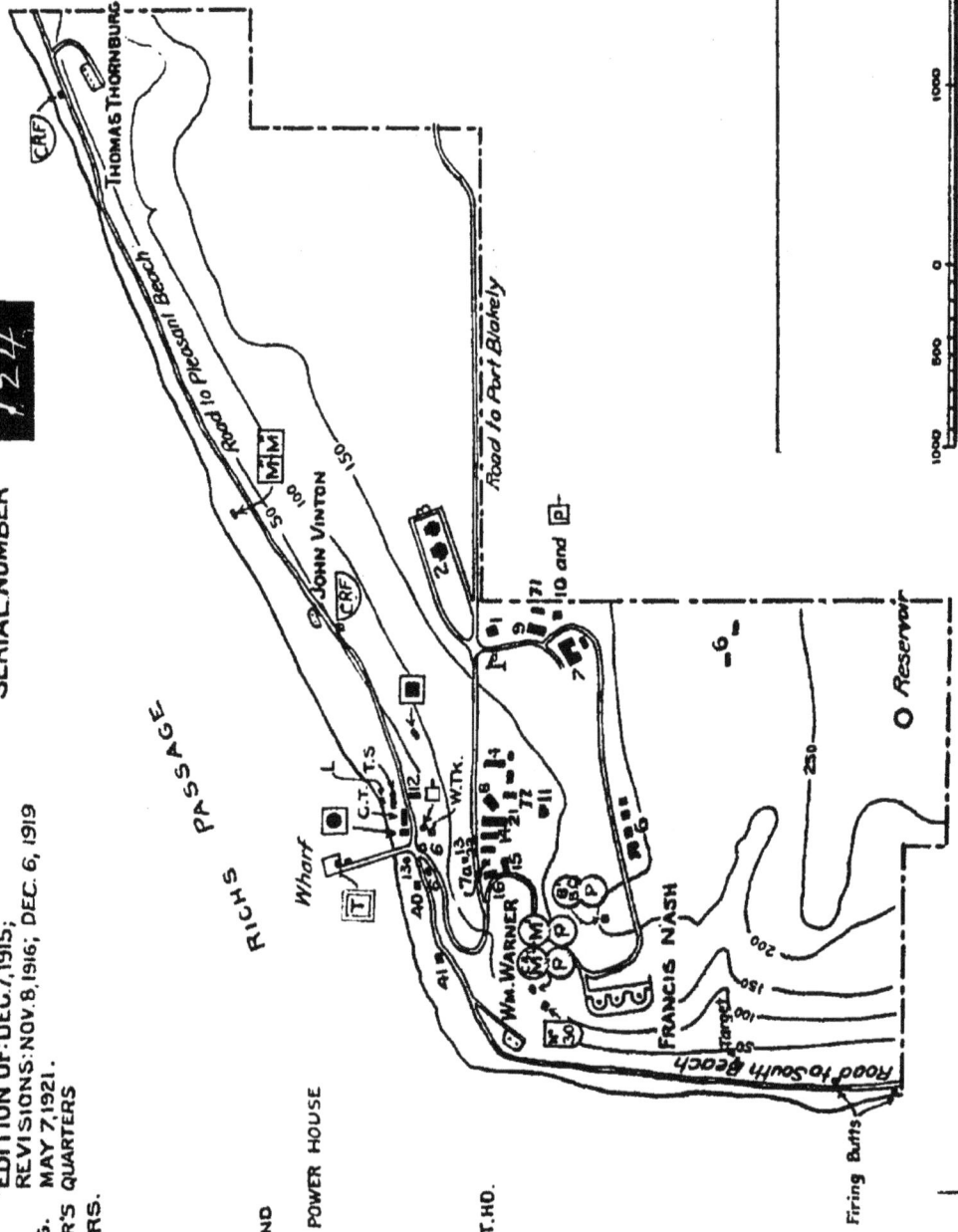

Road to Pleasant Beach

THOMAS THORNBURG

CRF

RICH'S PASSAGE

Wharf

JOHN VINTON

CRF

Wm. WARNER

FRANCIS NASH

Road to Port Blakely

Road to South Beach

Firing Butts

Reservor

N.

Fort Ward 1932 (NARA)

Fort Ward 1932 (NARA)

PUGET SOUND

MIDDLE POINT
AND
ORCHARD POINT.

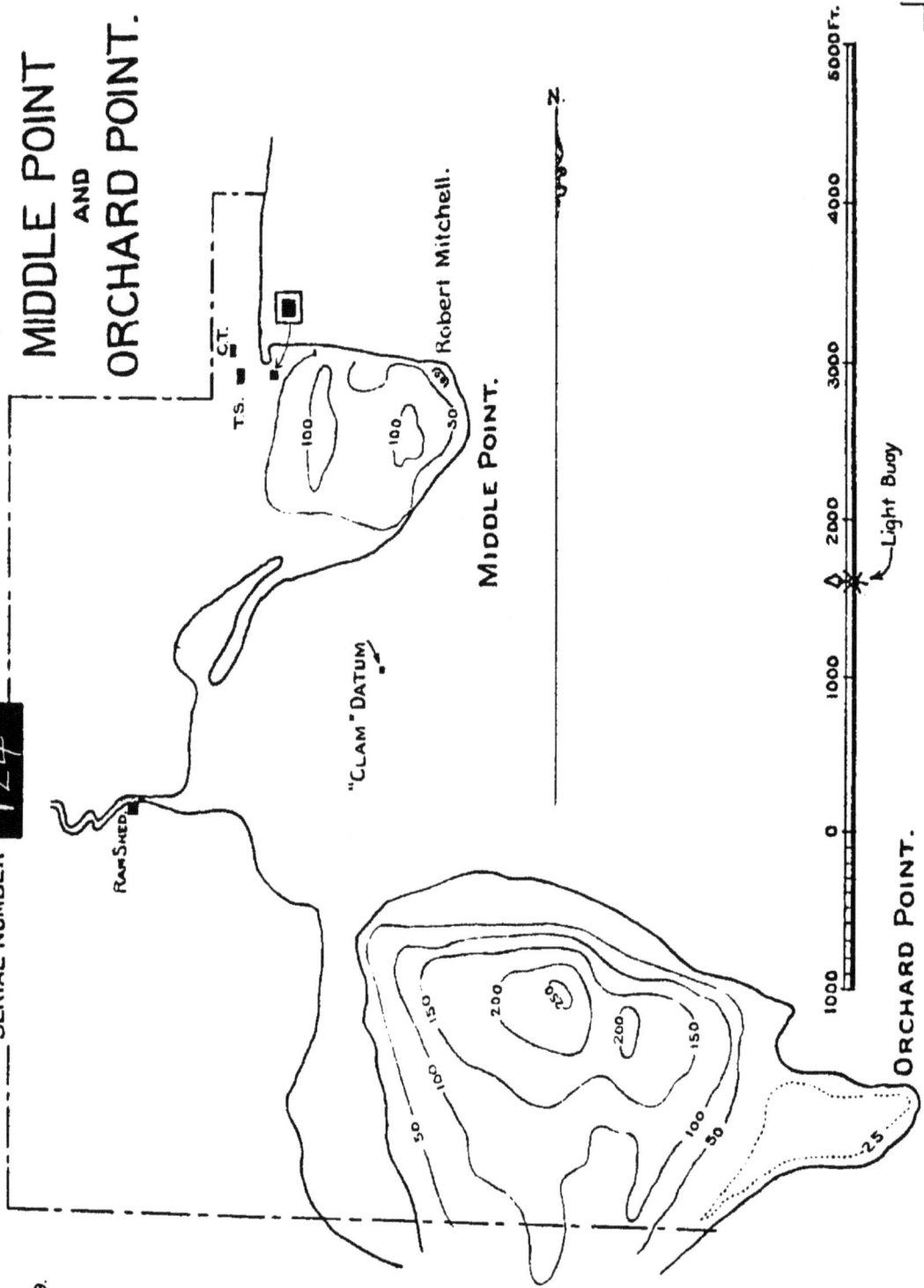

SERIAL NUMBER 124

EDITION OF APR 23, 1915.
REVISIONS: DEC. 7, 1915;
NOV. 8, 1916; DEC. 6, 1919.

C.T.

T.S.

Robert Mitchell.

100

100

50

50

RAM SHED.

"CLAM" DATUM

N.

MIDDLE POINT.

50
100
150
200
200
250
150
100
50
25

ORCHARD POINT.

Light Buoy

1000   0   1000   2000   3000   4000   5000 Ft.

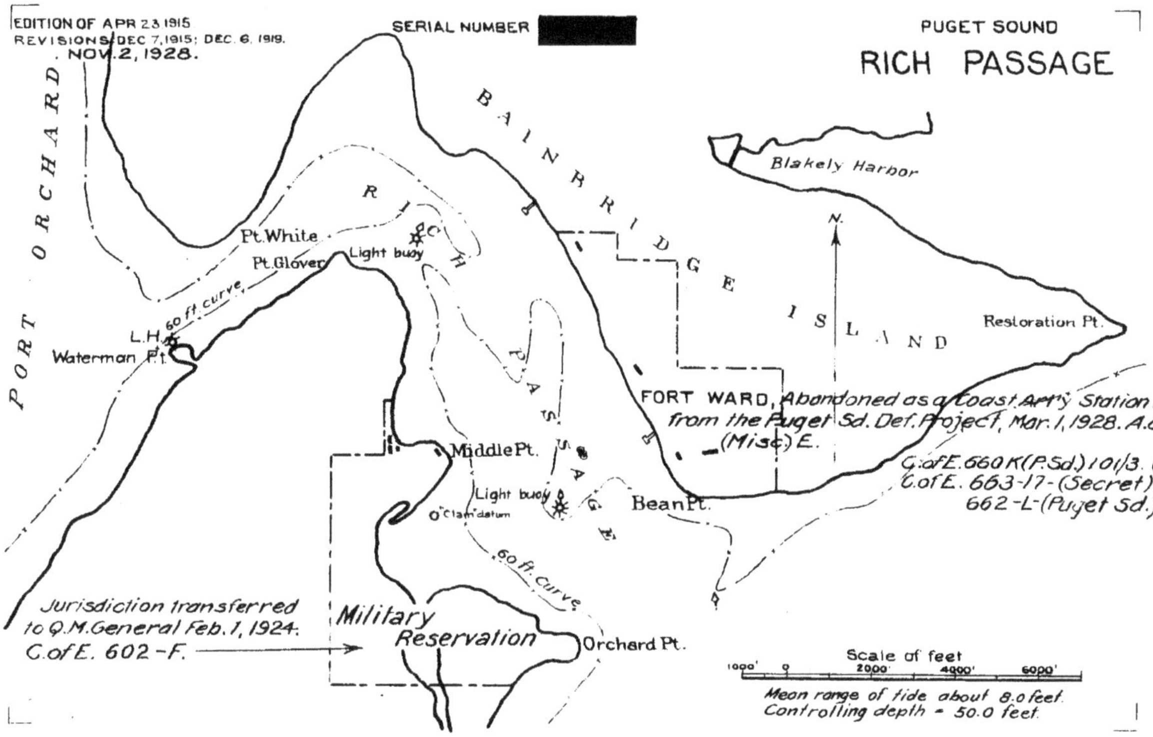

PUGET SOUND
RICH PASSAGE

EDITION OF APR. 23 1915
REVISIONS: DEC. 7, 1915; DEC. 6, 1919.
NOV. 2, 1928.

SERIAL NUMBER

PORT ORCHARD

BAINBRIDGE ISLAND

Blakely Harbor

Pt. White
Pt. Glover
Light buoy

L.H.
Waterman Pt.

60 ft. curve

RICH PASSAGE

Restoration Pt.

Middle Pt.
Light buoy
O'Clam datum
Bean Pt.

FORT WARD, Abandoned as a Coast Art'y Station and withdrawn from the Puget Sd. Def. Project, Mar. 1, 1928. A.G.660.3(11-30-27) (Misc.) E.

C. of E. 660 K (P.Sd.) 101/3. (Secret)
C. of E. 663-17 (Secret)
662-L-(Puget Sd.) 1.

Jurisdiction transferred to Q.M.General Feb. 1, 1924.
C.of E. 602-F.

Military Reservation

60 ft. curve

Orchard Pt.

Scale of feet

Mean range of tide about 8.0 feet.
Controlling depth = 50.0 feet.

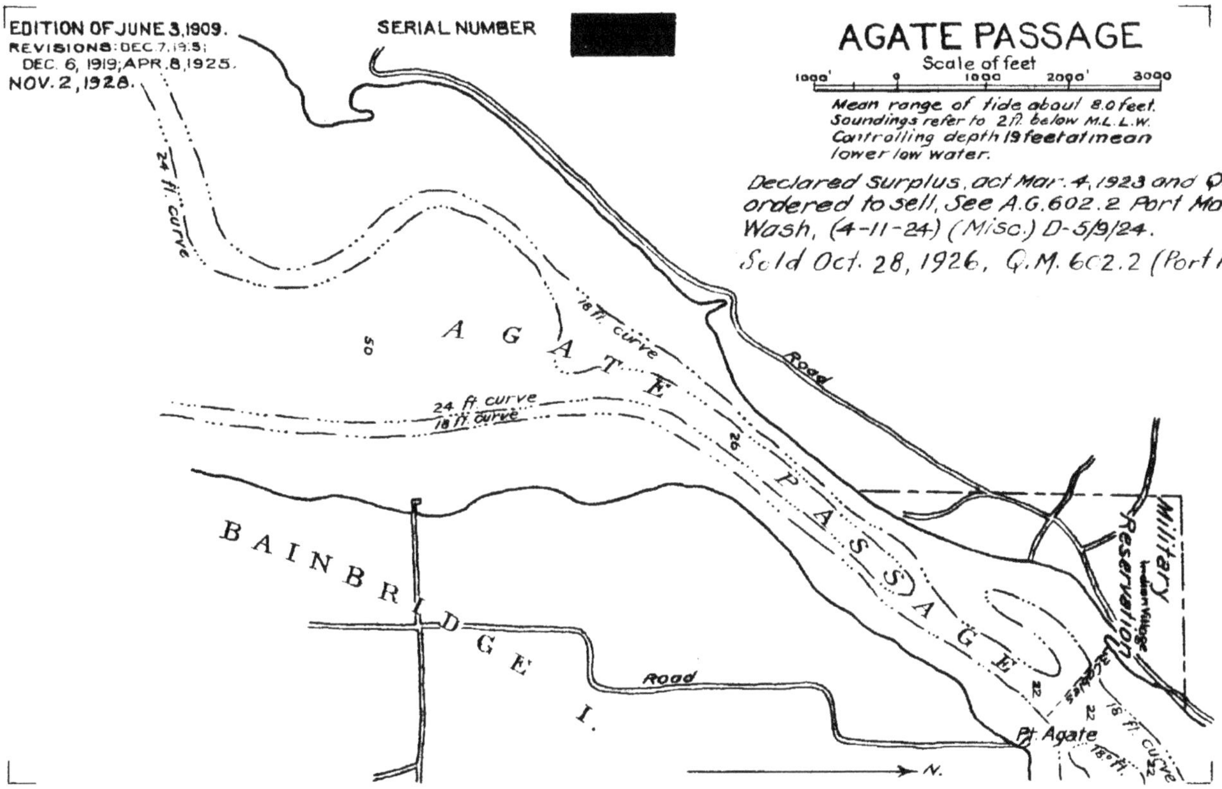

AGATE PASSAGE

EDITION OF JUNE 3, 1909.
REVISIONS: DEC. 7, 1915;
DEC. 6, 1919; APR. 8, 1925.
NOV. 2, 1928.

SERIAL NUMBER

Scale of feet

Mean range of tide about 8.0 feet.
Soundings refer to 2 ft. below M.L.L.W.
Controlling depth 19 feet at mean lower low water.

Declared Surplus, act Mar. 4, 1923 and Q.M.G. ordered to sell, See A.G. 602.2 Port Madison, Wash. (4-11-24) (Misc.) D-5/9/24.

Sold Oct. 28, 1926, Q.M. 602.2 (Port Madison)

24 ft. curve

AGATE PASSAGE

18 ft. curve
24 ft. curve
18 ft. curve

Road

BAINBRIDGE I.

Road

Military Reservation

Pt. Agate

N.

the height of the 8-inch disappearing battery behind it, and 750-feet away. It was planned for the 5-inch wire-wound type of gun that was never actually acquired. Original concrete work was done in 1900-1901. Completion was delayed; in 1903 the platform was finished for the Model 1903 pedestal mount, but the actual armament delayed until 1907. Transfer was made on January 18, 1904 at a cost of $24,935. It was named in General Orders No. 194 of December 27, 1904 for Brevet Captain William H. Warner who was killed in action with hostile Indians in the Sierra Nevada Mountains, CA in 1849. The two 5-inch Model 1900 Watervliet guns on Model 1903 pedestals were mounted on October 25, 1907 (#15/#12 and #19/#13). These were later dismounted by authority of July 18, 1918 but retained and then remounted in April 1919. They were authorized for final removal and scrapping on July 22, 1919. The emplacement still exists on private property, now fenced in and used as a patio for the house in front. The battery is closed to the public.

- **THORNBURGH**: A battery for four 3-inch, 15-pounder masking parapet guns erected at the northern end of the Fort Ward reservation along the eastern side of Rich Passage. The emplacements were in line, the battery firing to the southwest. Plans were submitted on July 19, 1900 and used funds made available by the Act of May 25, 1900. The plan followed recommended mimeographs. It had all internal platforms and gun centers of 29-feet, with magazine on lower left and ammunition service by hand. Work was done in early 1901. Transfer was made on February 17, 1904 for a cost of $21,884. It was named in General Orders No. 194 of December 27, 1904 for Major Thomas Thornburgh who was killed in action with the Ute Indians on the Milk River, CO in 1879. It was armed with four Model 1898 Driggs-Seabury guns on balanced pillar mounts (#9/#9, #15/#15, #40/#40, and #45/#45). They were authorized for conversion to M1898M1 pedestals on November 13, 1913. These served until authorized for removal on March 27, 1920, and removed by June. The emplacement is in a partially cleared state in the City of Bainbridge Island's Fort Ward Park and is open to the public.

- **VINTON**: A battery for two 3-inch masking parapet guns emplaced close to the shore in the center of the reservation along Rich Passage, firing to the southwest covering the mine fields in that passage. Plans were submitted on December 13, 1899. The plan featured internal platforms separated by 41-feet, with both service magazines located under the center traverse. Ammunition service was by hand. A small observation station was located on each flank. It was built in 1900. Transfer was made on February 1, 1904 at a cost of $10,966. It was named in General Orders No. 194 of December 27, 1904 for Captain John Vinton killed in action at Vera Cruz Mexico in 1847. It was armed with two 3-inch Model 1898 Driggs-Seabury guns and balanced pillar carriages (#2/#2 and #5/#5). In accordance with directives of July 24, 1916 the balanced pillars were altered to Model 1898M1 barbette pedestals. Authority to remove came on March 27, 1920, implemented that following June. The magazines have been filled in, but the emplacement is accessible by trail in the City of Bainbridge Island's Fort Ward Park and is open to the public.

- **MITCHELL**: A battery for two 3-inch masking parapet guns emplaced as the only armament at the Middle Point Military Reservation, opposite from Fort Ward on the western side of Rich Passage. It is sited to fire to the northeast. Submission was made on December 13, 1899. It followed the design type of having the platform gun centers separated by 41-feet, with both service magazines located under the center traverse. Ammunition service was by hand. It was planned as a sunken battery, with the slope behind the battery being at the crest height. There were plans submitted for a 6-inch disappearing battery to be built at the reservation, but its construction was never authorized. Work was done on the 3-inch battery beginning on July 1, 1901. It was named in General Orders No. 194 for 1st Lieutenant Robert Mitchell, Artillery Corps officer who died in

1904. The intended armament of two Model 1898 guns and masking parapet pillar mounts were never received or installed—the base rings were not even set. Apparently, there were real second thoughts about the utility of this very isolated post, the mine facilities located there were moved to Fort Ward in 1908. It was determined that the mine field in Rich Passage was adequately covered by the installed armament at Fort Ward. The Adjutant General authorized abandonment of the post and Battery Mitchell (still unarmed) on February 19, 1909. The emplacement still exists at the Manchester State Park, Washington. The battery is open to the public.

Fort Lawton (1900-1963) was originally reserved for the emplacement of seacoast artillery to defend the approach to Seattle. The planned defenses were never built. The reservation was then used an infantry post.

The Fort Ward Bakery was restored for use as a local community center (Mark Berhow)

Battery 131 at Salt Creek Recreation Area (Mark Berhow)

PUGET SOUND

FORT LAWTON

Magnolia Bluff

BALLARD

SALMON BAY

SALMON BAY WATER WAY

US Lock Reserve

G.N. R'Y

Locks

Dam

& I Wash Canal

G.N. Ry

SALMON BAY

Bascule Bridge

Seattle North Trunk Sewer

Target Range

Seattle Electric Cut Line

Seattle Electric

Cemetery

L.H. Reserve

West Pt.

L.H.

SOUND

PUGET SOUND

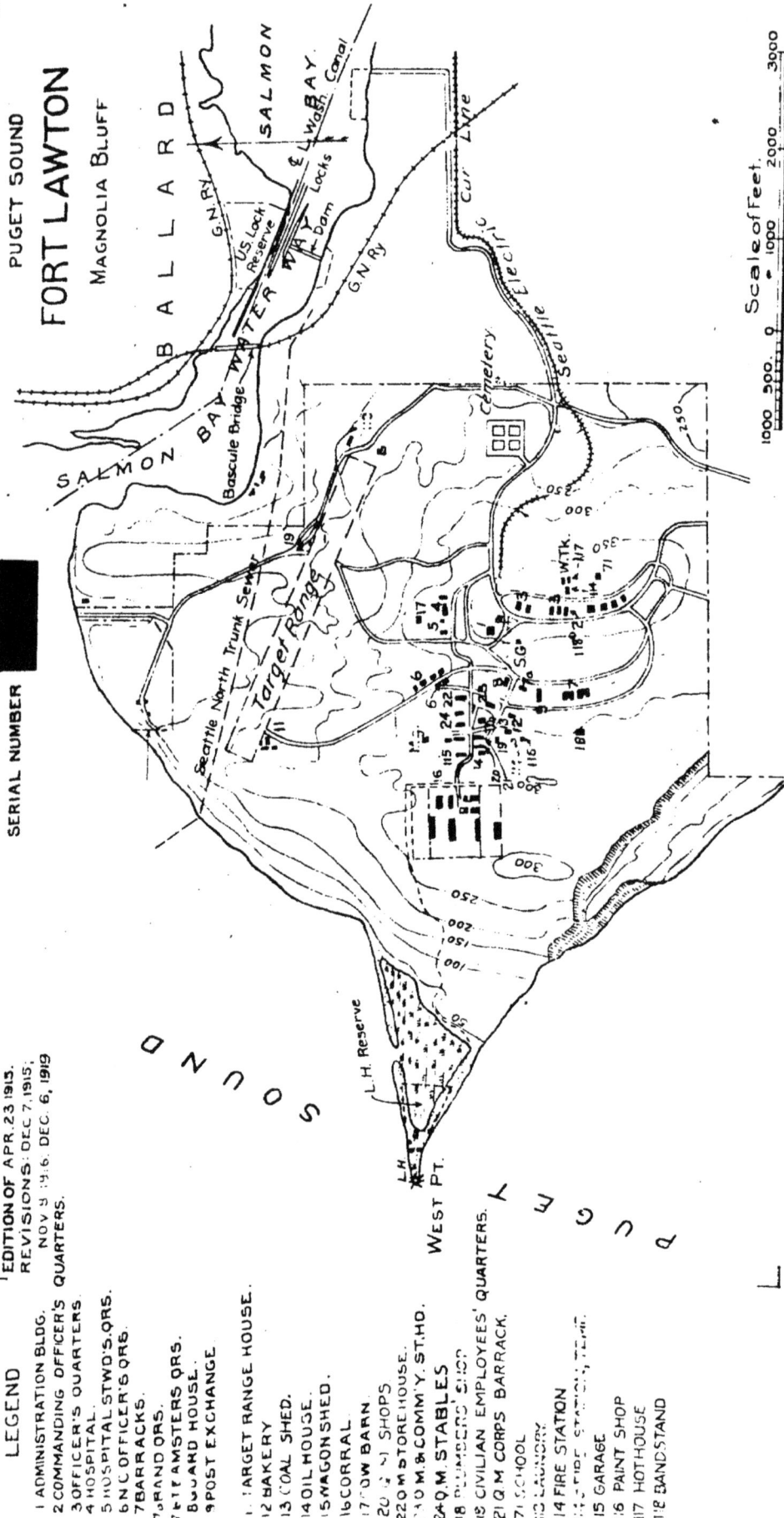

SERIAL NUMBER

EDITION OF APR. 23 1915.
REVISIONS: DEC 7, 1915;
NOV 9 1916. DEC. 6, 1919

Scale of Feet.
1000 500 0 1000 2000 3000

LEGEND

1 ADMINISTRATION BLDG.
2 COMMANDING OFFICER'S QUARTERS.
3 OFFICER'S QUARTERS.
4 HOSPITAL.
5 HOSPITAL STWD'S. QRS.
6 N.C. OFFICER'S QRS.
7 BARRACKS.
7ᵃ BAND QRS.
7ᵇ TEAMSTERS QRS.
8 GUARD HOUSE.
9 POST EXCHANGE.

11 TARGET RANGE HOUSE.
12 BAKERY.
13 COAL SHED.
14 OIL HOUSE.
15 WAGON SHED.
16 CORRAL
17 COW BARN.
20 Q.M. SHOPS.
22 Q.M. STOREHOUSE.
23 Q.M. & COMM'Y. ST. HO.
24 Q.M. STABLES.
18 PLUMBERS' SHOP.
18ᵃ CIVILIAN EMPLOYEES' QUARTERS.
21 Q.M. CORPS BARRACK.
7¹ SCHOOL.
1¹³ LAUNDRY.
1¹⁴ FIRE STATION.
1¹⁵ GARAGE.
1¹⁶ PAINT SHOP.
1¹⁷ HOTHOUSE.
1¹⁸ BANDSTAND.

Site obtained for seacoast artillery, but developed as a infantry post

Camp Hayden (1941-1949) is located at Tongue Point, near Striped Peak and Crescent Beach, facing north toward the Strait of Juan de Fuca, about twenty miles west of Port Angeles, Washington State. The military reservation (known at that time as Striped Peak Military Reservation) was established as part of the major effort to modernize the defense of Puget Sound for the coming of World War II. In 1942, the site was renamed as Camp Hayden in General Orders 27 of 1944 for Brig. Gen. John L. Hayden, U.S. Army CAC. and in 1944 was renamed again as Fort Hayden. The 1940 Program saw the addition of a #100 Series battery with two casemated 16-inch guns on long-range barbette carriages (Battery Construction #131) and a #200 Series battery with two 6-inch shielded barbette guns (Battery Construction #249). Several fire control stations were also constructed, along with a secondary HECP complex. Defenses were also planned at Cape Flattery (near Neah Bay) with two 16-inch batteries and two 6-inch batteries, but these were never constructed. In the 1950s the reservation was divided between Clallam County and the State of Washington. The main fortification area of Camp Hayden reservation is now Clallam County's Salt Creek Recreation Area, which requires a seasonal entry fee. A park road leads to Battery 131 with a parking area and an information kiosk. The casemated gun houses are open, but the interior of the battery is closed to the public. Hikers and mountain bikers access the State DNR's Striped Peak Recreation Area from the trailhead at the main gate. Battery 249 is just a short hike up the old military road leading to Striped Peak at a curve in the road on the property line between the county and the state. The interior is closed to the public and the battery is fairly overgrown. The battery commander's station is on the hillside just above the battery. Further up the road is a road gate blocking vehicle traffic into Salt Creek, which leads up to the Auxiliary HECP-Bn4-SCR 682 site on a bluff in front of the peak proper at the 900-foot level. The SCR 296A radar site for BCN 249 is further east along the road. The Striped Peak State Recreation Area is accessible by car on the Striped Peak Road from the east entrance at the Freshwater Bay County Park.

Camp Hayden Gun Batteries

- **Battery #131**: A 1940 Program dual 16-inch barbette battery emplaced as the primary defense to the entry of Puget Sound on the Strait of Juan de Fuca. Projects for heavy caliber weapons west of the Endicott defenses had long been contemplated. Planning began in the 1920s for a 16-inch battery at Middle Point west of Fort Worden, but it was not funded. However the new 1940 Program provided funding for three new 16-inch dual gun batteries in the Puget Sound. After much reshuffling of sites and priorities, two 16-inch projects were located at Cape Flattery in 1943 and ultimately not built, leaving this project as the sole unit constructed on the new Striped Peak Military Reservation above Crescent Beach west of Port Angeles. It was never named, during construction and service it was known as Battery Construction No. 131. It was designed to interdict the western end of the strait, in cooperation with the coverage by Canadian batteries on the opposite shore. Under the national priority list of September 11, 1940, it was initally assigned priority #11. It was funded in the FY-1942 Budget with $625,000. It was a fairly standard 100-series design, with the No. 1 gun house withdrawn in echelon, allowing the connecting gallery to run at an angled offset in order to fit the available land topography. Otherwise, the distance between gun centers and arrangement of internal traverse magazines and other rooms followed standard dimensions. The usual separate PSR room and numerous base-end station assignments complemented the battery construction. Work was done from October 30, 1942 to mid-1943. It was completed by May 10, 1945, and transfer was made on June 7, 1945 for a cost of $1,557,500. It was armed with two 16-inch navy MkIIM1 guns on Model 1919M4 barbette carriages (#51/#37 and #89/#38). It served in the Puget Sound defenses as local Tactical Battery No. 12. It was deleted and disarmed postwar, in 1948. The emplacement still exists at the Salt Creek Recreation Area and accessible by a paved road. The area is open to the public, but service gallery and rooms are sealed.

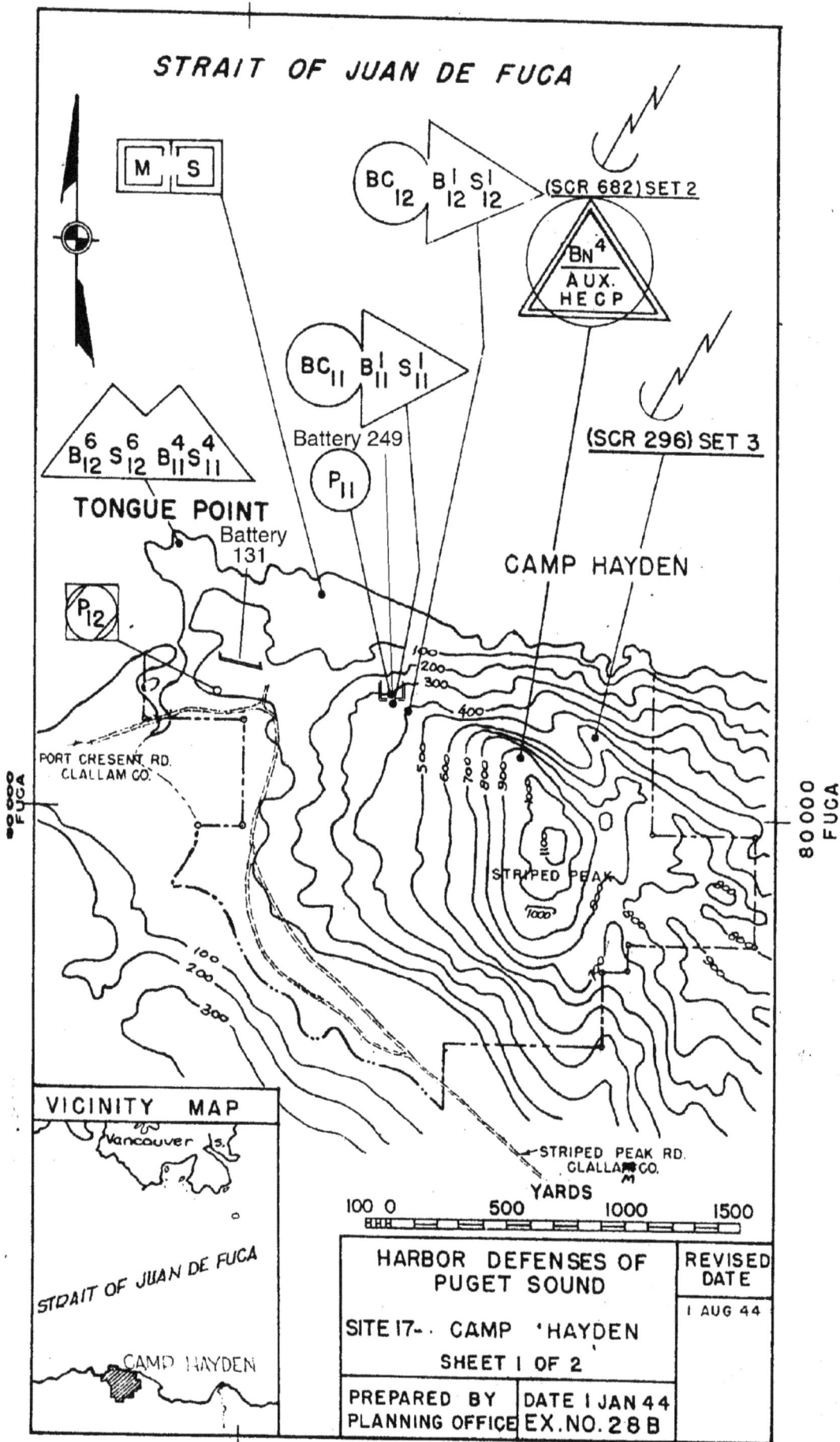

STRAIT OF JUAN DE FUCA

M S

BC₁₂ B¹₁₂ S¹₁₂ (SCR 682) SET 2

Bn⁴ / AUX. HECP

BC₁₁ B¹₁₁ S¹₁₁

Battery 249

P₁₁

(SCR 296) SET 3

B⁶₁₂ S⁶₁₂ B⁴₁₁ S⁴₁₁

TONGUE POINT

Battery 131

CAMP HAYDEN

P₁₂

PORT CRESENT RD. CLALLAM CO.

80000 FUCA

80000 FUCA

STRIPED PEAK

STRIPED PEAK RD. CLALLAM CO.

YARDS

VICINITY MAP

Vancouver I.s.

STRAIT OF JUAN DE FUCA

CAMP HAYDEN

100 0   500   1000   1500

HARBOR DEFENSES OF PUGET SOUND

SITE 17-- CAMP 'HAYDEN

SHEET 1 OF 2

REVISED DATE

1 AUG 44

PREPARED BY PLANNING OFFICE

DATE 1 JAN 44 EX.NO. 28 B

- **Battery #249**: A 1940 Program dual 6-inch barbette battery emplaced on a cliff shelf to the east of Battery #131, firing to the north into the Strait of Juan de Fuca. Of the original twenty-six batteries of the FY-1942 Budget, it was awarded priority #19. Work was not rapid, actually being put behind Battery #248 of the FY-1943 Budget at Fort Ebey. Construction was accomplished between April 12, 1943 and January 2, 1945. Transfer was made on January 5, 1945 for a construction cost of $350,000. It was of standard 200-series design but was one of the variants with the rear entry to the power room being folded to the left side. It was never named and was just referred to as Battery Construction No. 249 during building and service. It carried two 6-inch Model M1(T2) guns on Model M4 barbette carriages (#24/#5 and #25/#6), It served until deleted and disarmed in 1947 or 1948 at the end of the defenses for Puget Sound. The emplacement still exists at the Salt Creek Recreation Area. The battery is open to the public accessible by a short walk along the old military road, but the magazines are gated for use as a bat habitat.

Cape Flattery Military Reservation (1941-1943, 1950-1952) is located on tribal lands of Makah Indian Reservation on Cape Flattery, Clallam County, Washington. In 1942, the U.S. Army leased 4,024 acres of Makah Indian Reservation land to establish Cape Flattery Military Reservation as a coastal defense installation to protect the entrance to Puget Sound. The Military Reservation consisted of many non-contiguous sites that were to house two 6-inch and two 16-inch coastal gun batteries, approximately 25 fire control stations, a central magazine, numerous radar sites, and a combined HECP-HDCP.

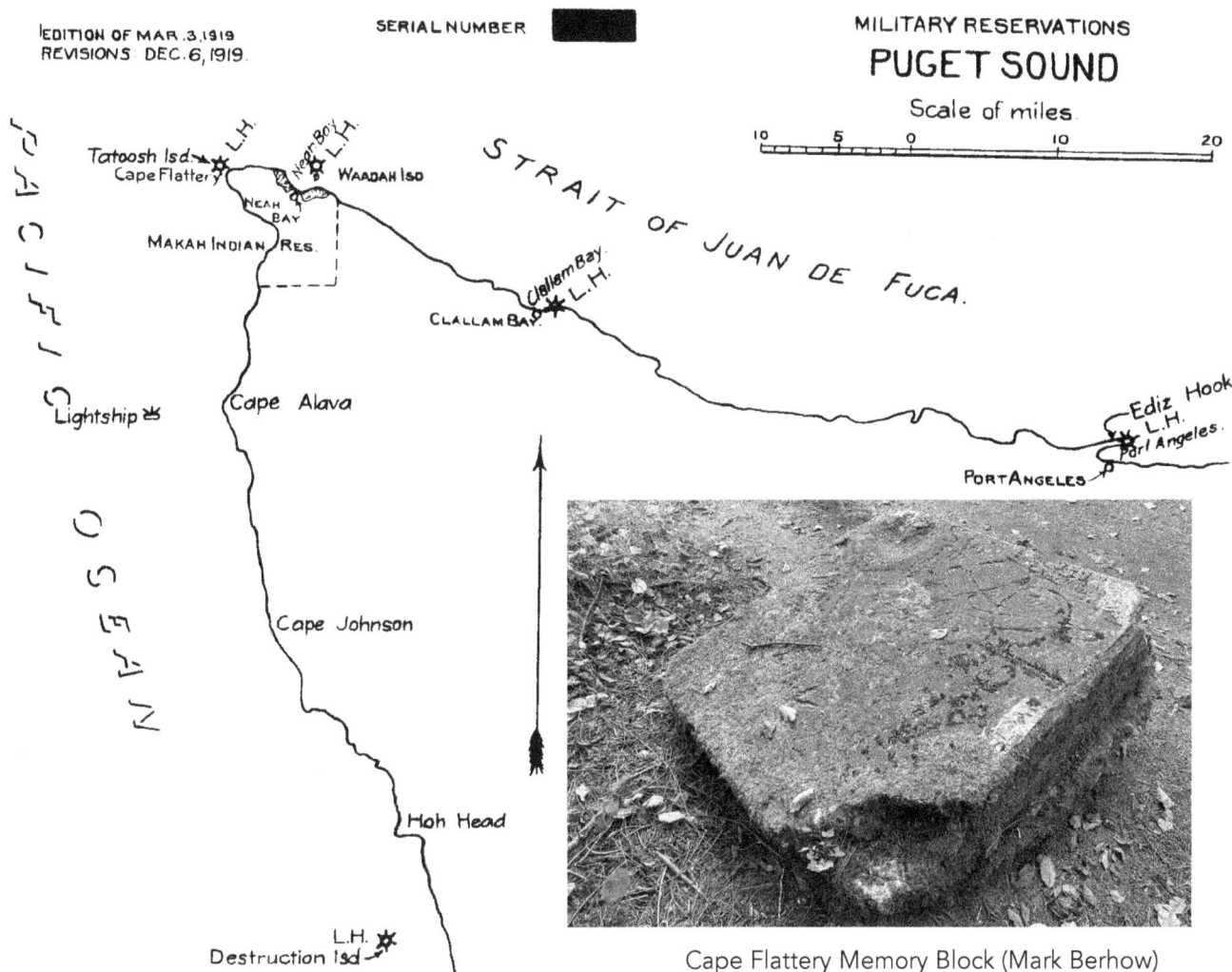

Cape Flattery Memory Block (Mark Berhow)

Roads were built and sites were cleared and graded but the project was terminated on 13 Oct 1943 before the batteries were built. Some support buildings and fire control stations were started, but very few were completed. The lease was terminated in 1945 and all the land, except a 10-acre site at Bahokus Peak, was transferred back to the Makah Tribe. Nine additional sites were established for fire-control stations and searchlight positions east and south of the main reservation (at Knob, Lower Agency, Upper Agency, Waadah Island, Wat, Portage Head (North), Portage Head (South), Point of Arches, and Duk). At least three of these fire-control stations were completed by 1943 (Wat, North and South Portage Head), and may still exist within the current Makah Indian Reservation lands.

Cape Flattery M.R. Gun Batteries (Gun Emplacements planned, not built)

- *Battery #132* (planned): A 1940 Program dual 16-inch casemated battery planned for Puget Sound. Under the original 1940 plan of September 11, this was the battery selected for Dungeness Point. It was assigned national priority #20 and an appropriation of $625,000 allocated from the FY-1942 Budget. It was soon relocated to a site with Battery #133 at Cape Flattery. Work was slow in starting, but some land clearing, excavation and laying of power conduits was done before work was cancelled in October of 1943. No actual concrete construction of the emplacement was done.

- *Battery #133* (planned): A 1940 Program dual 16-inch casemated battery planned for Puget Sound. Originally it was to be emplaced at Partridge Point, the site that eventually became Fort Ebey. It was assigned a national low priority of #32 of September 11, 1940. Funding was to be forthcoming in the FY-1943 Budget. By August 11, 1941 the site for the battery had been relocated to Cape Flattery, though carrying an even lower priority of #34 at that stage. The two 16-inch batteries were to be emplaced relatively close together, this one being the southern unit of the pair. While some of the site preparations were made, no actual construction was ever begun. All work was cancelled in October of 1943.

- *Battery #250* (planned): A 1940 Program emplacement for a dual 6-inch gun battery. It was intended to emplace this battery at the tip of Koitlah Point, on the eastern side of the new Cape Flattery Military Reservation, firing to the northeast. It was to be funded in the FY-1942 Budget and given the national high priority of #3. This priority was changed, with the emphasis being given the inner defenses of Puget Sound at Port Angeles and Fort Ebey. No construction was ever undertaken for the work, and it was cancelled in October of 1943.

- *Battery #251* (planned): A 1940 Program emplacement for a dual 6-inch gun on barbette battery. It was originally intended to erect the battery neat the opening of Ocean Creek, east of Battery #133 but west of the new HECP. It would have been funded under the FY-1943 Budget. No construction was undertaken, and it was cancelled in October of 1943.

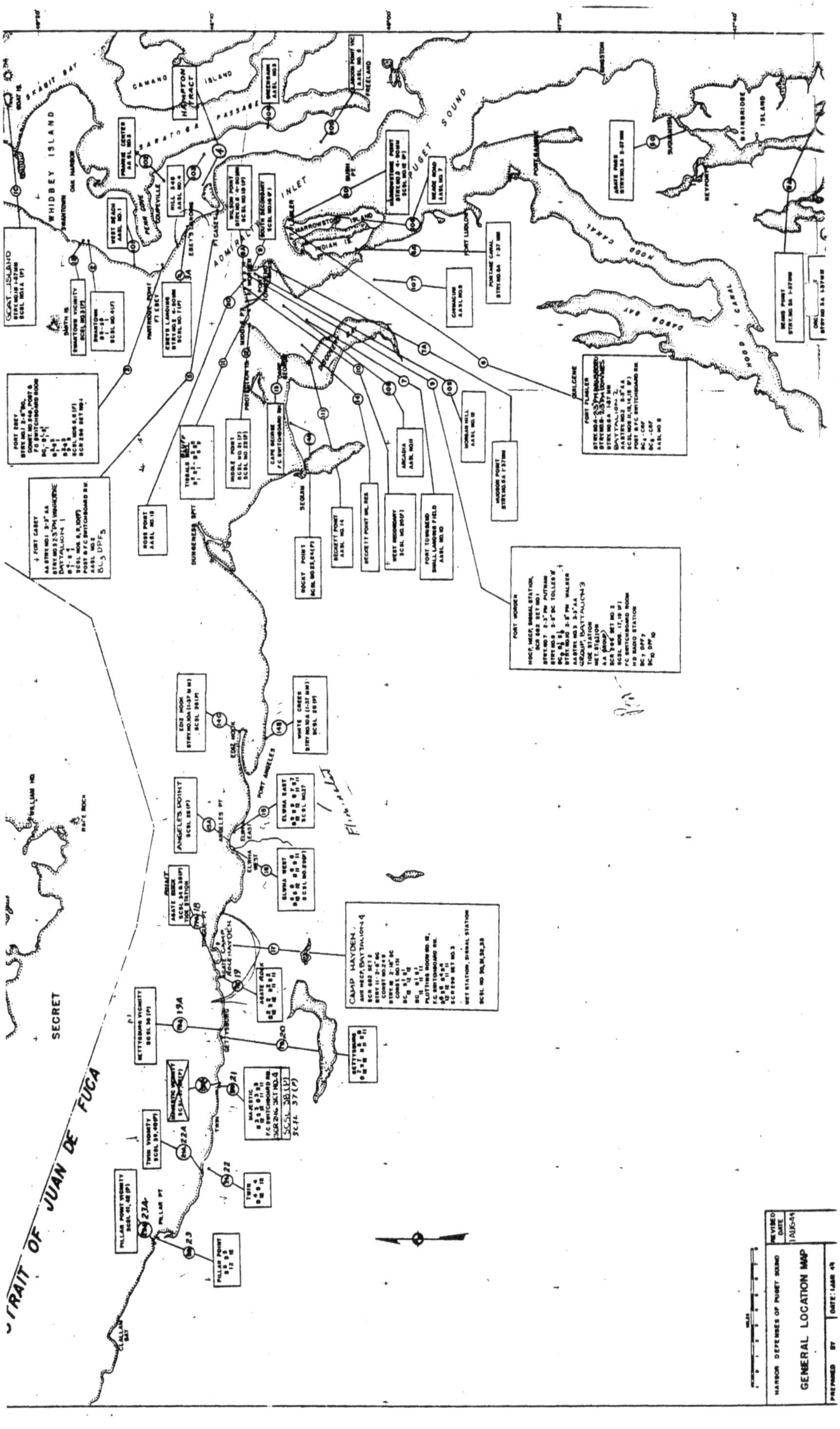

STRAIT OF JUAN DE FUCA

PUGET SOUND

HOOD CANAL

SECRET

GENERAL LOCATION MAP

HARBOR DEFENSES OF PUGET SOUND

PREPARED BY   DATE: 1 JAN 45

REVISED DATE 1 AUG 44

Puget Sound World War II-era Site Locations. Stations housed in a single structure are connected by dashes (-)

location	Loc#	Purpose
Deception Pass	1	Batt Tact. #1A AMTB, BC AMTB BC
Dugalla Bay	1A	searchlight AMTB
Swantown area	1B	searchlight
Goat Island/ Fort Whitman	1C	searchlight AMTB
Swantown	2	BS5/248
Fort Ebey	3	Batt Tact. #1 BCN 248, B3/248, B3/Tolles BC-B1/248 SCR296/248
Ebeys Landing	3A	Batt Tact. #2 AMTB, searchlight
Hampton Tract	4	not used
Fort Casey	5	Batt Tact. #3 Van Horne, Bn1, B4/248, BC Van Horne, SBR
Beans Point (Ft Ward)	5A	AMTB
Orchard Point	5B	AMTB
Agate Passage	5C	AMTB
Marrowsone Point	5D	Batt Tact. #5 AMTB, searchlight AMTB BC5
Fort Flagler	6	Batt Tact. #4 Wansboro, BC/Wansboro, SWB, Bn2, BC/Downes
Portage Canal	6A	AMTB
Fort Townsend	7	searchlight AMTB
Hudson Point	7A	AMTB
South Secondary	8	searchlight
Wilson Point	8A	Batt Tact. #8 AMTB, searchlight AMTB BC8
Fort Worden	9	Batt Tact. #7 Putnam, Batt Tact. #9 Tolles B, Batt Tact. #10 Walker, BC/Putnam, BC/Walker, Bn3, HECP/HDCP, HDOP, SCR582, BC-B1/Tolles B, SCR296-Tolles B
West Secondary	10	searchlight
Tibbels Bluff	11	BS2/248-BS2/Tolles B
Middle Point	12	searchlight
Cape George	13	SWB
Beckett Point	14	not used
Rocky Point	14A	searchlight
White Creek	14B	AMTB searchlight
Ediz Point	14C	searchlight AMTB
Elwa East	15	BS9/131-BS7/249, SL
Angeles Point	15A	searchlight
Elwa West	16	BS8/131-BS6/249
Camp Hayden	17	Batt Tact. #11 BCN 249, Batt Tact. #12 BCN 131, HDCP Aux, Bn4, SCR682, SCR296-249, BC-BS1/131, BC-BS1/249, PSR/131, BS6/131-BS4/249
Agate Point	18	searchlight, T
Agate Rock	19	BS2/131-BS2/249
Low Point	19A	searchlight
Gettysburg	20	BS7/131-BS5/249
Majestic	21	B3/131-B3/249, SCR296/131
Twin	22	BS4/131
Twin	22A	searchlight

Pillar Point   23   BS5/131
Pillar Point   23A   searchlight

Buchanan, Terry, *The Pictorial History of Fort Casey,* Serenity Ridge Press, 2016

Lyman P. *Fort Flagler over 100 years of History.* Friends of Fort Flagler. Nordland, WA, 2006.

Hansen, David M. *Battle Ready: The National Coast Defense System and the Fortification of Puget Sound, 1894-1925.* Washington State University Press, Pullman, WA 2014.

Gregory, V.J. *Keepers at the Gate.* Port Townsend Publishing Co. Port Townsend, WA, 1976.

Berhow, Mark A. Harbor Defenses of Puget Sound and the Modernization Program of 1940. *Coast Defense Journal* Vol. 35 No. 1, 2021 p. 37, Vol. 35, No. 2, p. 50.

Hansen, David M. "Fortress Without Guns" (Fort Ward, WA). *CDSG Journal* Vol. 9, No. 3, Aug. 1995, p. 4.

Blackburn, Marc K. "The Establishment of the Coastal Defenses of Puget Sound." *CDSG Journal* Vol. 14, No. 2, May 2000, p. 4.

Fort Casey State Park view (Steven Kobylk)

Fire Control Stations at Fort Casey State Park (Terry McGovern)

Army cemetary at Fort Worden (Mark Berhow)

www.ingramcontent.com/pod-product-compliance
Lightning Source LLC
Chambersburg PA
CBHW040259100426
42811CB00011B/1314